W0227898

John Older (Ed.)

Bone Implant Grafting

With 174 Figures

Springer-Verlag London Ltd.

M. W. J. Older, MBBS, BDS(Lond), FRCS(Ed)

Consultant in Orthopaedic and Traumatic Surgery,
Royal Surrey County Hospital, Guildford, Surrey, GU2 5XX, UK

Consultant Orthopaedic Surgeon, The John Charnley Hip Unit,
King Edward VII Hospital, Midhurst, West Sussex, GU29 0BL, UK

Honorary Senior Research Fellow, University of Surrey,
Guildford, Surrey, GU2 5XH, UK

ISBN 978-1-4471-1936-4 ISBN 978-1-4471-1934-0
DOI 10.1007/978-1-4471-1934-0

British Library Cataloguing in Publication Data
Bone Implant Grafting
I. Older, John
617.4
ISBN-13: 978-1-4471-1936-4

Library of Congress Cataloging-in-Publication Data
Bone Implant Grafting/John Older,(ed.).
p. cm.
Includes index.
ISBN-13: 978-1-4471-1936-4
1. Bone Grafting Congresses. 2. Orthopedic implants—Congresses. 3. Hip joint—Surgery —Congresses. 4. Knee—Surgery—Congresses.
I. Older, John, 1935– .
[DNLM: 1. Bone Transplantation—methods—congresses. WE 190 B7116]
RD123.B6465 1992
617.4'710592—dc20
DNLM/DLC 92–2327
for Library of Congress CIP

Apart from any fair dealing for the purposes of research or private study, or criticism or review, as permitted under the Copyright, Designs and Patents Act 1988, this publication may only be reproduced, stored or transmitted, in any form or by any means, with the prior permission in writing of the publishers, or in the case of reprographic reproduction in accordance with the terms of licences issued by the Copyright Licensing Agency. Enquiries concerning reproduction outside those terms should be sent to the publishers.

© Springer-Verlag London 1992
Originally published by Springer-Verlag London Limited 1992
Softcover reprint of the hardcover 1st edition 1992

The use of registered names, trademarks, etc. in this publication does not imply, even in the absence of a specific statement, that such names are exempt from the relevant laws and regulations and therefore free for general use.

Product liability: The publisher can give no guarantee for information about drug dosage and application thereof contained in this book. In every individual case the respective user must check its accuracy by consulting other pharmaceutical literature.

Typeset by Wilmaset Ltd, Birkenhead, Wirral
Printed by Henry Ling, The Dorset Press, Dorchester, UK
2128/3830–543210 Printed on acid-free paper

Preface

Today there is an increasing demand for major revision hip and knee surgery. For great defects in the skeleton which demand cure, bone grafting may be the only solution. Surgeons differ in their enthusiasm for autografts, allografts and metallic implants but all have their place.

The symposium on Bone Implant Grafting held in May 1991, on which this book is based, sought accurately to reflect the diversity of thinking world-wide on this important aspect of orthopaedic surgery.

Midhurst
February 1992

John Older

Acknowledgements

The Bone Implant Grafting symposium, on which this book is based, was made possible by the financial generosity of DePuy with assistance from Howmedica International, Johnson and Johnson, 3M and Straumann.

My sincere thanks go to King Edward VII Hospital, Midhurst, for providing the venue; Michael Phillips of Metaphor for organising the meeting; Robert Bohill Associates for providing sound and vision; and Ivor Williams and Barbara Bennett of Palantype for their verbatim reporting.

During this book's preparation, which was supported financially by the Charnley Trust, I have greatly valued the editorial assistance of Bobbin Baxter. My thanks and appreciation also go to Tracy Bilsland, who typed the manuscript, and my secretary, Christine Higgitt.

John Older

Contents

Part IV Bone Banks

Faculty

Dr P. Buma, MD
Department of Orthopaedics, Histomorphology Section, University Hospital of Nijmegen, PO Box 9101, 6500 HB Nijmegen, The Netherlands

Dr H.P. Chandler, MD
Assistant Professor of Orthopaedics, Harvard Medical School, ACC 531, Massachusetts General Hospital, Boston, MA 02114, USA

Dr A.A. Czitrom, PhD, MD, FRCS(C)
Clinical Associate Professor, Department of Orthopaedic Surgery, The University of Texas, Humana Advanced Surgical Institutes, Medical City Dallas, 7777 Forest Lane, Suite C-770, Dallas, TX 75230, USA

Dr D.M. Dall, MCh(Orth), FRCS
Professor of Clinical Orthopaedics, University of Southern California, Hospital of the Good Samaritans, 637 South Lucas Avenue, Los Angeles, CA 90017, USA

Dr C. Delloye, MD
Chef de Clinique Associé, Cliniques Universitaires St Luc, Service d'Orthopedie-Traumatologie, 53 Avenue Mounier, Brussels B1200, Belgium

Mr R.A. Elson, MBBChir, FRCS
Consultant Orthopaedic Surgeon, Northern General Hospital, Herries Road, Sheffield S5 7AU, UK

Mr G. Gie, MBChB, FRCS(Ed)
Consultant Orthopaedic Surgeon, Princess Elizabeth Orthopaedic Hospital, Wonford Road, Exeter, North Devon EX2 4LE, UK

Dr V.M. Goldberg, MD
Charles Herndon Professor and Chairman, Department of Orthopaedics, Case Western Reserve, University Hospital of Cleveland, 2074 Abington Road, Cleveland, OH 44106, USA

Dr A.E. Gross, MD, FRCS(C)
A.J. Latner Professor and Chairman, Division of Orthopaedic Surgery, University of Toronto, Mt Sinai Hospital, 600 University Avenue, Suite 476A, Toronto M56 1XS, Ontario, Canada

Dr A.K. Hedley, MD, FRCS
Institute for Bone and Joint Disorders, Osborn Plaza, Medical Office Building, 3320N 2nd Street, Phoenix, AZ 85012, USA

Dr A. Kocialkowski, MD
Orthopaedic Registrar, Department of Orthopaedic Surgery, Queen's Medical Centre, Nottingham NG7 2UH, UK

Dr B. Loty, MD
Hospital Cochin, Pavillon Ollier, 27 Rue du Faubourg St Jacques, Paris 75014, France

Dr W.G. Paprosky, MD, FACS
Clinical Associate Professor, Loyola University of Chicago, Chicago Medical Centre, 454 Pennsylvania Avenue, Glen Ellyn, IL 60137, USA

Miss H.G. Prince, MBChB, FRCS
Consultant Orthopaedic Surgeon, Harlow Wood Orthopaedic Hospital, Nottingham Road, Mansfield, Nottingham NG18 4TH, UK

Dr J.W. Schimmel, MD
Orthopaedic Resident, Department of Orthopaedics, University Hospital of Nijmegen, PO Box 9101, 6500 HB Nijmegen, The Netherlands

Dr B.W. Schreurs, MD
Orthopaedic Resident, Department of Orthopaedics, University Hospital of Nijmegen, PO Box 9101, 6500 HB Nijmegen, The Netherlands

Dr T.J.J.H. Slooff, PhD, MD
Professor of Orthopaedic Surgery, Department of Orthopaedics, University Hospital of Nijmegen, PO Box 9101, 6500 HB Nijmegen, The Netherlands

Mrs H. Stafford, BSc
Manager, Leicester Bone Bank, Department of Orthopaedic Surgery, Glenfield General Hospital, Groby Road, Leicester LE3 9QP, UK

Introduction: History and Early Research on Bone Transplantation

M.W.J. Older

Bone grafting has a long history and in legend goes far into antiquity. The first xenograft was possibly performed in Greek mythology when Demeter, the mother of Pelops, ate a portion of the young man's shoulder. To restore function, Zeus and the Gods constructed a shoulder joint out of ivory.

Was the first isograft performed in the Old Testament when Adam created Eve from a rib as an allograft?

The twin physicians, St Cosmas and St Damian, lived in the third century AD and were put to death in 255 AD by an angered Emperor Diocletian (Danilevicius 1967). In the fifth century, a faithful church retainer, exhausted by the pain from cancer of a limb, fell asleep in the temple of the Roman Forum, now known as the Basilica Cosmas et Damiano. In his dream the two saints came to him, removed the affected limb and implanted one from a Moor who had just died. This was the first allograft. Because the Moor had darker skin than the recipient, the event became known as "The Miracle of the Black Leg". The twins were beatified and their spectacular surgery was used as a subject by numerous Renaissance artists (Fig. I.1).

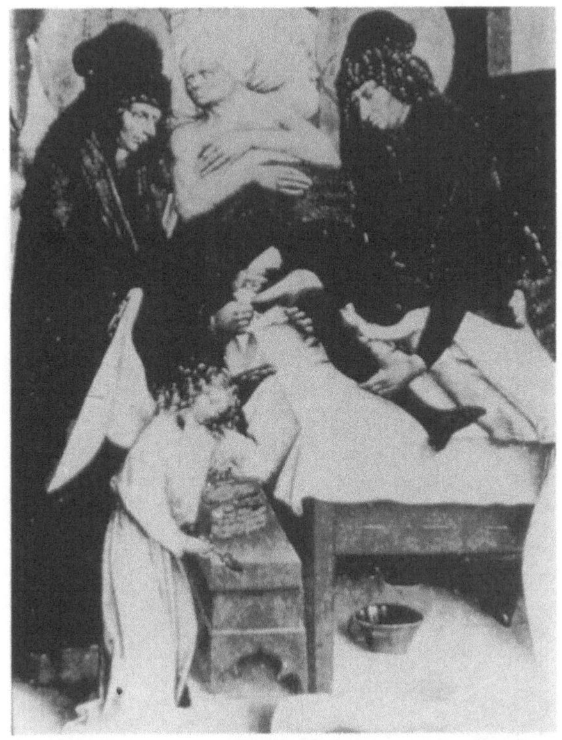

Fig. I.1. St Cosmas and St Damian, transplanting a limb from a Moor. The amputated leg lies on the floor in the foreground. Fifteenth Century painting, unknown artist. Wurttembergisches Landesmuseum, Stuttgart, Germany.

In 1668, Job van Meekeren, a Dutch surgeon, described the first bone graft procedure (Fig. I.2). Graft was taken from a dog's skull and used for the

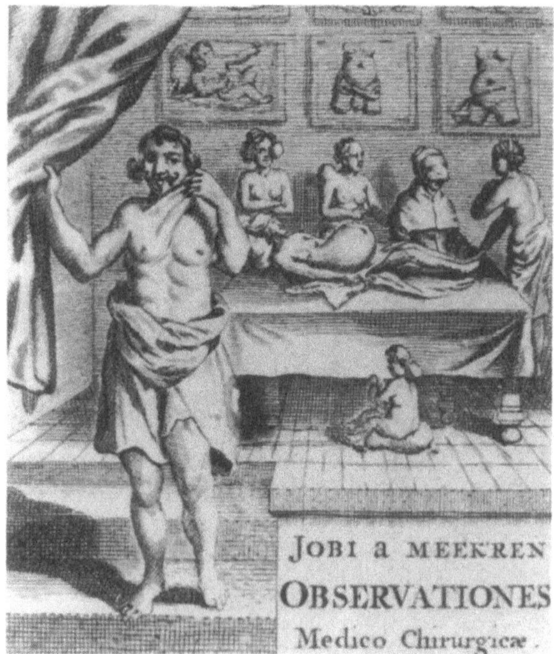

Fig. I.2. Job van Meekeren 1668, title page.

Fig. I.3. Anton van Leeuwenhoek 1674.

successful repair of a traumatic defect in a soldier's cranium, but the patient was excommunicated for this very barbaric treatment. The surgeon was asked to remove the graft so that the patient could be returned to the good grace of the Church, but the graft was found to have taken and he could not remove it.

Anton van Leeuwenhoek, a Dutch scientist, was the first to describe bone structure in 1674 (Fig. I.3). Ten years later, de Heyde, another Dutchman, experimented on frogs' legs and demonstrated, it is thought for the first time, the callus formation in blood clot around bone fractures. In 1742, the Frenchman Duhamel was responsible for the first scientific investigation with the problem of osteogenesis (Chase and Herndon 1955). The first clinical autograft was performed in Germany by von Walter who replaced parts of the skull surgically removed after trepanotomy.

In 1867, the great surgeon Ollier, experimenting with rabbits and young dogs in France, showed that autografts could be viable. In this classic surgical work he recognised that separate living bone fragments within the periosteum could live and grow in a suitable environment.

Macewen in Scotland (1880), performed the first allograft on a four-year-old boy, reconstructing an infected humerus with graft taken from the tibia of a child with rickets.

In 1893, Penski in Russia was the first to perform an allogeneic joint transplantation in animals (Aho 1973). Working independently in the same year – as so often happens in medicine throughout the world – Barth in Germany and Curtis in America published work on bone transplantation. They described the absorption of dead tissue in bone graft and the formation of new bone which grew into the graft from the surrounding living bones. In 1914, Phemister described this process as "creeping substitution".

Lexer described the use of joints in the reconstruction of traumatically injured knees in 1908. Seventeen years later, he reported that half the patients still retained their grafts.

Albee's work in America on bone graft surgery was published in 1915, which resulted in an increase in bone transplantation.

In the past 75 years, cadaveric allograft implantation has waxed and waned in popularity as a clinical activity (Groves 1917, Brooks and Hudson 1920, Levander 1934, Lacroix 1947, Mankin 1983). The concept of bone banking became a reality during and after World War II. It was used mostly for fractures and later fell into disfavour, probably on the basis of bad surgical experiences.

In 1953, Urist in America developed the theory of osseo-induction. This suggests that a chemical mediator from the bone graft could induce bone formation by the recruitment of cells with potential for bone formation.

In London, England, during the 1960s, Burwell, possibly more than anyone, established the histological and immunological events that ultimately defined the natural history of allo-implantation. Campbell and others tested and compared allo-implant systems against autografts and defined the fate of bone and cartilage grafts in experimental animals (Campbell et al. 1953). All this work set the stage for the establishment of reliable bone banks, especially in the USA, which provided a source of materials. Some clinicians around the world have reintroduced the concept of massive allografts as an orthotopic replacement for traumatically lost, tumour-ridden and diseased bones.

The system became relatively unused, however, apart from isolated centres, because of the complexity of the procedures; the unpredictability, in some eyes, of the results; the necessity to maintain a rigorously controlled bone bank; and the complications of infection, fracture and graft dissolution. Improvements have been made but allograft fracture and infection remain major issues.

So what of the future? The original views expressed by Barth and Curtis in their classic papers are very similar to the current understanding of the histological fate of non-vascularised auto- and allografts, except that the theory of induction developed by Urist and others in recent years gives promise for future investigation and use of bone grafts.There is a need for greater understanding of the cellular processes involved in bone vascularisation and cellular interaction at the host–donor interface. Cartilage preservation and joint physiology also need concentrated research to further our understanding of the biology of transplanted cartilage as a method of managing joint disease. The final breakthrough will be a safe method of altering the immune mechanism of the host or the immune composition of the donor so that the allograft is treated by the host in the same manner as an autogenic graft. Hopefully, research and development will lead to living joint transplants, microvascular anastomosis of host vessels to donor parts and transplants of viable epiphyseal growth plates.

References and Further Reading

Aebi M, Regazzoni P (Eds) (1989) Bone transplantation. Springer-Verlag, Berlin Heidelberg New York

Aho AJ (1973) Allogenic joint transplantation in the dog. Ann Chir Gynaecol Fenniae 62:226–233

Albee FH (1915) Bone Graft Surgery. Saunders, Philadelphia

Barth A (1893) Ueber histologische Befunde nach Knochenimplantationen. Arch Klin Chir 46:409–417

Brooks B, Hudson WA (1920) Studies in bone transplantations. An experimental study of the comparative success of autogenous and homogenous transplants of bone in dogs. Arch Surg 1:284–309

Burwell RG (1963) Studies in the transplantation of bone. The capacity of fresh and treated homografts of bone to evoke transplantation immunity. J Bone Joint Surg (Br) 45B:386–401

Burwell RG, Gowland G (1961) Studies in the transplantation of bone. I. Assessment of antigenicity. Serological studies. J Bone Joint Surg (Br) 43:814–819

Burwell RG, Gowland G (1962) Studies in the transplantation of bone. III. The immune responses of lymph nodes draining components of fresh homologous cancellous bone and homologous bone treated by different methods. J Bone Joint Surg (Br) 44:131–148

Campbell CJ, Brower T, Macfadden DG, Payne GB, Doherty J (1953) Experimental study of the fate of bone grafts. J Bone Joint Surg (Am) 35:332–346

Chalmers J (1959) Transplantation immunity in bone homografting. J Bone Joint Surg (Br) 41:160–179

Chandler HP, Penenberg BL (Eds) (1989) Bone Stock Deficiency in Total hip replacement. Slack Inc., Thorofare, NJ

Chase SW, Herndon CH (1955) The fate of autogenous and homogenous bone grafts. A historical review. J Bone Joint Surg (Am) 37:809–841

Curtis BF (1893) Cases of bone implantation and transplantation for cyst of tibia, osteomyelitic cavities and ununited fractures. Am J Med Sci 106:30–37

Czitrom AA, Gross AE (Eds) (1992) Allografts in orthopaedic practice. Williams and Wilkins, Baltimore

Danilevicius Z (1967) SS Cosmas and Damian. The patron saints of medicine in art. JAMA 201:1021–1025

Farrow RC (1948) Summary of the results of bone-grafting for war injuries. J Bone Joint Surg (Am) 30:31–39

Friedlander GE, Goldberg VM (Eds) (1989) Bone and Cartilage Allografts: Biology and Clinical Applications. Am Acad Orth Surg; National Inst Arth Musculoskeletal and Skin Diseases

Groves EWH (1917) Methods and results of transplantation of bone in the repair of defects caused by injury or disease. Br J Surg 5:185–242

de Heyde A (1684) Anatomia Mytuli, Subjecta centuria observatorium. Janssonio Waesbergios, Amsterdam

Lacroix P (1947) Organisers and the growth of bone. J Bone Joint Surg (Am) 29:292–296

van Leeuwenhoek A (1674) Microscopical observations about blood, milk, bones, the brain, spittle, cuticula, sweat, fat and tears. Philos Trans R Soc Lond 9:121–131

Levander G (1934) On the formation of new bone in bone transplantation. Acta Chir Scan 74:425–426

Lexer E (1908) Die Verwendung der freien Knochenplastik nebst Versuchen über Gelenkversteifung und Gelenktransplantation. Arch Klin Chir 86:939–954
Lexer E (1925) Joint transplantation and arthroplasty. Surg Gynec Obstet 40:782–809
van Meekeren J (1668) Heel–en geneeskonstige aanmerkingen. Commelijn, Amsterdam
Macewen W (1881) Observations concerning transplantations on bone. Proc R Soc Lond 32:232–247
Mankin HJ, Doppelt SH, Tomford WW (1983) Clinical experience with allograft implantation. The first ten years. Clin Orthop 174:69–86
Ollier XEL (1867) Traité expealrimental et clinique de la régénération des os et de la production artificielle du tissu osseux. 2 vols. Victor Masson et Fils, Paris
Phemister DB (1914) The fate of transplanted bone and regenerative power of its various constituents. Surg Gynecol Obstet 19:303–333
Urist MR (1953) The physiological basis of bone graft surgery, with special reference to the theory of induction. Clin Orthop 1:207–216
Von Walter P (1821) Wiedereinheilung der bei der Trapanation ausgebohrten Knochenscheibe. Journal der Chirurgie und Augen-Heilkunde 2:571

Part I
Basic Principles

1 Immunology of Bone Grafting

A.A. Czitrom

Experiments with bone cell or crushed bone fragment transplants placed under the kidney capsule in syngeneic and allogeneic combinations of mice demonstrate the transplantation reaction to bone allografts. At six weeks after transplantation in the syngeneic combination there is bone development under the kidney capsule. The bone is organised as an ossicle which forms marrow spaces populated by normal bone marrow in both the case of a bone cell transplant and a crushed bone transplant (Fig. 1.1a,b). In the allogeneic combination of mice, there is no bone present and a strong inflammatory response with many lymphocytes which can destroy the graft. This finding is seen in both the case of the bone cell transplant and the crushed bone allograft. (Fig. 1.1c,d).

These experiments show that an allograft bone graft put into a heterotopic site in a well vascularised bed such as under the kidney capsule will be destroyed. This represents the immunological transplant reaction against bone allografts. Although bone allografts are not vascular it is not surprising that they are rejected in the same way as vascular organ grafts. Many other avascular grafts such as skin allografts are rejected by the immune system. Thus, fresh bone allografts follow the general rules of recognition by the immune system.

When bone grafts are transplanted orthotopically, the transplantation reaction manifests itself by slower healing and remodelling of the allografts compared to autografts. This can be shown in a segmental graft model in dogs at 20 weeks (Czitrom 1989, 1992, Schwarzenbach et al. 1989). The autograft is well healed and remodelled while the allograft shows delayed healing and remodelling (Fig. 1.2a,b). This delay can be well shown and quantitated on fluorochrome labelled sections and no doubt represents the result of the immunological response against allograft bone.

It has been shown repeatedly in many laboratories that the outcome of bone allograft healing can be improved if there is histocompatible matching between donor and recipient (Goldberg et al. 1989, Stevenson 1987), and by immunosuppression of the host (Aebi et al. 1987, Schwarzenbach et al. 1989), hence autografts behave differently from allografts. In an orthotopic situation with an autograft, live cells that are transplanted will survive and produce bone, and the graft will be osteoconductive. In contrast, with an allograft the live cells will be rejected, the dead cells will not make bone but the graft will still be osteoconductive. These rules apply to fresh and not freeze-dried grafts, and explain the delay in healing of allografts by the lack of contribution of donor cells to the healing process.

The kinetics of bone formation by autografts and allografts has been well demonstrated (Chalmers 1959). New bone formation by the

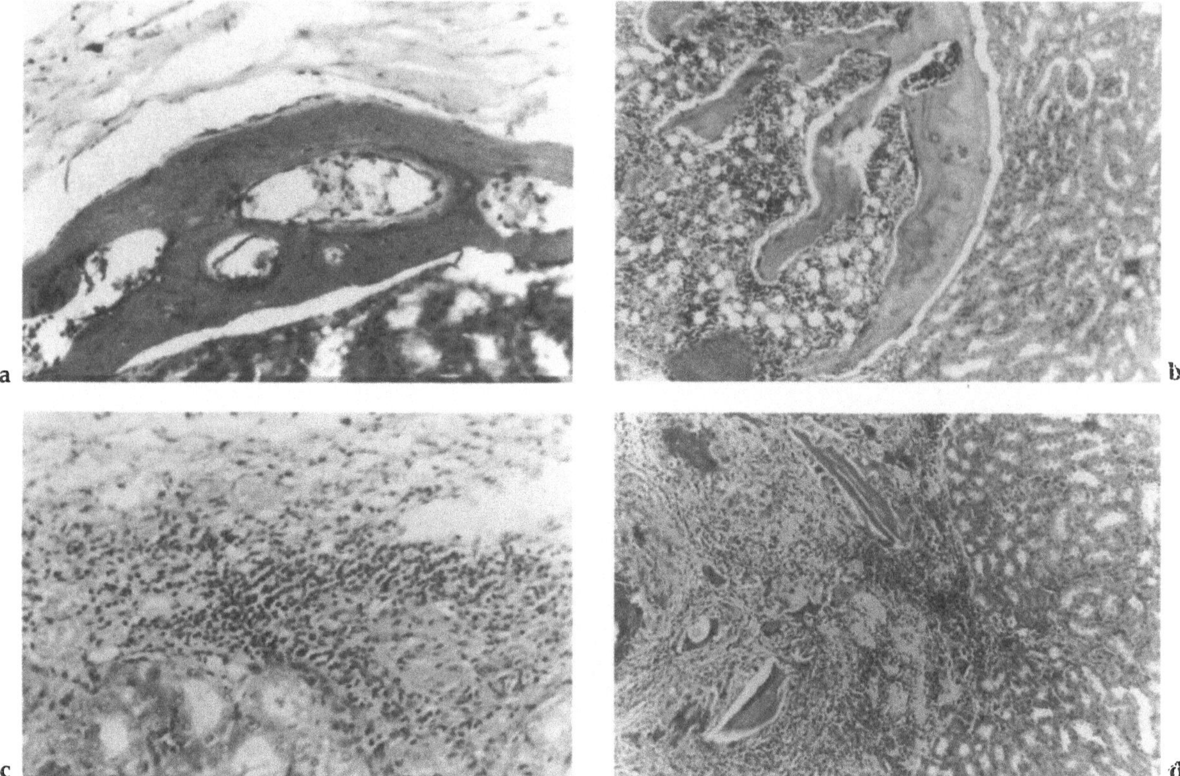

Fig. 1.1a–d. Transplantation reaction to heterotopic bone allografts. Clavaria-derived bone cells or crushed bone fragments were transplanted under the kidney capsule of syngeneic or allogeneic mice and assessed by histology at six weeks after grafting. **a** Photomicrograph of syngeneic bone cell transplant showing viable ossicle next to the normal kidney parenchyma (H & E,×325). **b** Photomicrograph of syngeneic crushed bone transplant showing viable bone with marrow spaces next to kidney tissue (H & E,×325). **c** Photomicrograph of allogeneic bone cell transplant showing no bone formation and an inflammatory response with lymphocytic infiltrate indicative of immune rejection (H & E,×325). **d** Photomicrograph of allogeneic crushed bone transplant showing fragmentation and destruction of bone with massive lymphocytic infiltrate. (Reproduced with permission from Czitrom (1992) and Czitrom (1989).)

allografts ceases after a short period of time, while the autografts form new bone for months. A late phase of bone formation by allografts does occur but it is inconsistent and not observed in all situations.The immune response which causes this delay in remodelling and rejection is both a cellular and a humoral response. The immune response is directed to cell surface allo-antigens. In the human, the major targets of the transplantation response are HL-A cell surface glycoprotein antigens (Tilney et al. 1989, Batchelor et al. 1990).

The cells that trigger and induce the transplantation response are antigen presenting cells. These cells trigger helper T cells which will amplify the cellular and humoral arms of the

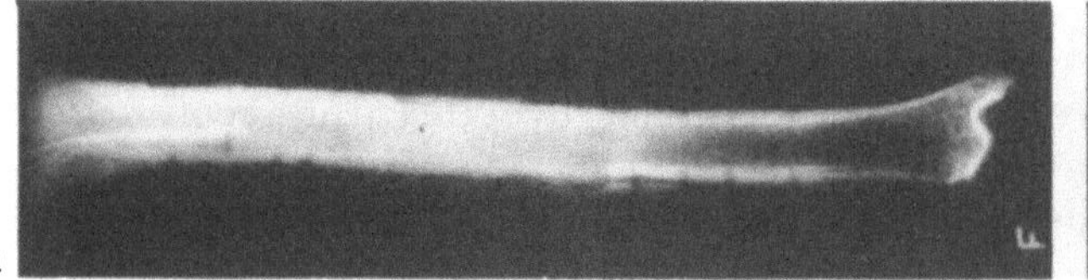

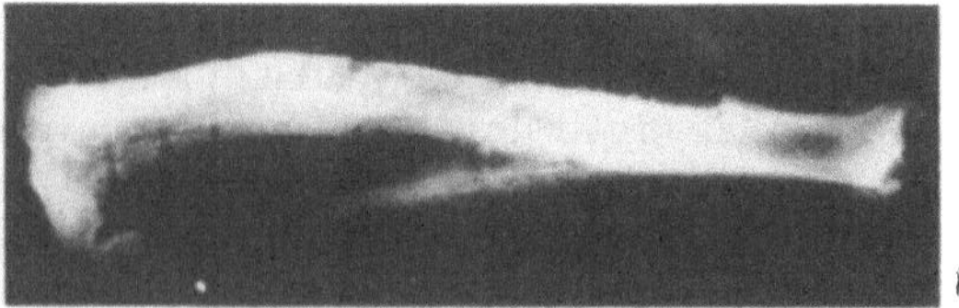

Fig. 1.2a,b. Outcome of orthotopic bone transplants. Diaphyseal autograft and allograft bone segments were transplanted in a dog model and assessed at 20 weeks. **a** Radiograph of autologous transplant showing complete healing and advanced remodelling. **b** Radiograph of allogeneic transplant showing incomplete healing and less remodelling. (Courtesy of Dr O. Schwarzenbach and Dr M. Aebi, University of Berne, Switzerland.)

immune system (Tilney et al. 1989). Examples of antigen presenting cells are dendritic cells and macrophages residing in haemopoietic organs.

The immunogenic cells in bone allografts are derived from bone marrow. This has been demonstrated by transplants carried out in chimeras (Esses et al. 1983). The experiment involves making a chimera where the donor is mixed so that the marrow will be type B and the bone will be type A. This can be done by irradiating an animal and replacing the entire marrow with marrow from a different animal. This chimera is used as a donor to bone graft a recipient type A. Rejection will occur with this combination. The reverse chimera with A type marrow and B type bone will not cause rejection when its bone is transplanted to an type A recipient. Thus, if the marrow is mismatched with the recipient the bone graft will be rejected. This demonstrates that it is the bone component of the bone allograft which is its immunogenic component. This experiment has been repeated recently with vascularised bone allografts and the outcome was exactly the same (Lee et al. 1989).

My laboratory has been involved in experiments aimed at the question of which cell in bone marrow is the one that initiates the immune response against bone allografts (Czitrom et al. 1985, 1988). The procedure used was to fractionate bone marrow by density gradients and enrich sequentially for cells that are the best stimulators of a standard in vitro T cell cytotoxicity assay.

The cells that were enriched sequentially as the best stimulators of T cell responses were a population of early granulocytic cells (Czitrom et al. 1988). These immunogenic cells had a large nucleus and the characteristic electron microscopy features of an early myeloid cell of the granulocyte lineage (Fig. 1.3).

Collectively, these immunogenic cells are termed passenger leucocytes and in various lymphoid organs they have been described as dendritic cells or macrophages. In the blood, B cells are very good antigen presenting cells; in the liver it is the Kupffer cells, in the skin it is Langerhans cells, in the brain it is oligodendrocytes, and in the bone marrow I think these cells are early granulocytes.

Can we reduce the immunogenicity of bone by eliminating some of these cells that are stimulating immune responses? It has been well shown that immunogenicity can be reduced by freezing or freeze-drying. The disadvantage of these methods is that they produce dead bone and therefore abolish osteogenesis from the bone graft. Selective removal of immunogenic cells without impairing the viability of osteogenic cells would be a better method to reduce the immunogenicity of allograft bone.

This can be done in various ways, one of which is organ cultures (Czitrom 1989). Preliminary experiments from my laboratory indicate that bone organ culture is effective in improving the incorporation of allogeneic bone grafts. The hypothesis is that this method is effective because it eliminates immunogenic cells while maintaining the viability of bone forming cells.

The immunogenic property of bone allografts can be demonstrated by their ability to induce

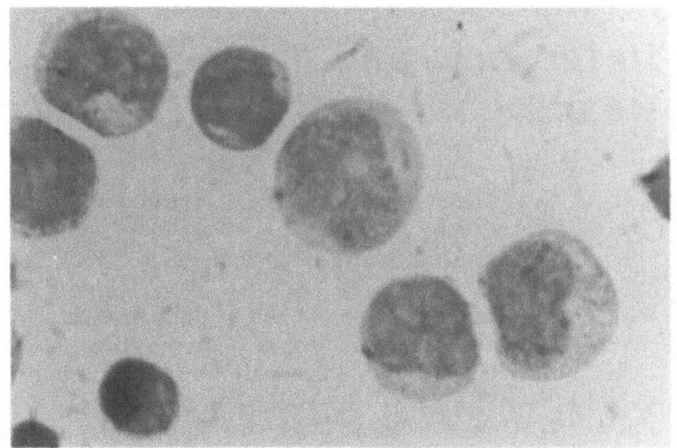
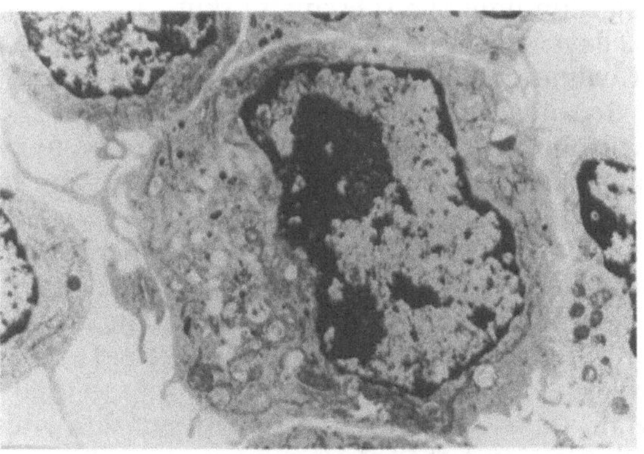

Fig. 1.3. The immunogenic cells in bone marrow. **a** Light micrograph of cells that were purified as the most immunogenic cells in bone marrow. They are large cells with irregular nuclei, a granular cytoplasm and a pale nucleolus (H & E,×650). **b** Electron micrograph of immunogenic cells in bone marrow showing surface projections, an indented nucleus, a small nucleolus, a prominent Golgi, numerous mitochondria, scanty lysosomes and many intracytoplasmic granules (×9940). The features are those of a myeloid cell of the granulocyte lineage. (Reproduced with permission from Czitrom (1992).)

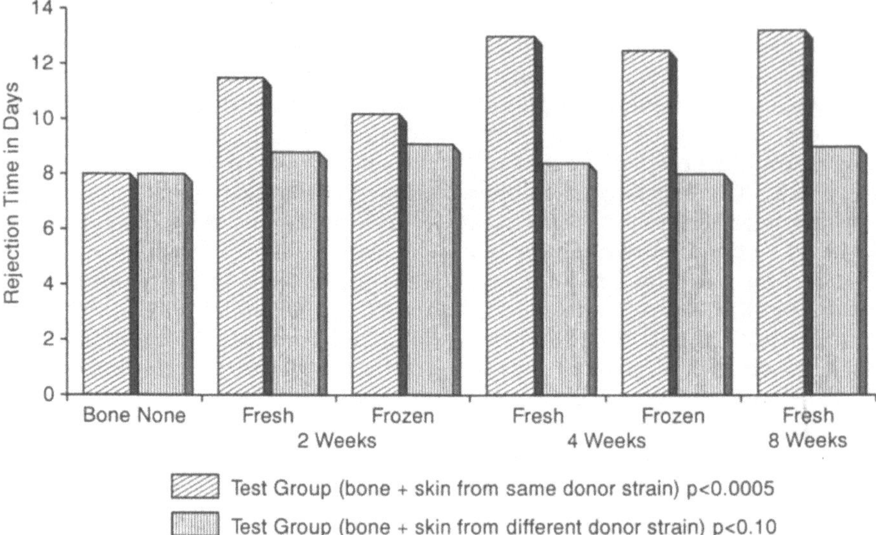

Fig. 1.4. Strain-specific immunosuppressive effect of bone allografts. Rejection time of skin allografts in normal rats without prior bone grafting is represented by the first pair of columns. Each pair of columns shows the rejection time of allogeneic skin grafted following the grafting of allogeneic bone derived from the same donor strain (first of two columns in each pair) versus the rejection time of a third-party skin graft derived from a different strain than the previous donor bone allograft (second of two columns in each pair). (Reproduced with permission from Langer et al.(1975).)

accelerated rejection of subsequent skin grafts (Chalmers 1959, Brooks et al. 1963). However, experimental evidence indicates that bone allografts can also suppress subsequent host immunity. Two sets of data will be reviewed briefly to demonstrate this point. First, Chalmers established that the rejection time of first set skin grafts in rats was 11 days, and that of second set skin homografts was six days. After grafting rats with a fresh bone allograft, a subsequent skin graft was rejected in an accelerated fashion. However, if the animal was grafted with freeze-dried bone, the rejection time of a subsequent skin allograft was 12 days, suggesting an immunosuppressive effect (Chalmers 1959).

Second, experiments carried out by Langer and Czitrom suggested that the fresh and frozen bone allografts induce specific immunosuppression to subsequent skin grafts (Fig. 1.4). The two columns illustrated represent the comparison of rejection times of skin grafts that are of the same strain as the bone grafts versus skin grafts that are of third-party origin. If bone was grafted at two weeks, four weeks or eight weeks prior to a skin graft, there was a delay in subsequent skin graft rejection and this phenomenon was observed with the fresh and frozen bone allografts (Langer et al. 1975).

The results indicate that bone allografts, in addition to being immunogenic, are able to provide specific suppression of the immune response under certain conditions that are not well understood. The future of bone transplantation is without doubt the vascularised joint and limb allograft. This will only be possible when the induction of specific immunosuppression becomes possible or when immunosuppressive drugs that have no side effects are found.

The eradication of AIDS remains a great challenge. Fresh grafts will probably not be widely used until this threat is totally eliminated.

References and Further Reading

Aebi M, Schwarzenbach O, Regazzoni P (1987) Long-term versus short term immunosuppression in experimental bone allotransplantation. Trans Orthop Res Soc 11(2):283–284

Batchelor JR, Kaminski E, Lombardi G, Goldman JM, Lechler RI (1990) Individual variation in alloresponsiveness and the molecular basis of allorecognition. Hum Immunol 28(2):96–103

Brooks DB, Heiple KG, Herndon AH, Powell AE (1963) Immunological factors in homogeneous bone transplantation. IV. The effect of various methods of preparation and irradiation on antigenicity. J Bone Joint Surg (Am) 45:1617–1626

Chalmers J (1959) Transplantation immunity in bone homografting. J Bone Joint Surg (Br) 41:160–179

Czitrom AA, Axelrod T, Fernandes B (1985) Antigen presenting cells and bone allotransplantation. Clin Orthop 197:27–31

Czitrom AA, Axelrod TS, Fernandes B (1988) Granulocyte precursors are the principal cells in bone marrow that stimulate allospecific cytolytic T-lymphocyte responses. Immunology 64(4):655–660

Czitrom AA (1989) Bone transplantation, passenger cells and the major histocompatibility complex. In: Bone Transplantation, Aebi M, Regazzoni P (eds). Springer-Verlag, Berlin Heidelberg New York, p103

Czitrom AA (1992) Immunology of bone grafting. In: Allografts in Orthopaedic Practice, Czitrom AA, Gross AE (eds). Williams and Wilkins, Baltimore

Esses SI, Halloran PF (1983) Donor marrow-derived cells as immunogens and targets for the immune response to bone and skin allografts. Transplantation 35(2):169–174

Goldberg VM, Powell A, Shaffer JW, Zika J, Stevenson S, Davy D, Heiple K (1989) The role of histocompatibility in bone allografting. In: Bone Transplantation, Aebi M, Regazzoni P (eds). Springer-Verlag, Berlin Heidelberg New York, p126

Langer F, Czitrom A, Pritzker KP, Gross AE (1975) The immunogenicity of fresh and frozen allogeneic bone. J Bone Joint Surg (Am) 57:216–220

Lee WPA, Yaremchuk M, Manfrini M, Pan YC, Randolph MA, Weiland AJ (1989) Prolonged survival of vascularised limb allografts from chimera donors. Trans Orthop Res Soc 13(2):412 Abstract

Schwarzenbach O, Regazzoni P, Aebi M (1989) Segmental vascularised and non-vascularised bone allografts. In: Bone Transplantation, Aebi M, Regazzoni P (eds). Springer-Verlag, Berlin Heidelberg New York, pp78–81

Stevenson S (1987) The immune response to osteochondral allografts in dogs. J Bone Joint Surg (Am) 69:573–582

Tilney NL, Kupiec-Weglinski JW (1989) Advances in the understanding of rejection mechanisms. Transplant Proc 21(1):10–13

2 Natural History of Autografts and Allografts

V.M. Goldberg

The increasing demand for bone graft material in revision arthroplasty and tumour reconstruction requires a thorough understanding of the natural history and biology of autografts and processed allografts. A number of reports in the literature indicate that bone grafts may have a failure rate approaching 25% (Mankin et al. 1987). The failure is a result of inadequate revascularisation and mineralisation of the bone graft. Immunological rejection of bone allografts still remains an important consideration in the failure of these grafts. The natural history of bone grafts has been defined from extensive laboratory studies and clinical experience (Goldberg and Stevenson 1987). This chapter will review the present knowledge in this field.

The Function of Bone Grafts

The functions of autogenous or allogeneic bone grafts are listed in Table 2.1. *Osteogenesis* is a process where the cells of the graft which survive produce new bone. When the graft is not immediately revascularised, only the superficial cells survive for a period of time supported by diffusion.

Table 2.1. Stages of bone graft incorporation (autograft or processed allograft)

Stage 1	Inflammation	Hours
	↓ ↑	
Stage 2	Pre-osteoblasts Osteoblasts	Days
	↓ ↑	
Stage 3	Osteoinduction	Weeks
	↓ ↑	
Stage 4	Osteoconduction	Months
	↓ ↑	
Stage 5	Mechanical support	Years

These cells usually undergo necrosis in time and the entire graft is replaced by a process of creeping substitution. *Osteoinduction* as described by Urist is a process where protein mediators present in the matrix of the bone graft induce mesenchymal stem cells of the host to migrate into the graft. These cells become osteoblasts which produce mineral to reconstitute the graft. There have been major advances recently in identifying these mediators, since bone is the major reservoir for these growth factors. *Osteoconduction* is a process where the graft acts as a trellis or roadway for the ingrowth of blood vessels and cells. Structural support is the ultimate function of bone grafts; this may be provided immediately by cortical bone or developed as a response to environmental

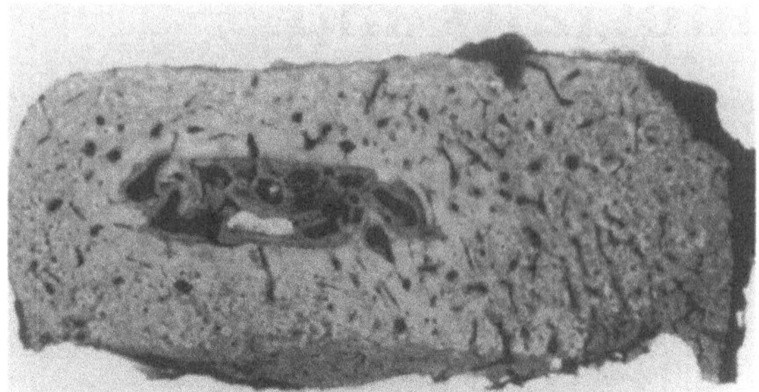

Fig. 2.1. Photomicrograph of the interface between graft and host demonstrating early union and creeping substitution (von Kossa stain, ×70).

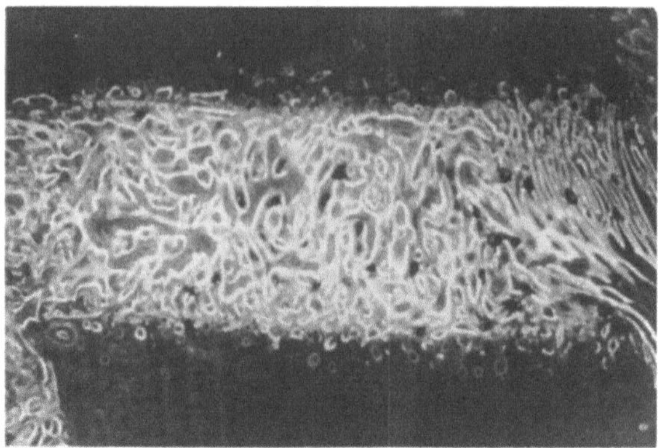

Fig. 2.3. Transverse section through the diaphysis of a cortical autograft demonstrating significant osteoporosis.

cues. Bone is unique compared to other organs as it has the capacity to regenerate, driven by the biological responses of cell recruitment and vascularity as well as the requirements mandated by environmental loads.

Stages of Bone Graft Incorporation

Bone graft incorporation is a continuum of five stages which are in dynamic equilibrium with each other (Table 2.1). The initial stage is an inflammatory process, with the ingrowth of mesenchymal stem cells which become osteoblastic cells. Next in sequence but also proceeding in equilibrium are osteoinduction, osteoconduction and the mechanical support stage. These stages are seen during the entire lifetime of the graft. Cortical grafts may remodel for years so bone scans may be active for a prolonged period of time.

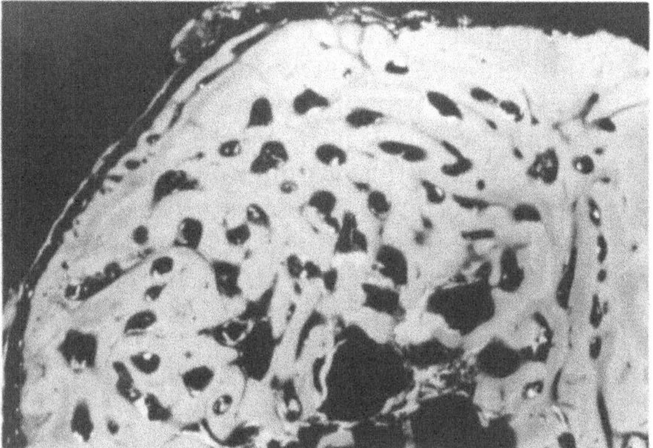

Fig. 2.2. Fluorochrome labelled photomicrograph of a diaphyseal cancellous autograft demonstrating complete replacement by viable bone (×40).

The incorporation of bone graft reflects its specific structural characteristics. Cancellous autograft has been classically considered the "gold" standard. The initial stage is characterised by inflammation, haemorrhage and neovascularisation. Osteoinductive proteins induce host preosteoblasts to migrate into the graft with the new blood vessels. Osteoblastic activity goes on concomitantly with osteoclastic resorption. Graft-host union occurs early and by the end of the first month after surgery significant new bone is deposited around the old necrotic trabeculae (Fig. 2.1). This creeping substitution ultimately replaces the entire cancellous autograft with new bone. The remodelling which occurs over many months is a response of the bone to its mechanical environment under the influence of Wolff's Law. A fluorochrome labelled cancellous autograft completely replaced by viable bone nine months after surgery is seen in Fig. 2.2. These grafts are highly efficient in filling defects and reconstituting structure.

The incorporation of cortical autografts parallels the sequence of incorporation of cancellous bone but because of their dense structure revascularisation is delayed. The initial process is characterised by bone resorption, an osteoclastic function which is brought into the graft by new blood vessels (Fig. 2.3). This resorption, seen histologically as extensive cortical porosity, significantly weakens the bone graft from approximately six weeks to six months after surgery. Remodelling and mineralisation of the graft is a

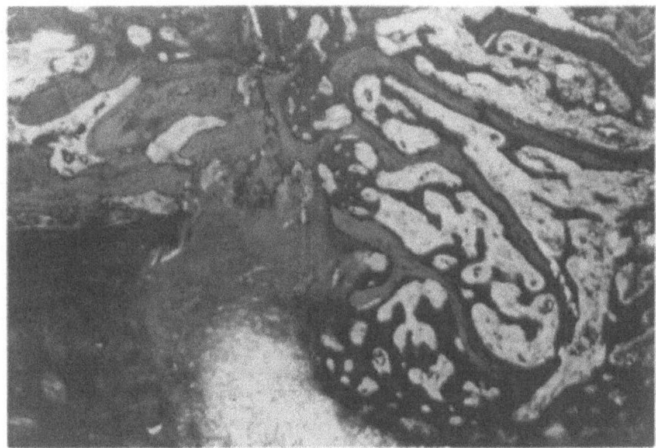

Fig. 2.4. Photomicrograph of a cortical autograft demonstrating a mixture of necrotic and viable bone six months after surgery (H & E ×12).

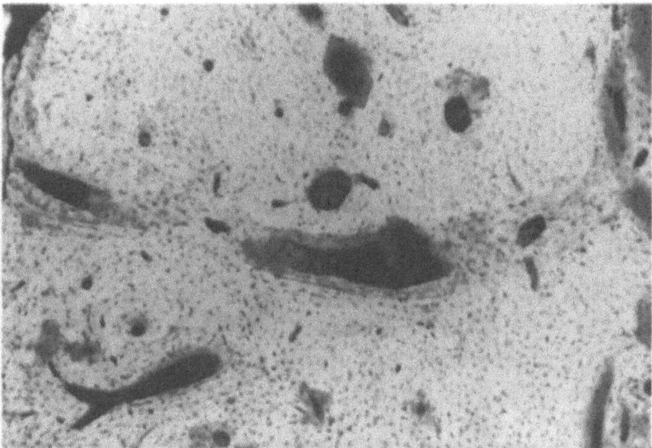

Fig. 2.5. High-power photomicrograph of a frozen cortical allograft nine months after surgery demonstrating a mixture of viable and old necrotic bone (H & E ×70).

prolonged process which results in a bone structure that is a mixture of necrotic and viable bone (Fig. 2.4).

Autografts do have difficulties which include limited supply, size and shape. Donor site morbidity also has been reported as significant. Finally, articular surfaces cannot be reconstructed. Because of these problems, allografts have been used since there is no donor site morbidity, no loss of normal structures, and no limits on size, shape and quantity. However, allografts have problems including no graft osteogenesis capability and, most important, the potential transmission of infectious agents. Additionally, an immune response may be induced in the host which will compromise the graft.

Fresh bone allografts are rarely used any more since they do induce a significant immunological response which compromises the procedure. Because of this, bone banking technology has used freezing or freeze-drying to process and preserve the bone. These procedures reduce the immune response without compromising the osteoinduction and/or osteoconductive properties of the bone. The incorporation of frozen or freeze-dried cancellous allografts parallels fresh autogenous material except the healing is slower and no initial osteogenesis occurs. Complete resorption and replacement is significantly slower and the allograft may never be completely remodelled. However, it is an effective material as a filler for bone defects.

Fresh frozen or freeze-dried cortical allografts incorporate by biological processes which parallel autogenous cortical grafts. Resorption is a significant event and the incorporation is a prolonged process with a mixture of old non-viable and new viable bone present almost throughout the lifetime of the individual (Fig. 2.5). Frozen cortical allografts have little if any osteoconductive material. Freeze-dried cortical allografts should not be used as structural immediate weight-bearing material as the preservation process destroys this capacity.

One of the major issues in bone banking is the transmission of infectious agents and most specifically the HIV virus. The role of irradiation in sterilising the graft remains controversial. Clearly it is an effective technique at high dosage to destroy transmissible agents, but at these levels the irradiation may destroy the biological functions of the bone graft. Additional experimental work is necessary to define the exact role of this method of bone graft processing.

Conclusions

The biological functions of bone grafts must be understood in order to use the best material for each specific clinical problem. Osteogenesis is of central importance in most clinical problems. Only successfully vascularised, fresh autografts provide this function. Fresh cancellous autografts provide direct osteogenesis but only for a short period of time while the superficial osteocytes survive by diffusion. Osteoconduction is a function provided by both cancellous autografts and allografts. Frozen cortical allografts have little or no osteoinductiveness, while freeze-dried bone appears to retain this function. Both freezing and freeze-drying inhibit the immunological reactivity of the allograft. Chemically processed bone, e.g. AAA bone, provides material high in inductive function with no immunological reactivity. Osteoconduction is preserved in most grafts and acts as a major source of new blood vessels and cells. Cortical bone grafts are best although cancellous bone influenced by Wolff's Law ultimately becomes a significant weight bearing structure.

The success or failure of a bone grafting procedure is dependent upon the biological issues of revascularisation and the concomitant mineralisation, the technical aspects of host bed preparation and fixation of the graft to the host bone. Finally, because allografts have significant clinical problems, alternative technologies are being explored. These include the use of tissue engineering, growth peptides e.g. the bone morphogenetic protein family, and alternative biomaterials such as ceramics. The future is bright for new predictable procedures to manage bone loss problems.

References and Further Reading

Goldberg VM, Stevenson S (1987) Natural history of autografts and allografts. Clin Orthop Rel Res 225:7–16

Mankin HJ, Gebhardt MC, Tomford WW (1987) The use of frozen cadaveric allografts in the management of patients with bone tumours of the extremities. Orthop Clin North Am 18(2): 275–289

3 Histological Evaluation of Allograft Incorporation after Cemented and Non-cemented Hip Arthroplasty in the Goat

P. Buma, W. Schreurs, D. Versleyen, R. Huiskes and T.J.J.H. Slooff

One of the major complications associated with the very successful operation of artificial hip replacement is the bone stock loss that may be found after aseptic loosening of the prosthesis. This complication hampers the successful fixation of a new prosthesis. In the past, orthopaedic surgeons tried to resolve this problem by the use of revision prostheses with longer stems and/or reinsertion of a standard prosthesis with more cement. However, these procedures are not very elegant, since in a second revision procedure further bone stock deficiencies will be found, and the problems for the orthopaedic surgeon will be increased.

The challenge to the orthopaedic surgeon is to restore the original bone stock during revision surgery. This goal can only be reached if the lost bone stock is supplemented with graft material, which has to be subsequently incorporated into a new bony structure. Different types of bone may be used in the grafting procedure: cortical bone, cancellous bone, chips or larger intact pieces of bone, fresh or deep frozen and lyophilised bone, and autograft or allograft bone (Oikarinen and Korhonen 1979, Mellonig et al. 1981, Kohler et al. 1986, Kakaiya and Jackson 1990, Alho et al. 1989, Wilson et al. 1989).

We present the first results of a study in the goat which investigated the incorporation of graft in combination with cemented and non-cemented prostheses. In our orthopaedic department, chip-like trabecular bone grafts were previously used with good results to restore the bone defects associated with acetabular protrusion (Slooff et al. 1984). The present study was designed to evaluate the feasibility of femoral revision surgery with morsellised allograft chips. Allograft bone was harvested from the sternum of donor goats. A straight stem cemented prosthesis or an experimental hydroxyapatite-coated titanium non-cemented (Osteonics) prosthesis was used. The operation technique and the first biomechanical results are described in Chapter 18.

Methods and Materials

In order to visualise new bone formation, the goats were labelled with different types of fluorochrome during the study. The animals were anaesthetised and fixed by perfusion. Thick sections of the femora were made with the

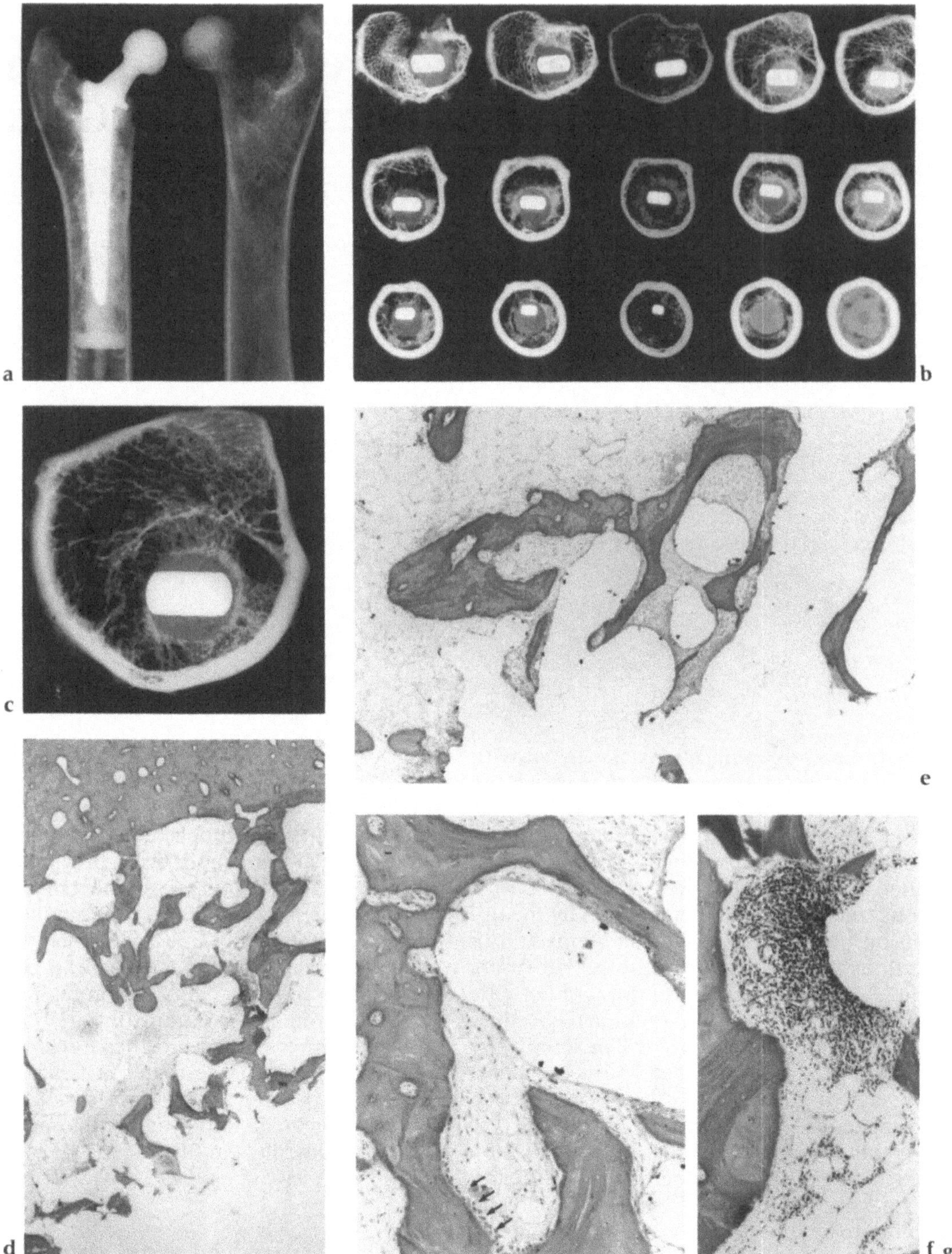

Fig. 3.1. a Roentgenograph of cemented prosthesis with intramedullary augmentation with allograft chips (×0.55). **b** Roentgenograph of thick sections from proximal (upper row, left) to distal levels (lower row, right) (×0.65). **c** Enlargement of fourth section shown in b. Note trabecular structure between the cement layer and the pre-existing trabecular bone (×1.7). **d** Low magnification haematoxylin–eosin stained section of mid-shaft level. Note trabecular system that connects the cement layer with the pre-existing compact cortical host bone (×18). **e** Interlock between cement and graft. The cement was resolved during histological preparation of the section but could be recognised by the barium sulphate (×35). **f** Detail of interlock between cement and graft. Dark areas in the bone are graft, with empty osteocyte lacunae, the lighter areas represent new bone. Note thin soft tissue interface between cement and graft, and area of active bone apposition (arrows) (×90). **g** Local accumulation of lymphocytes (×80).

prosthesis in situ. Roentgenographs of the sections were made, and the surfaces of the sections were stained with basic fuchsin. For the first evaluation of graft incorporation we applied fluorescence microscopy at low magnification. For the visualisation of the hydroxyapatite interface between prosthesis and new bone we used a Biorad confocal microscope. Subsequently, the sections were prepared for further histology. Undecalcified sections were stained with basic fuchsin and embedded in methylmethacrylate; thin sections were stained with haematoxylin–eosin. In the histological analysis, special attention was paid to graft lysis, incorporation of the graft, localisation of osteoclasts and osteoblasts, contact area of bone with cement and hydroxyapatite, the presence and extent of the periosteal reaction, immunological reactions to the graft, vascular invasion of the graft and the presence of infections.

Operations were performed on 12 goats; six received a cemented prosthesis and six a non-cemented prosthesis. Three goats in each group were sacrificed after six weeks and three after 12 weeks. Since not all histology was completed, only the results of the first two specimens can be presented here. Three specimens showed clear signs of infection after histological analysis: two cemented prostheses and one in the non-cemented group. Infected prostheses showed large numbers of polymorphonuclear leucocytes, lysis of the graft and bone loss. All infected goats were from the six week groups. These infected specimens were not analysed further so that of the six week cemented arthroplasties, only one remained for histological analysis. Therefore this chapter will focus on the results that were found after 12 weeks.

Cemented Arthroplasty

The sections showed that the prosthesis was surrounded by a homogeneous cement mantle and a layer of impacted pieces of trabecular bone graft 3mm thick. In the impacted graft, small pieces of original trabecular bone could be recognised. The spaces in between the graft were initially filled with a fibrin clot. After incorporation of the graft, there was a very good interlock between the pre-existing trabecular bone, the graft and the cement, as if a new trabecular structure had been formed (Fig. 3.1b,c). The basic fuchsin and the fluorescence microscopy (at low magnification) showed a trabecular structure in close contact with the cement mantle. Calcein green fluorescence labelling indicated that there was active bone turnover before sacrifice of the animal. At more distal levels both the roentgenographs and the fluorescence microscopy again showed that a good interlock existed between the cement, the graft and the compact cortical bone (Fig. 3.1b). Indeed, histology at low magnification showed trabecular bone interconnecting the cortical bone with the cement mantle (Fig. 3.1d). Although the cement was resolved during preparation of the histology, remnants of the barium sulphate remained in the sections. At all levels studied, the cement had penetrated into the graft (Fig. 3.1e,f). At high magnification, the original graft could be recognised by the empty osteocyte lacunae (Fig. 3.1g,d). It appeared that the trabecular bone was a mixture of graft and new bone which was characterised by the filled osteocyte lacunae. Rows of osteoblasts indicated that active bone formation took place (Fig. 3.1f). Changing numbers of lymphocytes were always seen in the graft, both with the cemented and non-cemented prostheses (Fig. 3.1g). No bone destruction was apparent in the direct vicinity of these cells which were probably a result of a mild immunological reaction to the allograft.

Occasionally, a relatively large protrusion of the cement mantle was seen in contact with the cortical bone. Locally, the new bone was contacting the cement mantle without a soft tissue interface. However, a thin soft tissue interface was generally present between the cement layer and the graft. Also macrophages were found at the interface between cement and graft indicating an immunological reaction to the cement mantle. The fluorescence microscopy at higher magnification showed active bone formation in the direct vicinity of the cement mantle.

The Non-cemented Arthroplasty

The roentgenograph of the femur after 12 weeks showed that the tip of the prosthesis was in very close contact with the cortex of the femur (Fig. 3.2a). Roentgenographs of the thick sections showed a very intimate contact between the graft and the prosthesis, particularly at the proximal levels (Fig. 3.2b). Fluorescence microscopy and basic fuchsin stained sections showed that the calcein green had been incorporated into the

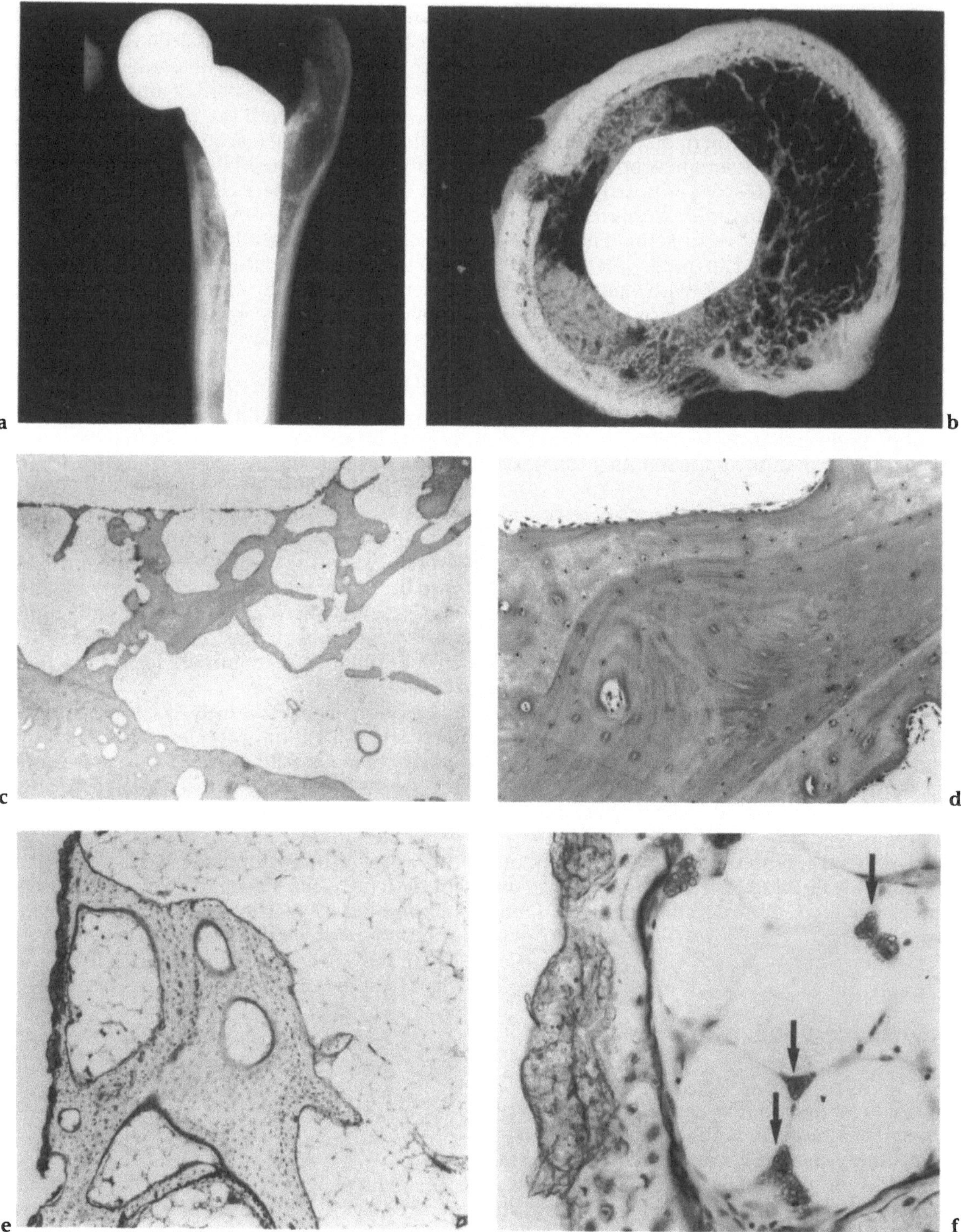

Fig. 3.2. a Roentgenograph of non-cemented prosthesis after insertion in the femur (×0.55). **b** Roentgenograph of thick section through proximal part of femur. Note very intimate contact between the trabecular bone and the corners of the prosthesis (×1.7). **c** Low magnification haematoxylin-eosin stained proximal section through lateral part of the femur. Note intimate contact between trabecular structure and the corner of the prosthesis (×18). **d** Detail of c, showing that the trabecular bone is a mixture of graft (empty osteocyte lacunae, darker stained bone) and new bone (filled osteocyte lacunae) (×120) **e** Interlock between the layer of hydroxyapatite and bone (×64). **f** Detail of **e**. Note isolated thydroxyapatite crystals between the medullary fat cells (arrows), and very intimate contact of new bone with the hydroxyapatite layer (×300).

trabecular bone. The new bone appeared to be in very close contact with the prosthesis (Fig. 3.2c,e,f). Confocal microscopy confirmed that there was a very good interlock between the new bone, the hydroxyapatite layer and the prosthesis. After removal of the prosthesis, this layer of hydroxyapatite stuck to the newly formed bone, which was present in between the layer of hydroxyapatite and the graft (Fig. 3.2e,f). With fluorescence microscopy of the same area, active bone apposition was still seen. Trabeculae connected the prosthesis with the host cortical bone, particularly at the lateral side of the prosthesis (Fig. 3.2c). At high magnification it became clear that this trabecular bone was again a mixture of graft and newly formed bone (Fig. 3.2d).

Roentgenographs of thick sections indicated that at midshaft and distal levels the graft was not in contact with the prosthesis and cortical bone. Ingrowth of new bone only appeared to have taken place in the region where the tip of the prosthesis was in close contact with the cortex of the bone. The graft had resolved and young trabecular woven bone was growing locally into the space between prosthesis and cortical bone. A bony bridge between prosthesis and cortical bone had only been formed where the tip of the prosthesis was in close contact with the lateral cortex of the femur.

Conclusions

The histological protocol used in the present study allowed a good evaluation of the graft incorporation. With this procedure of intramedullary augmentation with morsellised allograft chips a very good homogeneous wall of chips can be made. Signs of incorporation were found both in the graft in combination with the cemented arthroplasty, and in the graft around the non-cemented prosthesis.

References and Further Reading

Alho A, Karaharju EO, Korkala O, Laasonen EM, Homstrom T, Muller C (1989) Allogenic grafts for bone tumour. 21 cases of osteoarticular and segmental grafts. Acta Orthop Scand 60(2):143–153

Kakaiya RM, Jackson BJ (1990) Regional programs for surgical bone banking. Clin Orthop 251:290–294

Kohler P, Kreicbergs A, Stromberg L (1986) Physical properties of autoclaved bone. Acta Orthop Scand 57: 141–145

Mellonig JT, Bowers GM, Cotton WR (1981) Comparison of bone graft materials. Part II. New bone formation with autografts and allografts: A histological evaluation. J Periodontol 52:297–302

Nelson IW, Bulstrode DJK, Mowat AG (1990) Femoral allografts in revision of hip replacement. J Bone Joint Surg (Br) 72:151–152

Oikarinen J, Korhonen L (1979) The bone inductive capacity of various bone transplanting materials used for treatment of experimental bone defects. Clin Orthop 140:208–215

Slooff TJJH, Huiskes R, van Horn J, Lemmens AJ (1984) Bone grafting for total hip replacement for acetabular protrusion. Acta Orthop Scand 55:593–594

Wilson MG, Nikpoor N, Aliabadi P, Poss R, Weissman BN (1989) The fate of acetabular allografts after bipolar revision arthroplasty of the hip. A radiographic review. J Bone Joint Surg (Am) 71:1469–1479

4 A New Artificial Bone Graft Material

A. Kocialkowski

Three prospective international studies were carried out in 64 patients to determine the efficacy, benefits and safety of an artificial bone graft material. Several orthopaedic centres in France, Germany, Belgium and the United Kingdom participated in the studies. The material used was Collagraft, a mixture of granular porous calcium phosphate ceramic (Zimmer Inc., Warsaw, Indiana) with the addition of bovine fibrillar collagen (Collagen Corp., Palo Alto, California) and autogenous bone marrow. This mixture has been used as a bone graft in 42 patients with delayed union and non-union of long bones and in 22 acute fractures of long bones. In summary 10 out of 64 patients (16%) treated with the ceramic/collagen/marrow mixture failed to unite at one year and eight of them required re-operation. No adverse short-term reactions such as wound drainage, erythema and inflammation, or long-term graft site complications such as infection were seen. The laboratory tests performed in all patients six months and one year after grafting did show some minor abnormalities, none of which were considered to be clinically significant.

Materials and Methods

The ceramic used was a biphasic formulation of 65% tricalcium phosphate in a granular form with granules 0.5 to 1.0mm in diameter (Zimmer Inc., Warsaw, Indiana). The fibrillar collagen was a highly purified collagen obtained from bovine dermis, supplied in a gel form and contained in a syringe. The collagen and ceramic were thoroughly mixed in a bowl and autologous bone marrow aspirated either from the iliac crest or from the fracture site was added in a ratio of 5 to 7ml of bone marrow to 7ml of ceramic/collagen mixture.

To be eligible for the prospective delayed union study (Study 1), patients had to have a non-union or delayed union of a long bone (or a failed arthrodesis), be aged between 18 and 70 years and in good general health. Patients with a recent history of osteomyelitis, malignancy, metabolic bone disease and those using corticosteroids or immunosuppressive agents were excluded.

In this prospective randomised study (including centres in France, Germany, Belgium and the United Kingdom) 43 delayed union cases were included. Twenty-four fractures were treated with Collagraft and 19 received autogenous bone. In a second, prospective non-randomised multicentre study (Study 2), 18 delayed union cases were included, all treated with Collagraft. In a third, open multicentre study (Study 3), 22 acute fractures (treated within 30 days) were treated with Collagraft.

The follow-up assessment was performed at six weeks, three, six and 12 months, and finally when union was solid or the fracture had failed to unite. Each assessment included two radiographs

Table 4.1. Radiographic assessment

	Points
Appearance of graft site	
Total resorption (>90%)	0
Mostly resorbed (50–90%)	1
Largely intact with minimal bone ingrowth	2
Largely intact with bone ingrowth	3
Complete incorporation	4
Quality of union at proximal end	
Non-union/total resorption (>90%)	0
Mostly resorbed (50–90%)	1
Largely intact with minimal bone ingrowth	2
Largely intact with bone ingrowth	3
Radiographic early union	4
Radiographic solid union	5
Total obliteration of interface	6
Quality of union at distal end	
Non-union/total resorption (>90%)	0
Mostly resorbed (50–90%)	1
Largely intact with minimal bone ingrowth	2
Largely intact with bone ingrowth	3
Radiographic solid union	4
Radiographic solid union	5
Total obliteration of interface	6
Total score	16

Table 4.2. Assessment of activities of daily living (ADL)

Non-weight-bearing fractures
Washing face and hair
Brushing teeth
Tying shoelaces
Putting on a belt
Wringing out a washcloth
Throwing a ball
Making a bed
Reaching a high shelf
Pushing open a revolving door
Personal hygiene
Lifting weight

Weight-bearing fractures
Walking
Running
Getting up from a chair
Climbing stairs
Putting on trousers while standing
Getting in/out of a car
Sexual functions
Picking up items from the floor
Lifting weight

Slight impairment: 1 to 3 activities impaired
Moderate impairment: 4 to 7 activities impaired
Severe impairment: 8 or more activities impaired

Table 4.3. Pain assessment

1. None
2. Slight – no analgesics
3. Moderate – occasional analgesics
4. Severe – daily analgesics
5. Disabling

of the graft site at each visit (Table 4.1), an assessment of the activities of daily living (Table 4.2), a clinical assessment of post-operative pain and a record of any analgesics required for pain (Table 4.3). Any signs of inflammation, a hypersensitivity reaction to the graft, or of delayed wound healing were also recorded. The haematological and biochemical blood tests were performed at six and 12 months after the grafting procedure.

Preliminary Results

Study 1

In study 1 (the prospective randomised delayed union study), 24 patients were grafted with Collagraft and 19 with autogenous bone. The mean age of the patients was 35 years (range 21 to 68) in the Collagraft group, and 42 years (range 23 to 59) in the autogenous bone group. The mean time from the injury to the grafting operation was 25 weeks (range 12 to 36) in the Collagraft group, and 29 weeks (range 12 to 32) in the autogenous group. The mean follow-up period was 15 months (range 12 to 24) in the Collagraft group, and 13 months (range 12 to 24) in the autogenous bone group. Stabilisation of the fracture was performed with a plate in fourteen (59%), with an external fixator in seven (29%) and an intramedullary nail in three (12%) patients in the Collagraft group; in the autogenous group with a plate in eleven (60%), an external fixator in six (30%) and a nail in two (10%) patients. The Collagraft recipient site included the tibia in ten (42%), the femur in eight (33%) and the humerus in six (25%). The cancellous autograft site included the tibia in seven (37%), the femur in four (21%), the humerus in four (21%) and the forearm in four (21%).

An analysis of the outcome measures shows that at six months, 12 out of 24 (50%) patients implanted with Collagraft, and three out of 19 (16%) patients grafted with autogenous bone had slight to moderate pain at the fracture site. At one year, eight (33%) patients grafted with Collagraft and two patients (10%) grafted with autogenous bone had slight to moderate pain (Table 4.4).1 The assessment of activities of daily living (ADL) showed that at six months, 12 patients (50%) grafted with Collagraft and nine patients (47%) grafted with autogenous bone had slight to

Table 4.4. Randomised delayed union study: pain scoring

Graft	Months	Pain score			
		None	Slight	Moderate	Severe
Collagraft	6	12 (50%)	11 (45%)	1	0
(n = 24)	12	16 (66%)	8 (33%)	0	0
Autograft	6	16 (84%)	1 (5%)	2	0
(n = 19)	12	17 (89%)	1 (5%)	1	0

Table 4.5. Randomised delayed union study: activities of daily living (ADL) score

Graft	Months	ADL score			
		None	Slight	Moderate	Severe
Collagraft	6	11 (45%)	7 (29%)	5 (21%)	1
(n = 24)	12	14 (58%)	9 (37%)	1	0
Autograft	6	10 (53%)	9 (47%)	0	0
(n = 19)	12	17 (89%)	2	0	0

moderate impairment; at one year, ten patients (41%) grafted with Collagraft and two patients (10%) grafted with autogenous bone had slight impairment of ADL (Table 4.5). A mean radiographic score of 11.6 at six months and 13.7 at one year was obtained in the Collagraft group, and a mean of 12.9 at six months and 14.2 at one year in the autogenous bone group (0 = total resorption, 12 = united, 16 = perfect bony union) (Table 4.6). At one year, four out of 24 delayed union cases (16%) which had been grafted with ceramic/collagen/marrow failed to unite and required re-operation: two femoral delayed unions (one

Table 4.6. Randomised delayed union study: radiographic score

Graft	Months	Radiographic score (mean with standard deviation)
Collagraft	6	11.6±3.3
(n = 24)	12	13.7±3.0
Autograft	6	12.9±2.6
(n = 19)	12	14.2±1.2

treated with an external fixator and one with a dynamic hip screw) required further operation and were stabilised with an intra-medullary nail (Fig. 4.1). Two tibial delayed unions required re-operation: one after high tibial osteotomy was treated with a knee replacement, and one tibial fracture originally treated with an external fixator was plated (Fig. 4.2). Only one out of 19 delayed unions (5%) grafted with autogenous bone failed to unite. This was a delayed union of a humerus treated with a K-nail which required a re-operation and was subsequently plated. The follow-up radiographs of one delayed-union and one non-union of a humerus successfully grafted with the ceramic/collagen/marrow mixture are shown (Figs. 4.3a,b,c, 4.4a,b,c).

Study 2

In study 2 (the non-randomised prospective delayed union study), 18 patients with delayed unions were grafted using only Collagraft. The mean age of the patients was 48 years (range 18 to

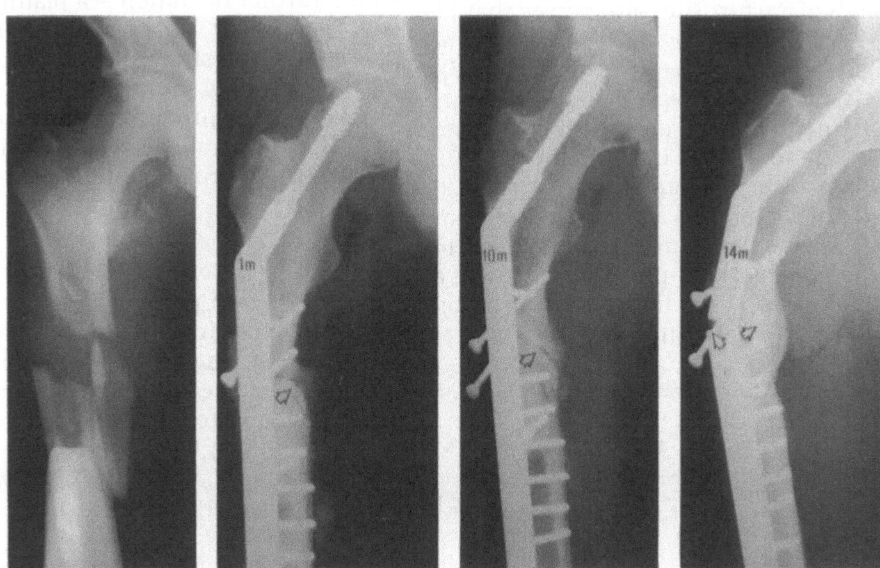

Fig. 4.1. Comminuted fracture of the proximal femur treated with Collagraft and dynamic hip screw. At the 14 months stage the non-union had established and the plate fractured.

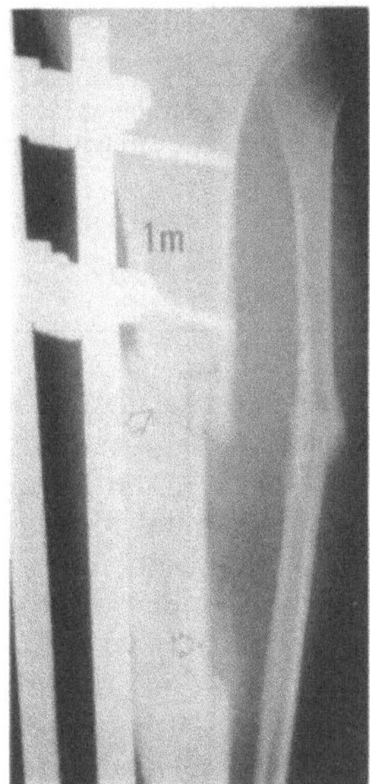

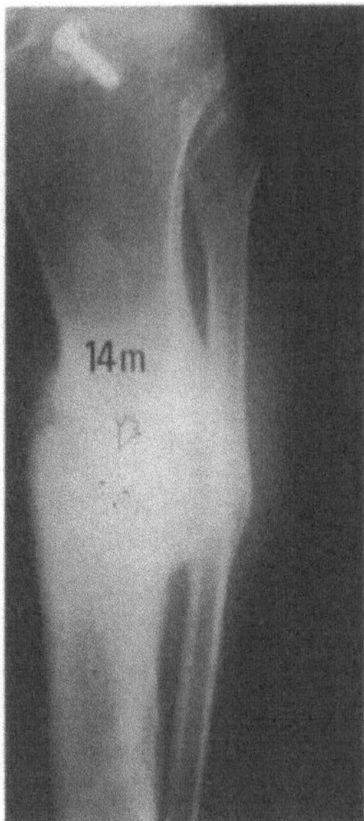

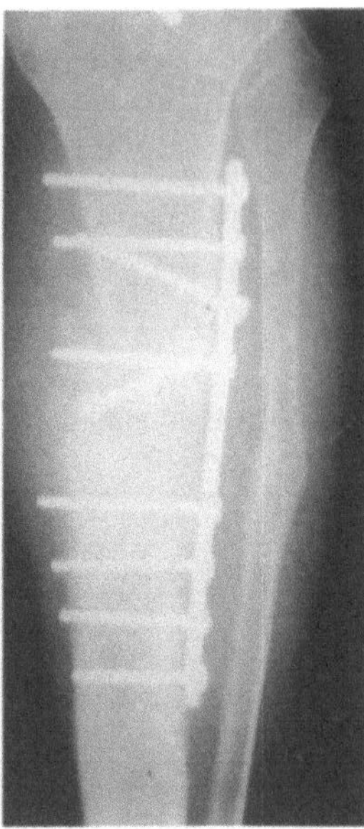

Fig. 4.2. Segmental fracture of a tibia treated with Collagraft and an external fixator. At the 14 months stage a hypertrophic non-union had developed and required re-operation.

70), the mean time from the injury to the grafting operation was 25 weeks (range 12 to 33), and the mean follow-up period was 18 months (range 12 to 34 months). The graft recipient site was the tibia in five (28%), a previously attempted knee arthrodesis in four (22%), the femur in four (22%), the humerus in two (12%) and the fore-arm in three (16%). Stabilisation was performed using a plate in eight (44%), an external fixator in five (28%), an intra-medullary nail in four (22%), and a plaster cast only in one case (6%).

The analysis of fracture site pain showed that 11 patients (61%) at six months and eight patients at one year had slight to moderate pain (Table 4.7). The analysis of ADL showed that ten patients (55%) at six months and seven patients at one year had slight to severe impairment of ADL (Table 4.7). A mean radiographic score of 11.0 was achieved at six months and 13.0 at one year (Table 4.8). At one year three patients failed to unite: one femoral delayed-union treated with a dynamic hip screw, one tibial delayed-union treated with an external fixator, and one knee re-

Table 4.7. Non-randomised delayed union cases treated with Collagraft: analysis of outcome measures

Outcome	Months	Score			
		None	Slight	Moderate	Severe
Pain	6	7 (39%)	6	5	0
(n = 18)	12	10 (55%)	6	2	0
ADL	6	8 (44%)	8	1	1
(n = 18)	12	11 (61%)	4	2	1

Table 4.8. Non-randomised delayed union cases treated with Collagraft: radiographic score

Graft	Months	Radiographic score (mean with standard deviation)
Collagraft	6	11.0±2.9
(n = 18)	12	13.0±2.6

arthrodesis treated with an intra-medullary nail. The tibial non–union was successfully treated with a further operation – a plating and grafting with cancellous autogenous bone. Biopsy at revision surgery showed fibrous tissue surrounding the ceramic granules with no bone formation. The knee re-arthrodesis was treated in a brace, and the femoral delayed-union is awaiting further surgical treatment.

Study 3

In study 3 (open prospective study of acute fractures) 22 acute fractures were treated with Collagraft at the time of primary stabilisation of the fracture. The mean age of the patients was 42 years (range 18 to 66); the graft site was the tibia in twelve (55%), the femur in eight (36%) and intra–medullary nail in six (28%) cases. The follow-up was a minimum of one year for all patients.

The analysis of the fracture site pain showed eleven patients (50%) at six months and six patients at one year with slight to moderate pain

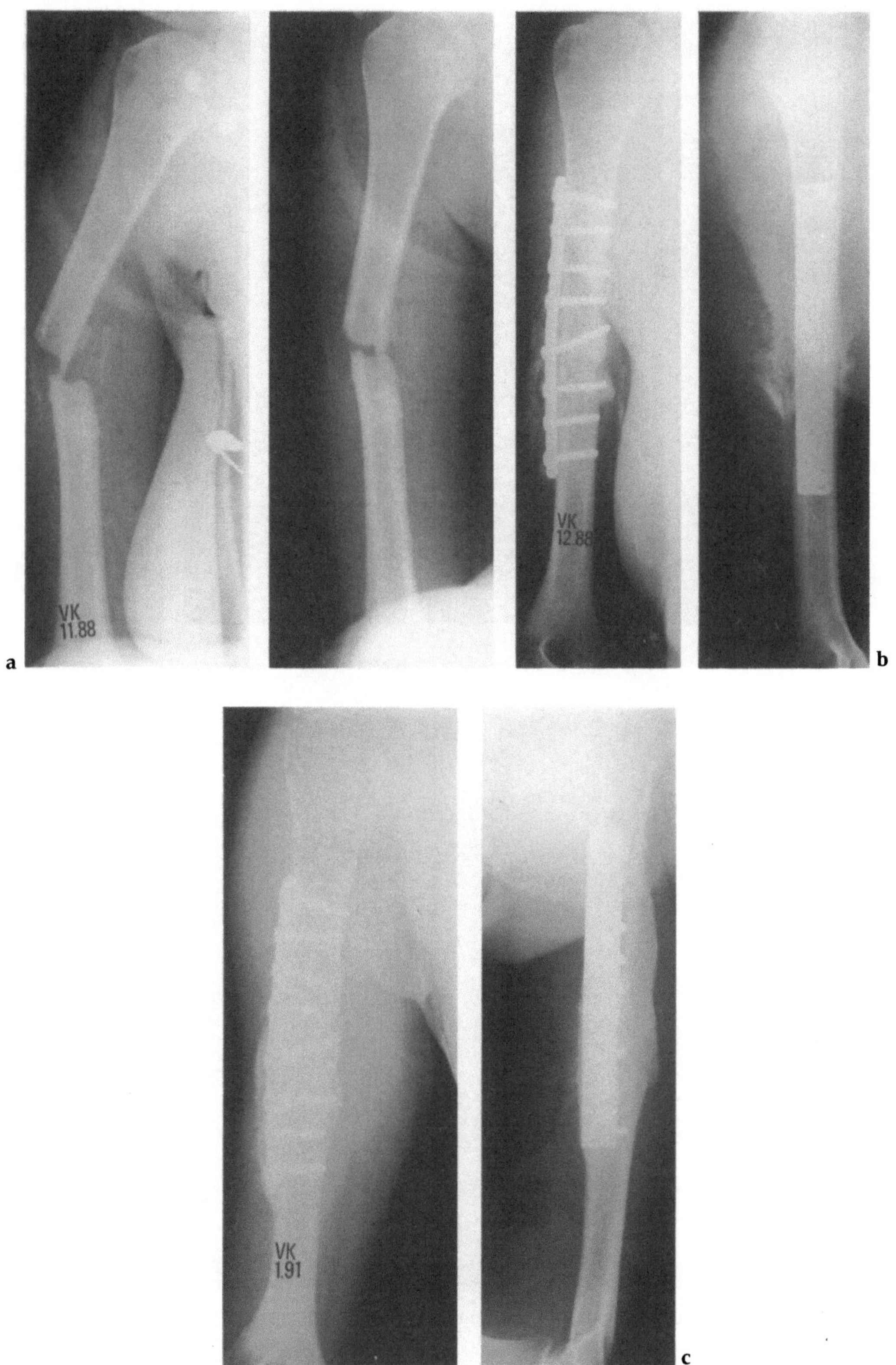

Fig. 4.3. a A delayed union of a humerus at 12 weeks after fracture. **b** Plating and grafting with Collagraft. **c** Full union at 24 months.

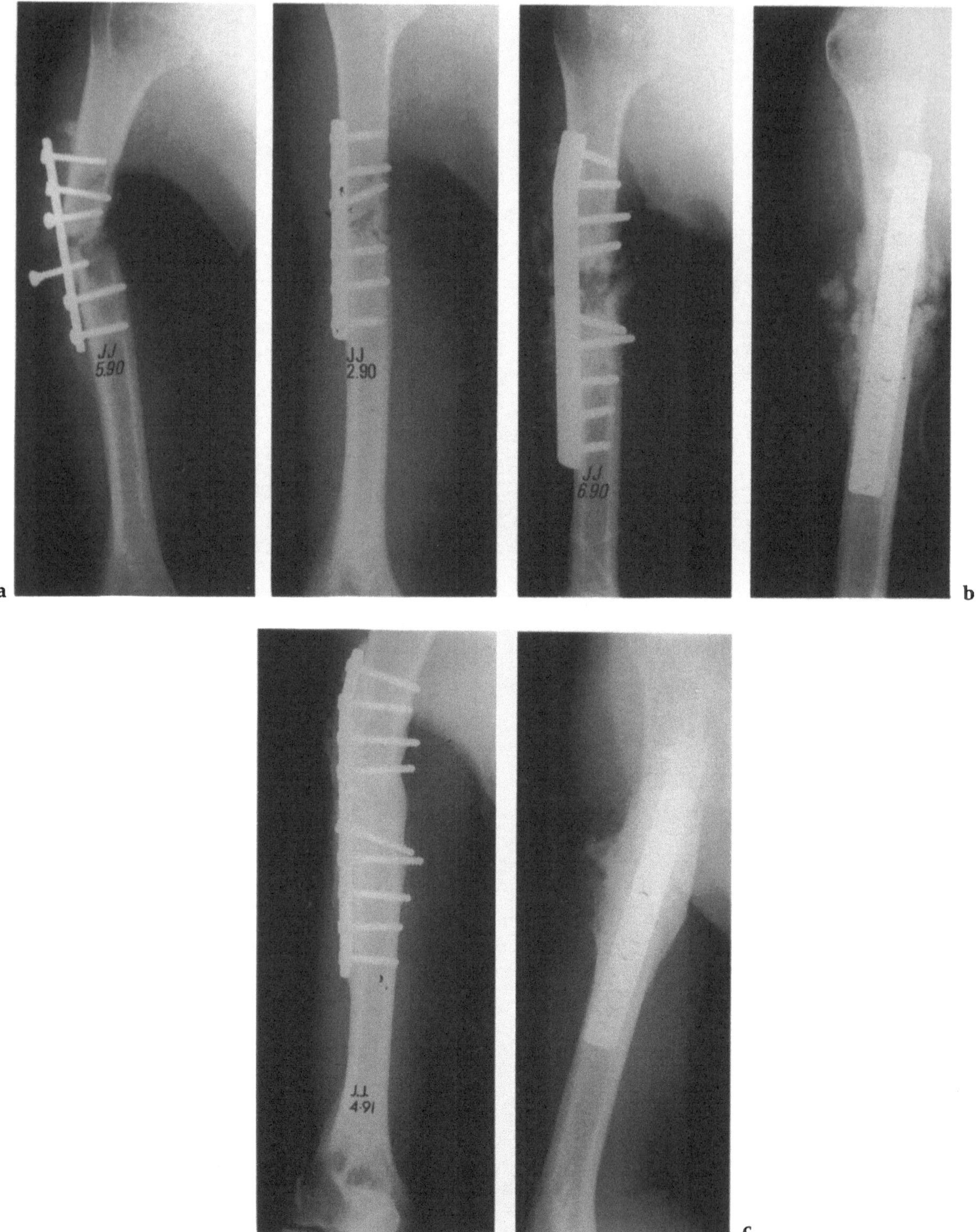

Fig. 4.4. **a** A failed fixation of a humerus. **b** Re-plating and grafting with Collagraft. **c** Full union at ten months.

Table 4.9. Acute fracture cases treated with Collagraft: analysis of outcome measures

Outcome	Months	Score			
		None	Slight	Moderate	Severe
Pain	6	11 (50%)	8	3	0
(n = 22)	12	16 (72%)	4	2	0
ADL	6	8 (36%)	12	2	0
(n = 22)	12	13 (59%)	8	0	1

Table 4.10. Acute fracture cases treated with Collagraft: radiographic score

Graft	Months	Radiographic score (mean with standard deviation)
Collagraft	6	12.4±3.8
(n = 22)	12	13.3±3.3

(Table 4.9). The analysis of ADL showed that 14 patients (63%) at six months and nine patients (41%) at one year had slight to severe impairment of ADL (Table 4.9). A mean radiographic score of 12.4 at six months and of 13.3 at one year was obtained in the study (Table 4.10). At one year three out of 22 patients (14%) failed to unite and required further operations: two tibial delayed-unions, one treated with an external fixator and one with an intra-medullary nail, and one femoral delayed union treated with a dynamic hip screw.

In summary ten out of 64 patients (16%) treated with the ceramic/collagen/marrow mixture failed to unite at one year and eight of them required re-operation. No adverse short-term reactions such as wound drainage, erythema and inflammation, or long-term graft site complications such as infection were seen. The laboratory tests performed in all patients six months and one year after grafting showed minor abnormalities in a few cases but these were not clinically significant and were not related to any particular test. Biopsies from the two delayed unions previously grafted with ceramic/collagen/marrow mixture which failed to unite after one year showed some foreign body reactions associated with resorption of the ceramic granules and fibrous tissue surrounding the ceramic material. No malignancy or genuine inflammatory changes were observed.

Discussion

The clinical studies show that an artificial bone grafting material (ceramic/collagen/marrow) does not do any harm to the host tissues and was seen to promote the union of fractured bones in these patients, but it was not as effective as autogenous cancellous bone. A major difficulty was its application to the fracture site; the mixture occasionally became very liquid as it absorbed blood and then started to flow out of the wound (Kocialkowski et al. 1989, 1990). Hence, the spoon filling technique was inappropriate as it was difficult to fill the bone defect thoroughly. Further escape of the mixture was stopped only by carefully suturing the soft tissues over it.

Despite the application problems, these reported results in 64 patients at one year (16% failure rate) are encouraging. The ceramic/collagen/marrow graft is inserted in a paste-like consistency around the outer fracture margins and into the fracture itself, simulating the spontaneous callus formation around the fracture (Kocialkowski and Wallace 1991, 1992). The advantages of artificial graft substitutes are an increase of graft surface and contact area, and healing may take place simultaneously over the whole graft area. In addition there is minimal discomfort from an iliac crest autogenous bone donor site. By comparison with solid ceramic implants or blocks of allo- or autografts this ceramic/collagen/marrow paste-like material offers advantages justifying its further evaluation as a potential graft material in the future.

Conclusions

In an American study, reported by Cornell et al. (1990), 232 patients with acute fractures were entered into the study, 117 receiving cancellous autograft and 115 receiving ceramic/collagen/marrow graft. At six months and at one year the artificial graft appeared to function as well as autogenous graft when used in the treatment of acute long-bone fractures.

The extensive work of Urist on bone induction and the identification of bone morphogenetic protein (BMP) has opened up new experimental approaches for the grafting of skeletal tissues. Recently, Johnson et al. (1990), reported four patients with severely deformed non-unions of distal tibia which united after grafting with human BMP.

In future, it may no longer be necessary to procure a bone autograft from another site but to use a bone substitute (ceramic/collagen)

composite with stromal marrow cells and an inductive agent (BMP). This goal has already been partially achieved.

Acknowledgement. The author would like to thank all investigators from the ten study centres in Europe who participated in the study: J. Barsotti, P. Bonnevialle, F. Lechner, A. Le Rebeller, Y. Masse, M. Saleh, P.G. Schneider, L. Solomon, J. Vidal, W.A. Wallace.

References and Further Reading

Cornell CN, Lane JM, Chapman MW, Seligson D, Henry S, Merkow R, Gustilo R, Vincent K (1990) Multicenter trial of Collagraft as bone graft substitute. Transactions of the 57th annual meeting of American Academy of Orthopaedic Surgeons, p.59

Johnson EE, Urist MR, Finerman GAM (1990) Distal metaphyseal tibial non-union. Clin Orthop 250:234–240

Kocialkowski A, Wallace WA (1989) Granular ceramic, fibrillar collagen and bone marrow contributions to bone formation. Transactions of the SICOT European Trainees Teaching Meeting. London, Oct 12–14 Book of Abstracts, p.9

Kocialkowski A, Wallace WA (1992) In search of a magic bone graft. Internat J Orthop Trauma 2(1) (In press)

Kocialkowski A, Wallace WA, Burwell RG (1990) The Nottingham experience with Colhap. Transactions of a Symposium on the Surgical Use of Bone Grafts and Bone Substitutes. Oswestry, UK, March 29–31

Kocialkowski A, Wallace WA, Prince HG (1990) Clinical experience with a new artificial bone graft: preliminary results of a prospective study. Injury 21:142–144

Kocialkowski A, Wallace WA, Burwell RG, Hardy JG (1990) Collagen and ceramic as an osteoconductive matrix for heterotopic bone formation. J Bone Joint Surg (Br) 72:163

Kocialkowski A, Wallace WA, Burwell RG (1990) Bone marrow and granular ceramic contributions to bone formation. Clin Anatomy 1:65

Kocialkowski A, Wallace WA, Burwell RG, Hardy JG (1990) Cancellous bone versus ceramic/collagen/marrow mixture in an experimental non-union. J Bone Joint Surg (Br) 72:735

Kocialkowski A, Wallace WA, Burwell RG (1990) Bone formation within ceramic implants. Clin Anatomy 3:327

Kocialkowski A, Wallace WA, Burwell RG (1990) The efficiency of a ceramic/collagen/marrow artificial graft in a tibial non-union in the rabbit. Clin Anatomy 3:237

Kocialkowski A, Wallace WA, Burwell RG, Hardy JG (1990) How to improve the osteogenicity of the ceramic implants. J Bone Joint Surg (Br) 72:1107

Kocialkowski A, Wallace WA, Hardy JG (1990) The use of bone scanning to predict non-union in an experimental model. Nucl Med Commun 11:90–91

Kocialkowski A, Wallace WA, Burwell RG, Hardy J (1990) Correlation between scintigraphy and histomorphometry in new bone detection. Eur J Nucl Med 12:259

Kocialkowski A, Wallace WA, Burwell RG (1991) Correlation between scintigraphy, clinical assay and radiography in the assessment of osteogenicity of different bone grafting materials. Clin Anatomy 4:74

Kocialkowski A, Wallace WA, Petty-Saphon S (1992) Experience with Collagraft in Europe. In: Lindholm S (ed) New trends in bone grafting. University of Tampere, Finland

Discussion

Dr Delloye: Did you try to inject bone marrow only within the non-unions because it has been shown that bone marrow alone is effective as osteogenic material?

Dr Kocialkowski: Professor Burwell did it in the 1960s and found that bone marrow was very effective. Lane in America has treated 12 non-unions and with the injection of bone marrow a high incidence of union was achieved. The bone marrow is very effective on its own but the problem when you apply bone marrow is that it drains the old bone marrow from the grafted site, so that at the end you may have no bone marrow left at the fracture site.

Dr Czitrom: I would caution you with regard to the conclusions derived from this study which, in my opinion, is not a soundly designed scientific study. You cannot randomise different kinds of fractures in different bones throughout the skeleton. It would require thousands and thousands of patients to do a proper study. Also, there was no group which did not have bone graft. Some of these patients could heal without a bone graft. That may not be possible to do clinically, because you cannot fail to bone-graft someone who needs it. Another major criticism is that a grade 2 fracture of the femur or a grade 1 fracture of the humerus differ, and that was not taken into account.

Dr Kocialkowski: I entirely agree with your comment. We have serious doubts about the study in the US as only fresh fractures were included and these can unite without any grafting. In the prospective randomised study there must have been at least eight weeks since the fracture. Our assessment of the patients was independent of the company producing Collagraft and each centre had the same protocol. I honestly reported what I found. I agree that you need thousands of patients, but it is difficult to get so many.

Dr Czitrom: You described some delayed unions at two months. That is not very delayed; it could still unite in three or four months. In these cases you put in the Collagraft together with internal fixation, so how can you tell that it is not the internal fixation alone that made it unite?

Dr Kocialkowski: I entirely agree that you cannot tell, therefore you have to wait for up to three years to find out if the fracture has really united. I know that some of the fractures failed at 14 months because the plate broke. Sometimes it is difficult to prove on the X-rays that it has united because, unfortunately, the ceramic stays at the fracture line and you cannot see this. The ceramic is visible but you cannot be sure that it has united. You have to wait between two and three years to be absolutely sure. These are early results and so far they are encouraging.

I have stressed that the autogenous bone is the better graft to use, but sometimes we cannot do that, and it is better to use the Collagraft than the allogeneic bone, especially when there is a possibility of infecting the patient with HIV. Unfortunately, HIV is becoming more common, and I predict that in ten years the use of allograft will be banned because of it. I do not think that a vaccine will solve the problem.

Mr Older: What you are saying is that we are really here under false pretences, because there will be so much HIV that nobody will be bone grafting, and we will all have to use your equipment?

Dr Kocialkowski: That is right!

Mr Older: From what you have seen in your work, how do you assess its place in orthopaedic surgery in the future?

Dr Kocialkowski: It can certainly have a place as a morsellised allogen graft, which is commonly used. With the increase in HIV, the morsellised allogeneic graft will probably be banned, so it is an alternative to use this type of ceramic, collagen and marrow as a morsellised graft instead of using the allogeneic morsellised graft.

Mr Older: Dr Czitrom, how do you regard what is being done both here and in America in relation to the future of orthopaedics?

Dr Czitrom: I think the idea of using collagen and ceramic mixed together is a good one. My criticism is of the way the study was designed. The idea of using a carrier for bone marrow cells is a good one and worth researching. I would probably do it in an experimental animal in the kind of defect which does not heal under normal conditions, and then compare this with autologous bone and with nothing. Then it should be possible to say whether it works or not.

Part IIA
Hip: The Acetabulum

5 Autografting in the Acetabulum

M.W.J. Older

Autografting of bone in the acetabulum is the "gold standard" in revision hip surgery. A series of case histories will best illustrate my personal experience in the past ten years. Until recently, when revising a socket, however large the bone defect I have simply filled the hole with more cement. In some cases this has been surprisingly successful.

Case Reports

Case 1

In 1971 this lady had her arthrosic left hip replaced with a metal-to-metal McKee–Farrar prosthesis, and the following year the right hip was replaced with a similar implant. Ten years later, at the age of 77, she presented with pain and disability in both hips. Radiographs showed extensive upward migration of the left socket with considerable leg length discrepancy. There was also some migration of the right socket (Fig. 5.1a). I revised the left hip and managed to manoeuvre the cup into a more anatomical position, but this required an enormous amount of cement which I used with considerable trepidation. In 1984, I replaced the right hip (Fig. 5.1b). She had an excellent result with legs of almost equal length and walked free of pain, without a stick and only a slight limp. She was very active and played bowls two or three times a week (Fig. 5.1c). In 1990 she died, for reasons quite unrelated to her hips. These two hips had given her good service for eight and six years respectively with no sign of clinical or radiological loosening. Despite this success, I was still concerned about the excessive use of cement.

Case 2

This patient with congenital dislocation of the hip had an autogenous bone graft from what was left of the femoral head. This hip is now 14 years old and is working well (Fig. 5.2).

Case 3

This patient with protrusio acetabuli had had a femoral osteotomy ten years previously (Fig. 5.3a). At hip arthroplasty I used layers of femoral head bone packed onto the medial wall of the socket instead of large quantities of cement. At seven years the hip is doing remarkably well (Fig. 5.3b).

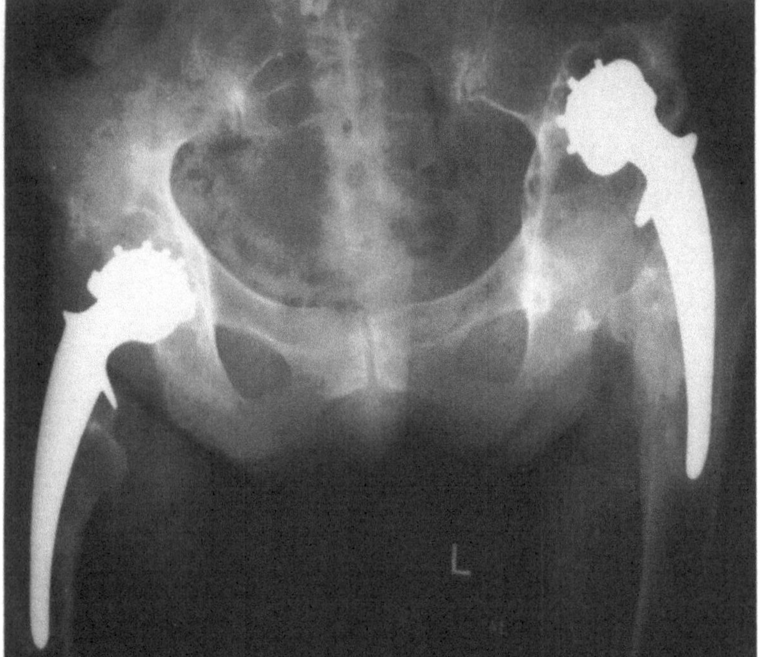

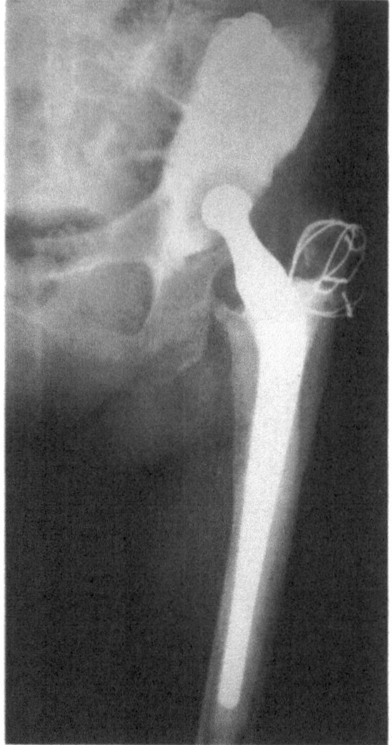

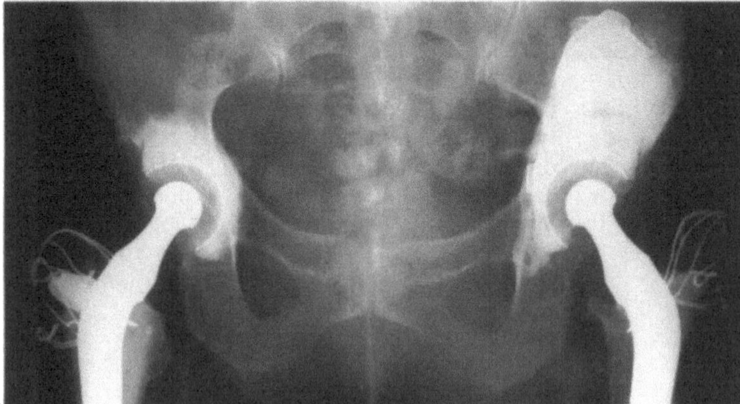

Fig. 5.1. **a** Pre-operative radiograph. **b** Post-operative radiograph. **c** Post-operative radiograph–left hip at eight years.

Fig. 5.2. Pre- and post-operative radiographs.
▼

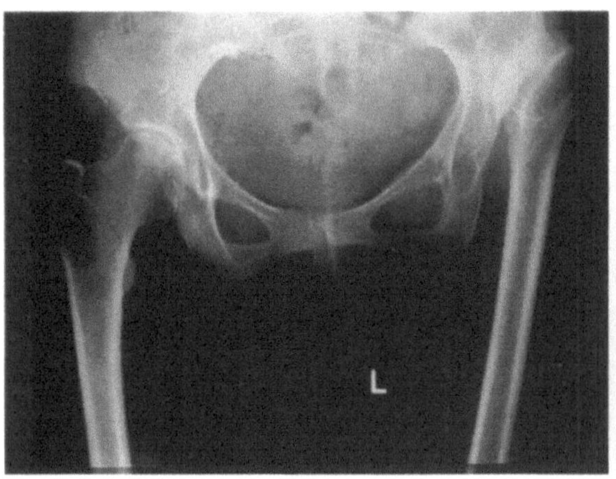

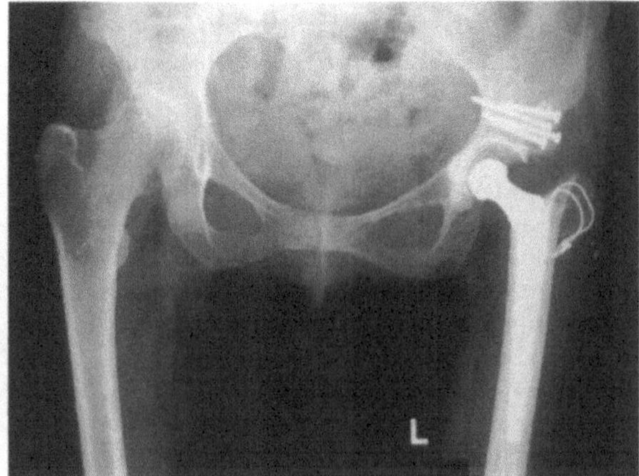

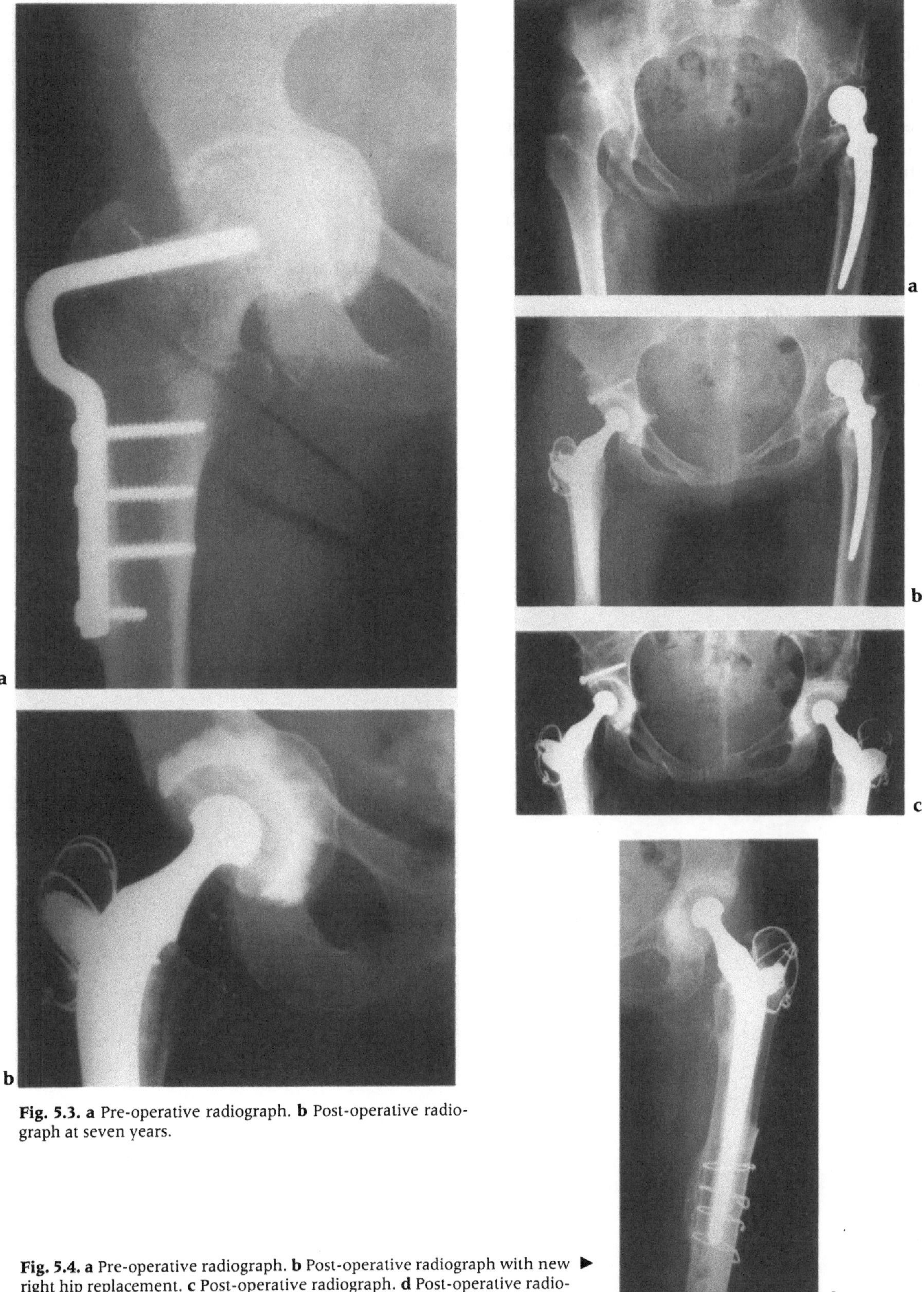

Fig. 5.3. **a** Pre-operative radiograph. **b** Post-operative radiograph at seven years.

Fig. 5.4. **a** Pre-operative radiograph. **b** Post-operative radiograph with new right hip replacement. **c** Post-operative radiograph. **d** Post-operative radiograph–left hip at three years. ▶

Case 4

This patient, aged 48, presented in 1988 with an osteo-arthrosic right hip associated with a very shallow socket. The left hip, which had been replaced ten years previously, showed radiological pictures of extensive upward migration of the socket. The left leg was very short. Interpretation of the radiological features was made difficult by the radiolucent cement that had been used (Fig. 5.4a).

I replaced the arthrosic right hip using the femoral head to reconstruct the roof of the acetabulum (Fig. 5.4b). A year later, I reconstituted the socket of the left hip into a good anatomical position with bone from both the left and right iliac crests (Fig. 5.4c). Three years later, this lady is now walking free of pain, without a stick and only a trace of a limp. Radiographs show sound union and consolidation of the bone grafts (Fig. 5.4d). There have been no unpleasant sequelae following the bilateral iliac crest bone donor sites.

Discussion

Following the swing away from cement to autogenous bone I have performed 20 revision procedures on the acetabular socket using the patient's own bone and the clinical and radiological results in the past three years have been excellent.

6 Banked Allograft Bone for Pelvic Defects

A.E. Gross

The primary goal in a revision arthroplasty is to implant uncemented components supported by host bone restoring the anatomy and equalising leg lengths. At our hospital, cemented components are acceptable for the lower demand patient in whom another revision is highly unlikely.

The secondary goal is to implant uncemented or cemented components supported primarily by host bone, with secondary support by morsellised allograft bone but still restoring the anatomy and leg lengths.

The tertiary goal is to implant uncemented components supported by host bone or primarily by host bone, with secondary support by morsellised bone but with some sacrifice of anatomy and leg lengths. This is where the controversy lies. How much leg length and anatomy should be sacrificed to avoid the quaternary goal, implanting components supported primarily by bulk allograft bone, in order to restore the anatomy and the leg lengths?

The other surgical principles are that we do not use cement in the host except in the lower demand patient – where we do not hesitate to use it. We do not sacrifice host bone and try not to devascularise it, and always autograft junctions generously between allograft and the host. If it is an easy revision, we use the transgluteal or a straight lateral approach. In a difficult revision, we do not hesitate to perform a trochanteric osteotomy.

We use a simple classification of bone defects. On the pelvic side, we have protrusio, a shelf or minor column defect and an acetabular or major column defect. Minor and major column defects require bulk grafts. On the femoral side, we have intramedullary and cortical defects which are either circumferential or non-circumferential. If they are circumferential, we have the calcar or the proximal femoral defect depending on the length.

Protrusio Grafts

The protrusio graft is used for a contained cavity defect. It can be a medial or a supramedial defect but the acetabular walls are intact. We use morsellised bone, but if an intact femoral head or intact head segments are used, it is very important to ensure that your bone bank distinguishes between the male and the female femoral head. Post-menopausal female femoral heads should not be used for a structural graft of any kind.

Our protrusio reconstruction in a high demand patient is the metal-backed, large diameter,

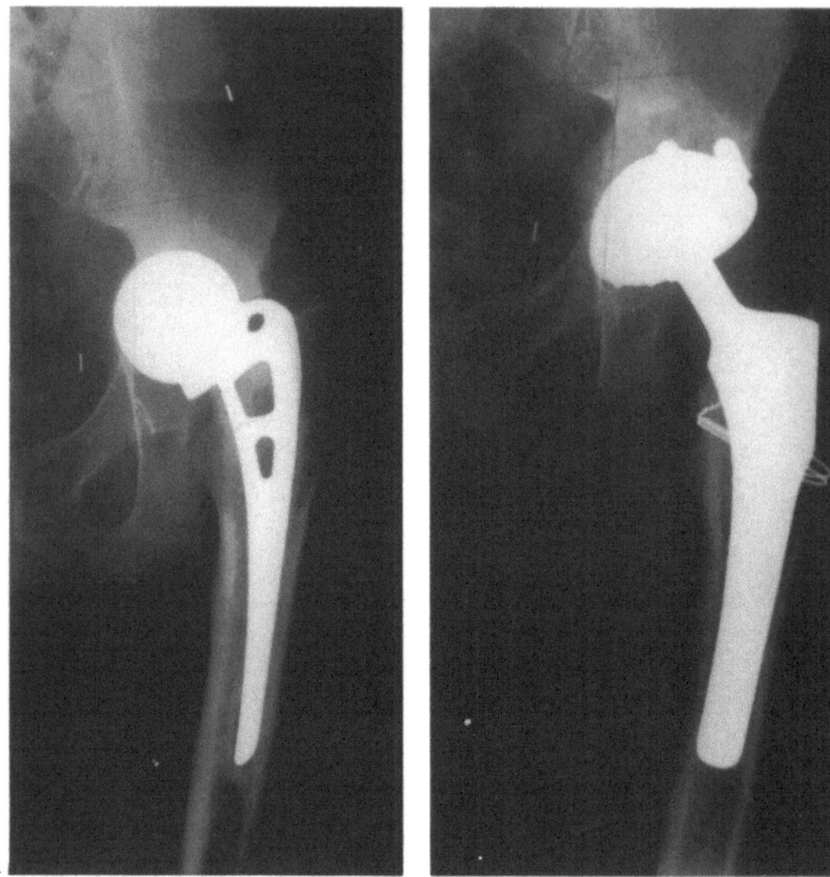

Fig. 6.1. a Pre-operative radiograph showing severe superior medial protrusio several years following an Austin Moore arthroplasty. **b** Post-operative radiograph five years after reconstruction with morsellised allograft bone and an uncemented cup.

uncemented cup positioned as horizontally as possible. In the lower demand patient we are using exactly the same reconstruction as Slooff.

Case Report

A young man who was a high demand patient had a kidney transplant and developed an osteonecrosis. He was in his thirties and had had an arthroplasty several years previously. We used morsellised bone and a large diameter, metal-backed cup placed as horizontally as possible to obtain good rim contact (Fig. 6.1a,b).

In the lower demand patient, we use morsellised bone, a ring support and a cemented cup. We no longer advocate reconstruction with bipolar prostheses and morsellised bone. We only use this in a salvage situation when a more sophisticated reconstruction is not feasible, if there is excessive blood loss or if we do not have the appropriate tissue.

Bulk Acetabular Allografts (Minor and Major Column)

The shelf graft – a small bulk allograft – is used where the defect involves the rim plus the acetabular wall but less than 50% of the acetabulum. If 50% of the host acetabulum remains, an uncemented cup can be used. If it is more than 50%, a major column graft is necessary and the cup is cemented. We tend to use a true acetabular allograft or a segment of a male femoral head.

Case Report

We used true acetabular allograft with a cemented cup for this patient. The cup may be fixed with large cancellous screws, a neutralisation plate or large protrusio ring. It is very important that, if possible, there is some form of

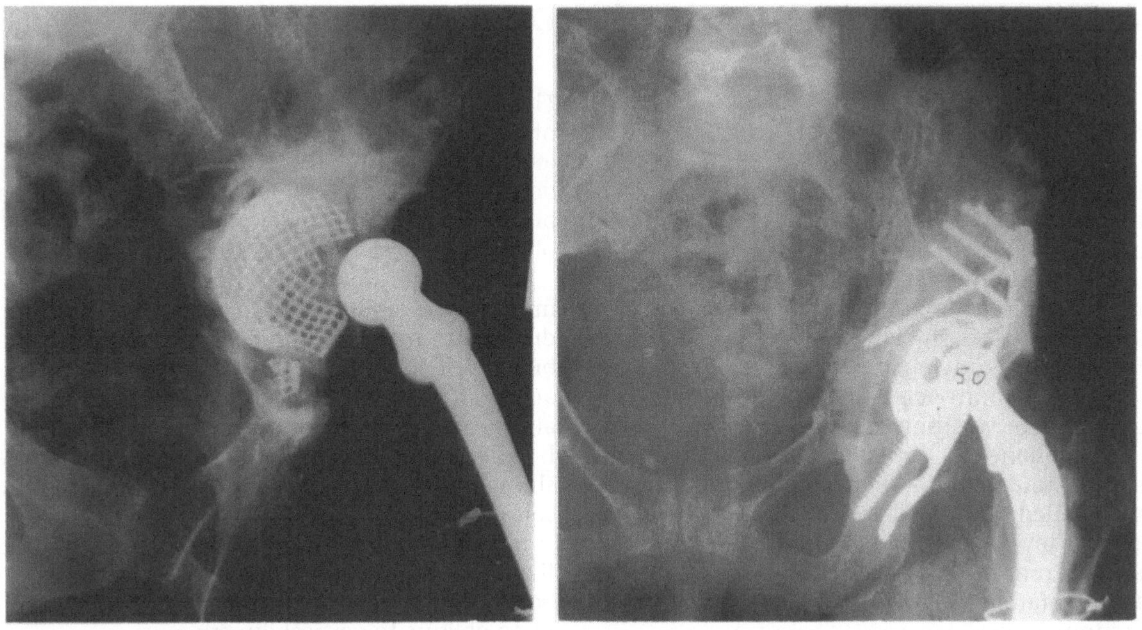

Fig. 6.2. a Judet view radiograph showing anterior column defect (major column). **b** Radiograph four years following reconstruction with major column, allograft.

fixation from host bone to host bone. These grafts are far from perfect and the more they are protected, the better (Fig. 6.2a,b).

Results

Protrusio Grafts

In January 1991, our total allograft series was 374 patients of whom 127 were protrusio grafts for contained cavity defects. Most of the defects you will see in your practice should fall into this classification, unless you happen to be a bone transplant centre.

In a prospective clinical and radiographic review and an average follow-up of 40 months of 55 patients with contained cavity defects, we looked very closely at the radiographic parameters, and tried to determine the most effective bone implant reconstruction for the protrusio or the contained type grafts. Some of these patients may do well clinically but the X-rays tell a different story.

We looked at the three most common reconstructions used: the uncemented fixed cup, the cemented cup with the ring and the bipolar device, which became very popular for a while and is no longer so. As far as migration was concerned, the uncemented and cemented cup did very well. The bipolar cup continued to migrate, particularly in a superior direction with this type of reconstruction. Loss of bone stock was minimal with both the uncemented and cemented cups but maximal with the bipolar cup.

The most graft resorption occurred with the bipolar device. Significant lucent lines were found around the cemented cup with the ring. Even though this is an excellent reconstruction with good results, we tend therefore to reserve it for the lower demand patient. Clinically the groups that did best were, as predicted, the ring and the cemented cup and the fixed uncemented cups.

Morsellised allograft bone incorporates and remodels well in the contained cavity defect. The fixed cemented or uncemented cups were clinically successful although there was a high incidence of lucent lines with the cemented cup. We tend, therefore, to reserve that procedure for the lower demand patient.

The uncemented cups are the preferred method for the high demand patient. The radiographic and clinical success rates are high. The reinforcement rings had a 100% success rate, but we were worried about the incidence of lucent lines. We are reserving the bipolar or bicentric device only for the extreme salvage situation.

Bulk Acetabular Allografts

In the bulk acetabular allografts, there were 34 of the minor and 32 of the major defects with an average follow-up of 36 months.

Minor column grafts did very well clinically and radiographically, except for some evidence of so-called "stress-shielding" which seemed to be limited to the unweighted part of the graft.

With the controversial major column grafts, the success rate was only 62%, but because of the restored bone stock we were able to increase the success rate with another revision. Even though they failed, we could put in a new cup at the correct level on further revision, increasing the success rate in that series to 82%. Fracture and fragmentation occurred in 47% of the major column grafts.

Discussion

The advantage of bulk allograft is the restoration of leg length; the cup can be placed at the right level and, if another revision is necessary, this can be done at the correct level. The problems with a bulk allograft are the high rate of revision because of resorption and fracture.

Better bone and better internal fixation will improve the results. It is absolutely necessary to distinguish male from female femoral heads if a bulk graft is to be used. On occasions in the autograft situation, it may be better to use a good allograft than an osteoporotic autograft femoral head with a dysplastic hip. Cancellous surfaces should only be exposed to the cup and not to the soft granulation tissue.

In our opinion, it is essential for good internal fixation that some form of plate or ring from host bone to host bone should be used with the screws in an oblique to vertical direction.

7 Classification and Treatment of the Failed Acetabulum

W.G. Paprosky

A systematic approach has been developed in the new field of reconstructive surgery. This is a presentation about allograft reconstruction in porous-coated implants. Cement has not been used since 1982.

We have found it difficult to reconstruct with cement in the face of massive osteolysis, especially of the acetabulum. There have been a number of studies to show that the long-term results of re-implantation with cement, especially in the face of lysis, have been relatively poor. Although for the short-term, re-implantation with a large amount of cement using various types of rings and other devices may have given a reasonable result, in our hands and others in the United States, the results have not been good.

Defect Classification

A simple defect classification system has been developed, based on whether or not the rim of the acetabulum is sufficiently stable and intact to provide a rigid support for the acetabular component. Defects are classified by types.

Type 1

The rim is supportive and regular. The anterior and posterior columns are totally supportive. Bleeding bone is present with potential for ingrowth into the acetabulum. Grafting is not necessary. Unfortunately, these simple types are few and far between (Fig. 7.1).

Type 2A

This defect is most common, with the rim being regular and supportive but with a cavity defect present. The largest implant is utilised, and cancellous filler is added to ensure rim contact. The entire approach requires the rim to be supportive. If the rim will not support the implant, supportive allograft reconstruction must be done (Fig. 7.2a).

Type 2B

If the tear drop is still intact, with less than 2cm superior migration but no massive superior bone loss, then allograft is unnecessary. The need for support allograft can be eliminated if you can put the component on the rim and fill in the rest of the defect with cancellous bone.

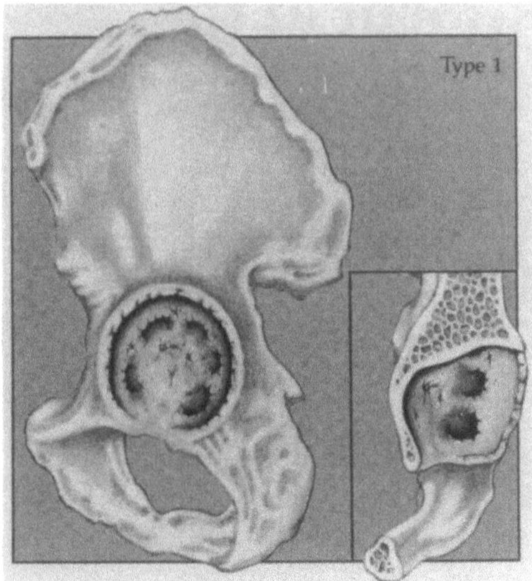

Fig. 7.1. Acetabular defect: Type 1.

There are also instances in which the rim is intact but irregular. There is less of a cavity defect but more superior migration of the implant with the tear drop still intact. Very often, the superior rim of the acetabulum will be missing, but support may still be obtained from the anterior and posterior columns. This is absolutely critical and mandatory in cementless reconstruction (Fig.7.2b).

Type 2C

In the protrusio type of defect, the rim is supportive; there is medial wall compromise with a cavity defect and bone loss medially. Rather than bottoming out the acetabular component, try to lateralise the component so that there is rim fixation. If you load the medial wall with a cementless implant, migration of the implant will occur with resulting failure. By attempting to lateralise the component, stability is gained on the rim. Grafts should incorporate when you apply a bending movement indirectly to the medial wall through loading of the anterior and posterior columns (Fig. 7.2c).

Type 3A

In this defect, the rim is non-supportive. There is bone loss involving 30–50% of the rim. Preoperative and intra-operative assessment of the problem must be made so the appropriate structural graft technique may be employed. In our experience, there is really no other way to achieve stability and restore the rim and bone stock other than using a structural support allograft (Fig. 7.3**a**).

Type 3B

There is dissociation of the anterior and posterior columns present. The bone loss involves 50–70% of the rim. These require a different type of grafting pattern, but again they must be reconstructed with plates in large allografts. In some cases, a whole acetabular transplant must be done if primary rim support cannot be achieved with the host–allograft construct.

To summarise, the principles involved are restoration of the bone and the peripheral rim so support is gained, and attempting to restore the acetabular level: a proximal relocation of the hip centre of up to 2cm is acceptable. The acetabular component must be reorientated so that there is post-reduction stability (Fig. 7.3b).

Materials and Methods

From 1982 to the end of 1988, 177 operative procedures were performed in 165 patients (Table 7.1). Revisions were performed almost exclusively for aseptic loosening of a prosthesis (Table 7.2) initially inserted for primary osteo-arthrosis (Table 7.3). The average number of previous operations was nearly four. There was an equal distribution of males and females. The age range was 24–89 with a mean of 57 years. The mean follow-up was just over five years (Table 7.4).

Patients were clinically evaluated with respect to pain, limp and overall satisfaction.

Table 7.1. Patients and operative procedures

	n
Patients	165
Surgical procedures	177
Components	
Porous	134
Threaded	43

Table 7.2. Diagnosis for revision surgery

	n
Aseptic loosening	168
Infection	5
Recurrent dislocation	3
Limb length discrepancy	1

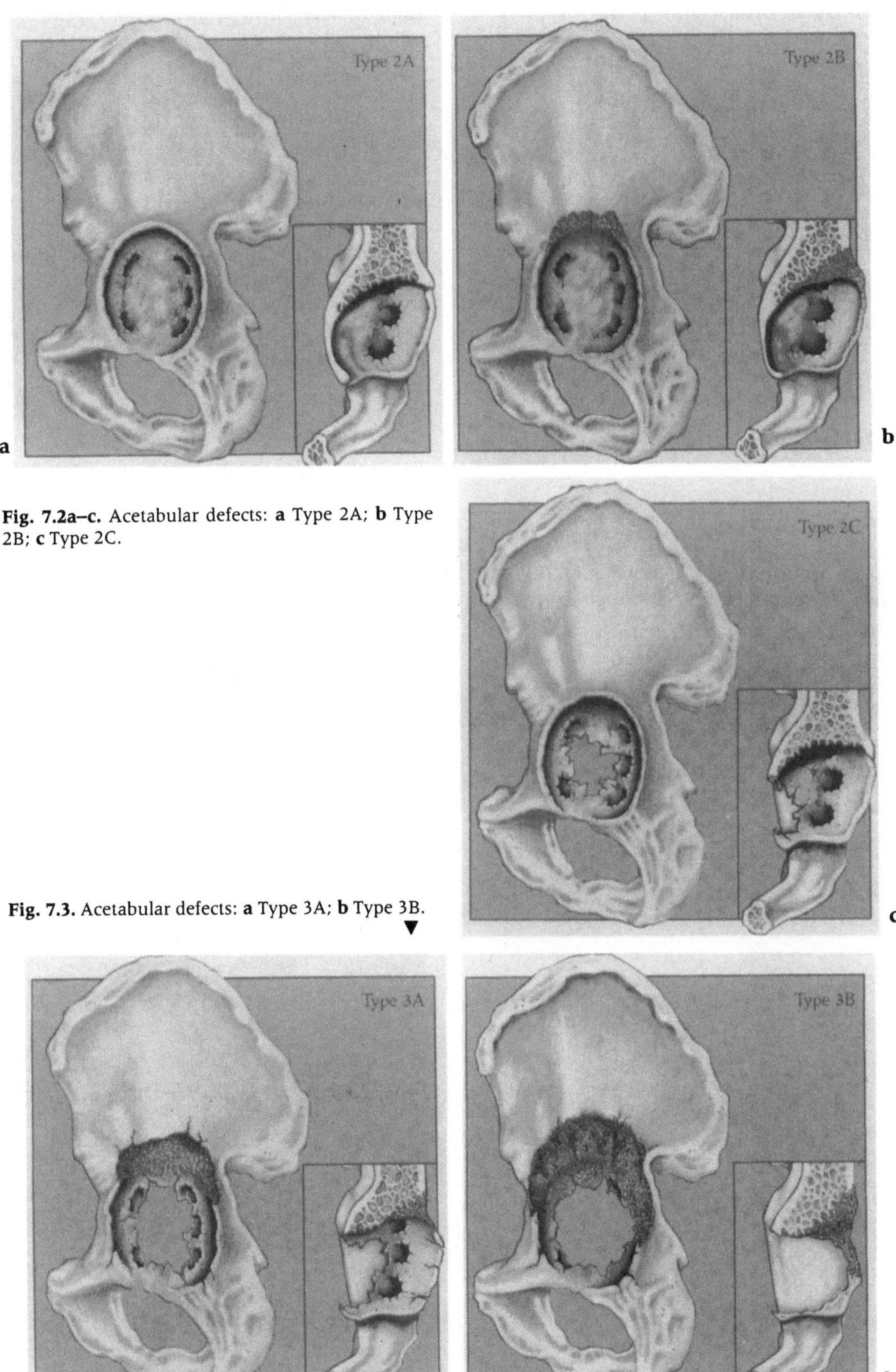

Fig. 7.2a–c. Acetabular defects: **a** Type 2A; **b** Type 2B; **c** Type 2C.

Fig. 7.3. Acetabular defects: **a** Type 3A; **b** Type 3B. ▼

Table 7.3. Diagnosis for initial arthroplasty

	n
Osteoarthritis	106
Congenital hip dislocation	16
Infection	5
Fracture	22
Avascular necrosis	7
Rheumatoid arthritis	15
Slipped femoral epiphysis	5
Achondroplasia	1

Table 7.4. Age, follow-up and previous operations

Age at operation (years)	57.4 (range 24–91)
Follow-up (years)	4.7 (range 2–8)
Previous surgery (no. of operations)	2.6 (range 1–6)
Previous infections (%)	7.0

graphic evaluation was made for component loosening, graft resorption and union.

Technique

Type 1

The acetabulum is relatively uncomplicated. Structural support grafting is not required and a porous implant is used. The surgeon should ream until there is contact with the anterior and posterior columns. In order to obtain optimum rim fixation an implant 1mm, 2mm, even 3mm larger than the original reamed number is inserted. Defects are filled with particulate cancellous graft material (Fig. 7.4).

Type 2A

Proximal migration is present with loss of the hemispherical shape of the acetabulum, and a certain amount of graft has to be employed. There is some thinning of the anterior and posterior columns. Rim fixation can be obtained and defects are filled with cancellous graft. With a

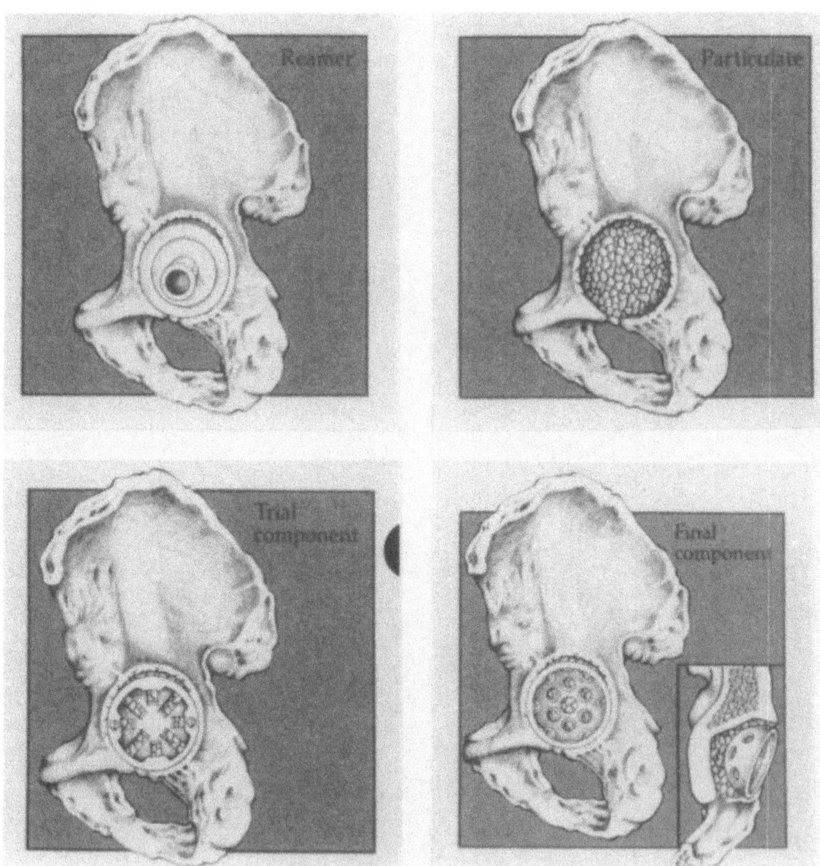

Fig. 7.4. Technique for Type 1 defects.

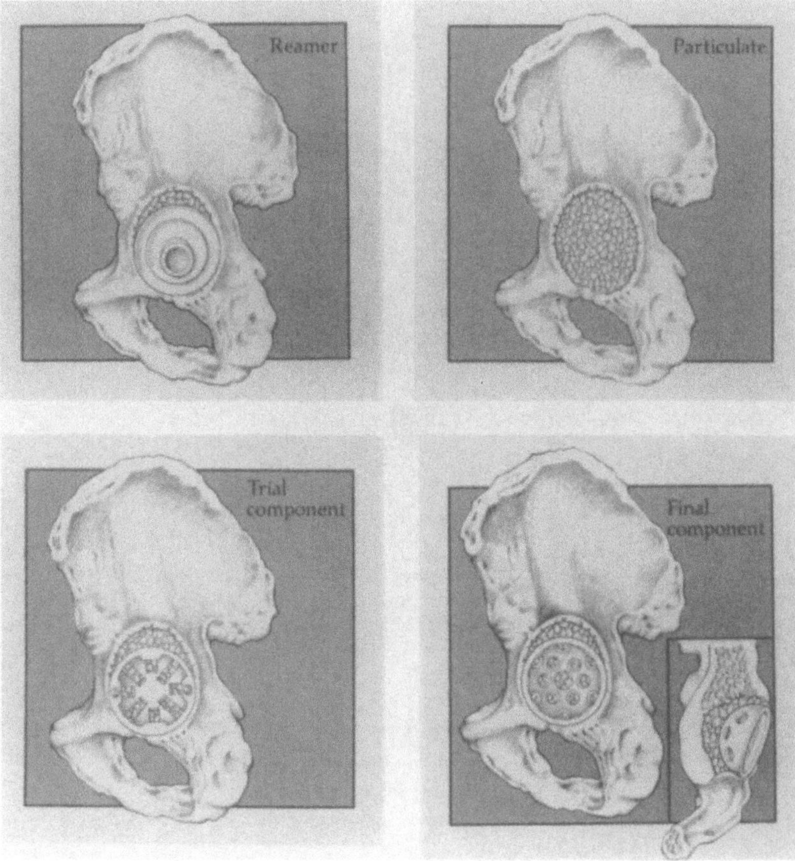

Fig. 7.5. Technique for Type 2A defects.

cementless device if stability of the rim is not obtained, the implant will migrate and failure of the graft will result. Anterior and posterior rim fit, provided primarily to the component, is mandatory (Fig. 7.5).

Type 2B

The superior rim of the acetabulum is absent. However, the anterior and posterior rim is still supportive. Placing a femoral head allograft within the acetabulum when the superior rim is absent makes internal fixation with screws very difficult. Instead, a "Number 7" graft is used. The femoral head is cut in the shape of a number 7 so that the long portion of the 7 is outside the acetabulum and the short portion is inside. This allows placement of screws on the outside of the acetabulum away from the surface which is to be reamed. The graft is stressed by the implant superiorly so that incorporation can occur. The graft trabeculae are also oriented in a vertical direction to permit maximum strength of the graft (Figs. 7.6, 7.7).

Type 2C

Where there is protrusio defect, we prefer to use cancellous chips because incorporation is faster. The support and stability of the acetabular component are achieved by rim support. The protrusio portion of the acetabulum which has been grafted should not be relied upon for any significant support. If the cancellous graft medially is directly stressed, medial migration of the component will occur (Fig. 7.8).

Type 3A

There is loss of bone and loss of rim fixation and support; the tear drop is no longer present. There is severe component migration both superiorly and medially with marked bone loss, so structural supportive allograft is necessary. We like to use a

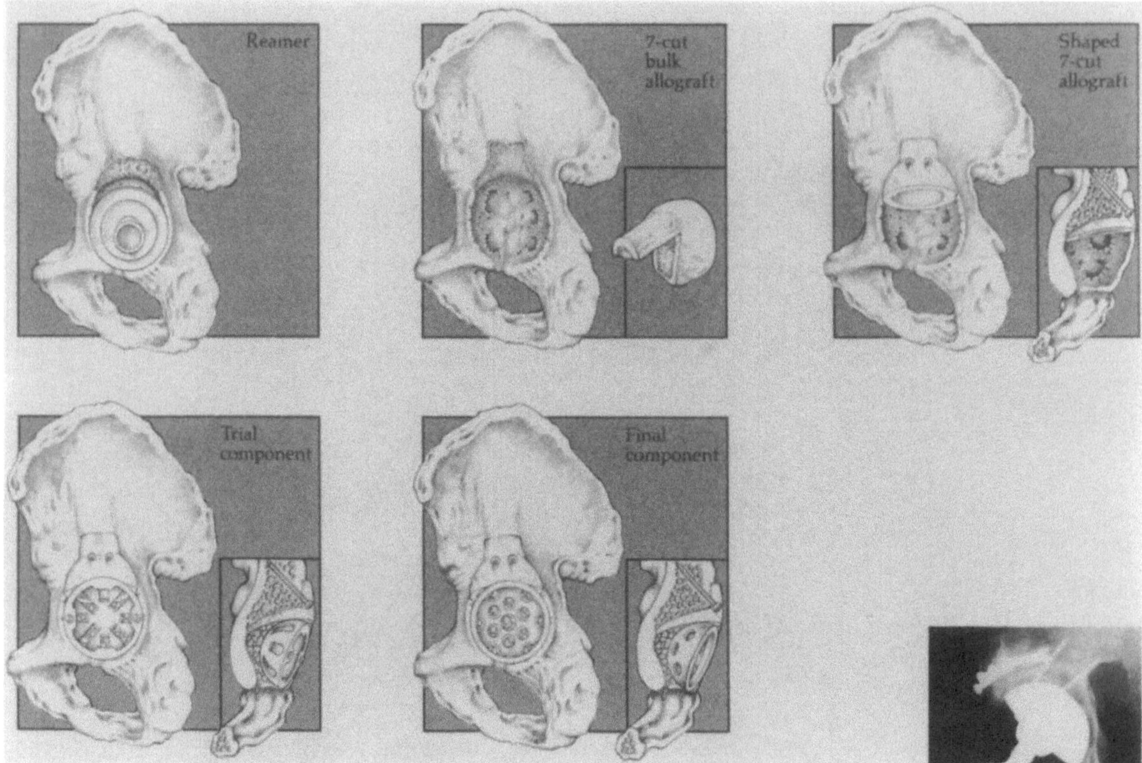

Fig. 7.6. Technique for Type 2B defects.

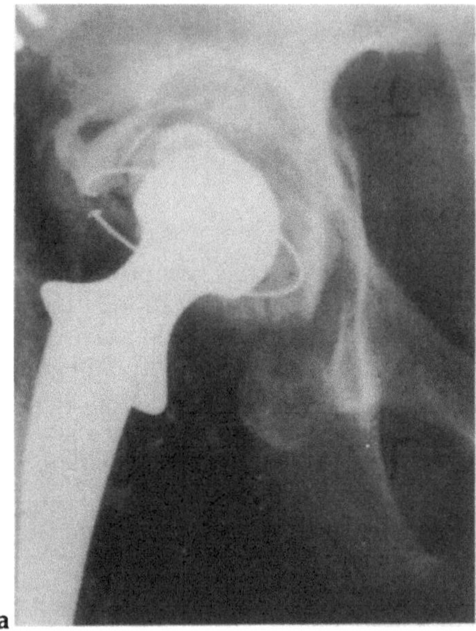

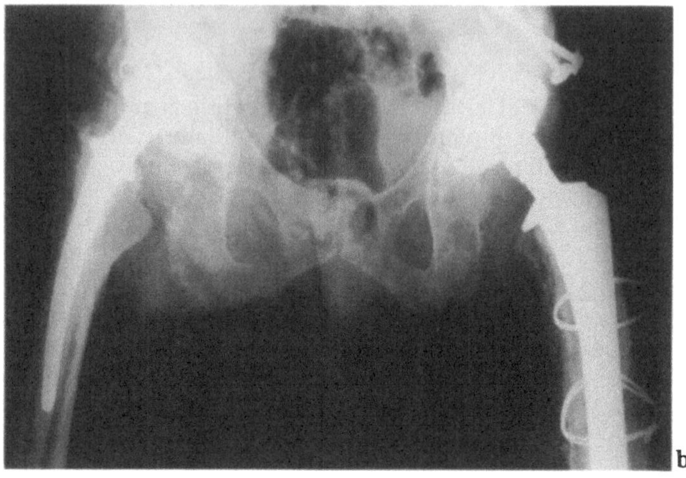

Fig. 7.7. A Type 2B acetabular defect reconstructed using a "Number 7" femoral head bone graft: **a** Pre-operation; **b** Two years post-operation.

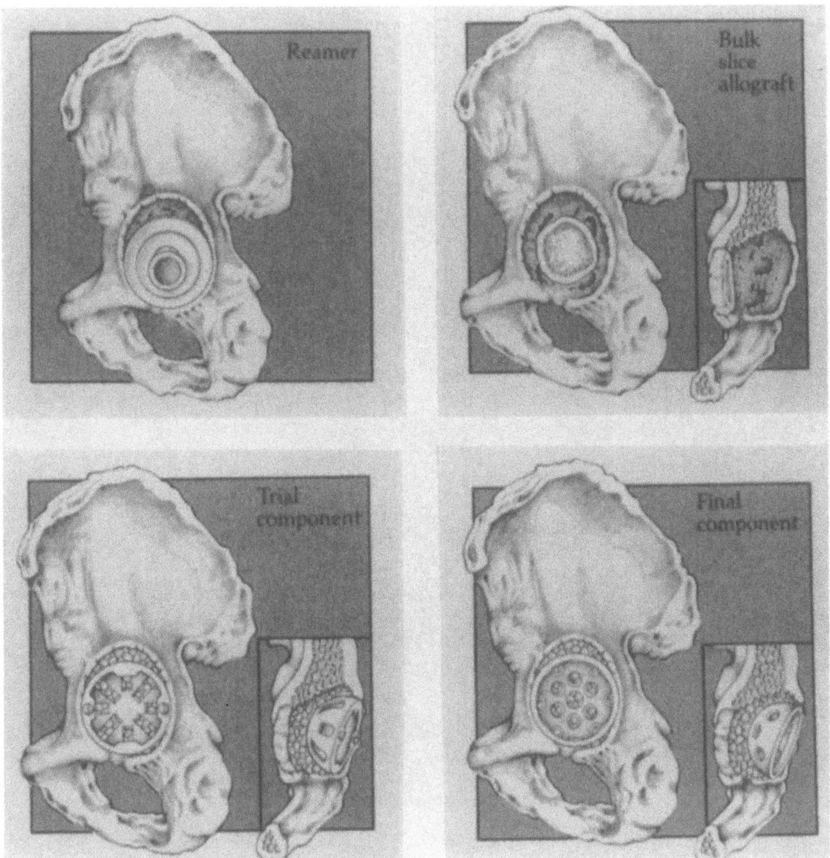

Fig. 7.8. Technique for Type 2C defects.

distal femur. We no longer use femoral heads for the large defects; we believe we can achieve a larger surface area for incorporation into the host as well as the ability to move the graft around and have solid bone to avoid early failure of the graft. The distal femoral allograft is cut in the shape of a number 7 and fixed to the ilium in the same manner as Type 2B (Figs. 7.9, 7.10).

There is a large surface area to achieve rapid bone incorporation. A portion of the graft is placed on the ilium; a portion is inside the acetabulum so there is a buttressing effect and the graft does not migrate superiorly. This technique is employed when 50% or more of the rim remains.

Type 3B

There is bone loss of the acetabulum from the "nine o'clock to five o'clock" position, or more than 50% of the acetabular rim is absent. Preoperative radiographs often show massive bone loss or pelvic dissociation. Structural support allograft is necessary. Rather than a column reconstruction, we prefer to span the defect with an "arc graft" using a proximal femur. This means circumferential reconstruction over a large area using one solid piece of rigidly fixed bone graft. The graft is fixed by plates and screws onto the ilium and ischium. The construct is then reamed. Cancellous bone is used to fill the protusio defect. In all Type 3 reconstructions, peripheral or superior large diameter bone screws should be used to give initial stability to the acetabular component–allograft construct (Figs. 7.11, 7.12). If primary stability of the acetabular component cannot be achieved using this method, then a whole acetabular transplant must be done and a cup cemented into the allograft.

Results

The 177 acetabular revisions were divided by type using the intra-operative assessment as the final

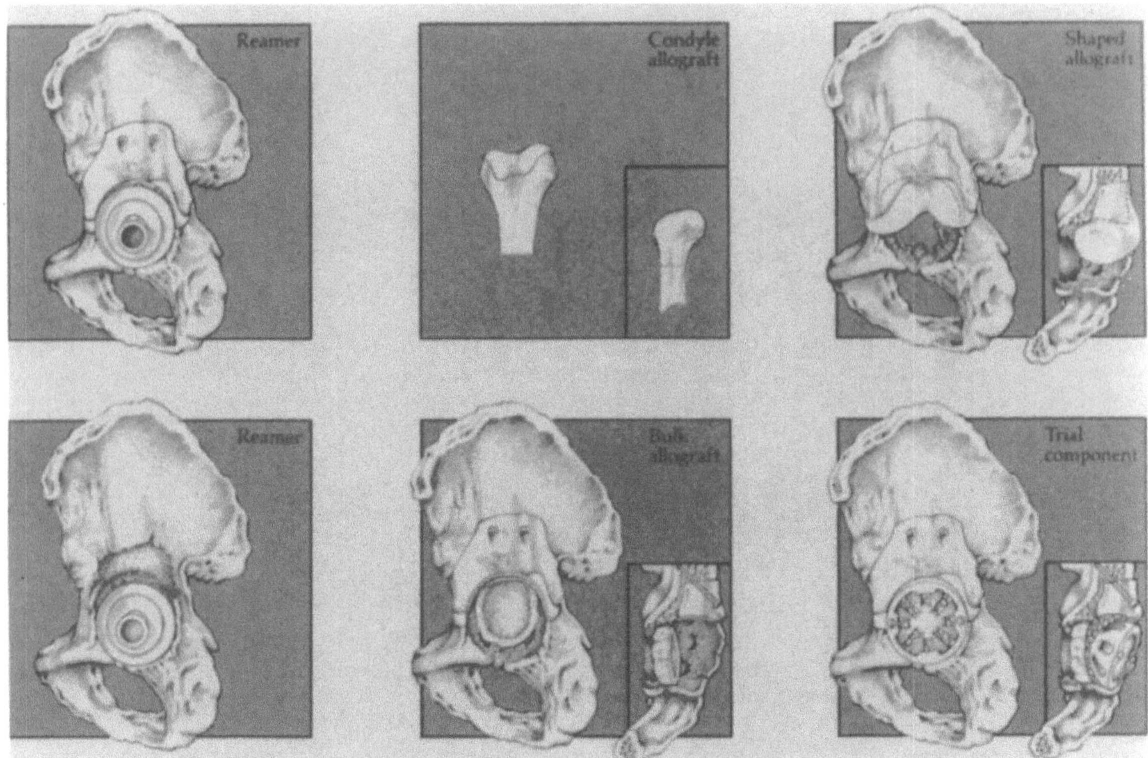

Fig. 7.9. Technique for Type 3A defects.

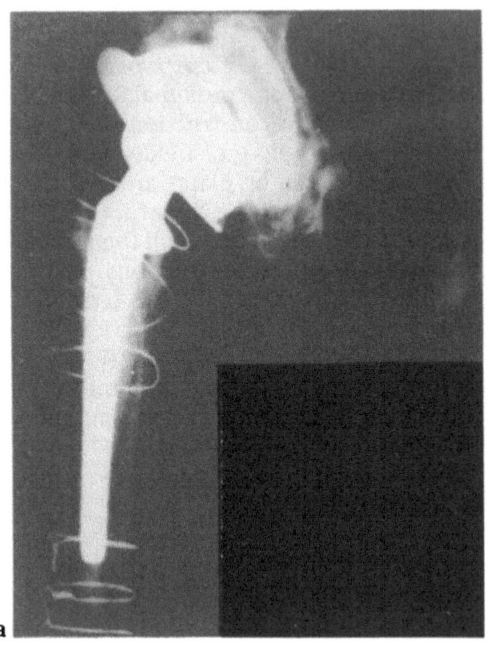

a

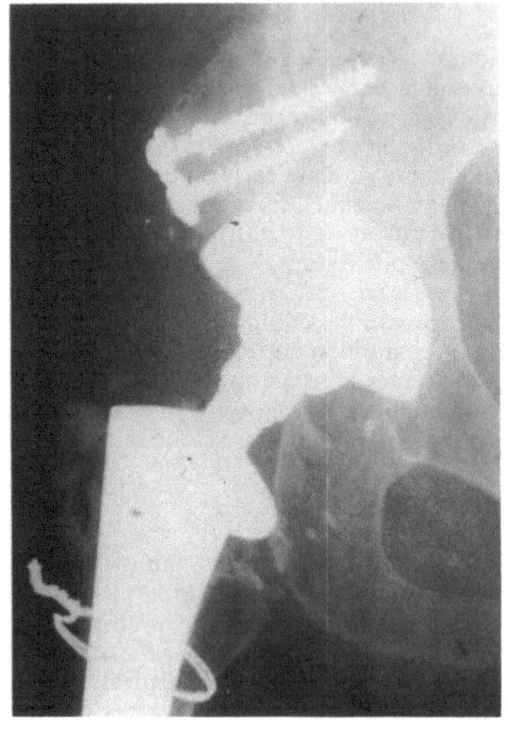

b

Fig. 7.10. A Type 3A acetabular defect reconstructed with a distal femur "Figure 7" graft: **a** Pre-operation; **b** Four years post-operation.

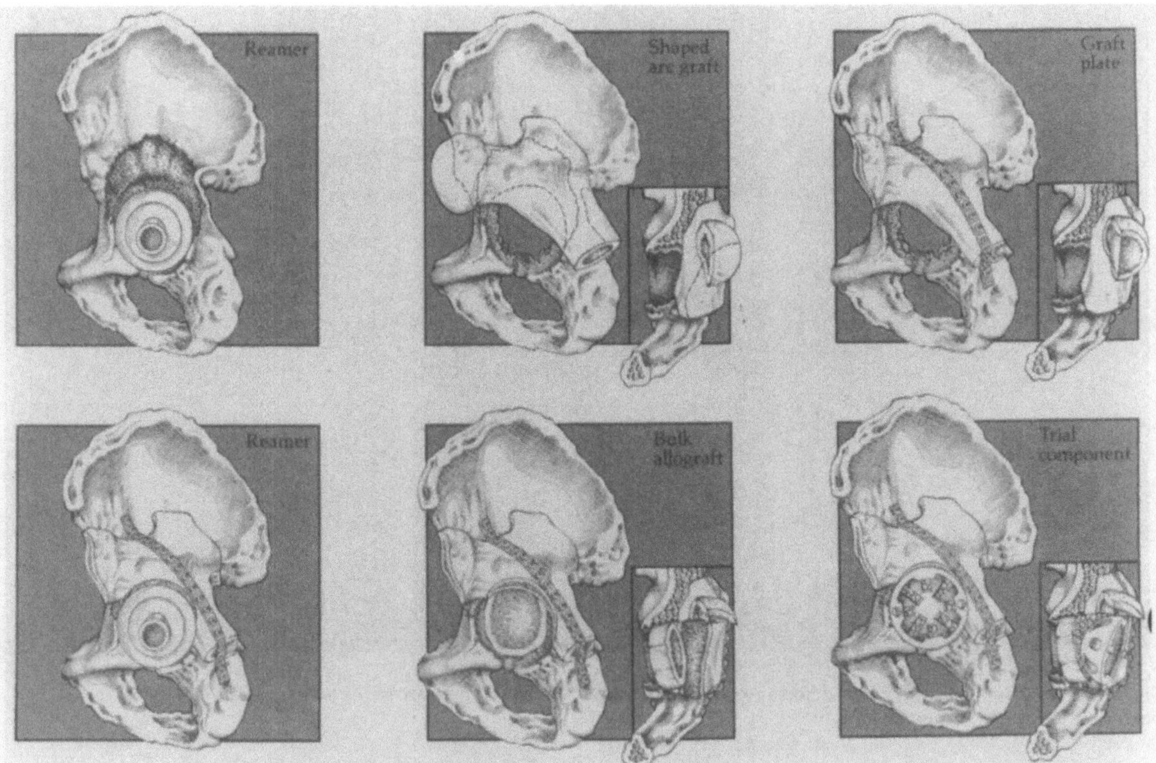

Fig. 7.11. Technique for Type 3B defects.

Table 7.5. The 177 acetabular revisions divided by type

Type	*n*
1	44
2	98
3	35

Table 7.6. Patients subdivided into type of grafting pattern

Type	*n*
1	44
2A	29
2B	63
2C	6
3	35

criteria (Table 7.5). Patients were further subdivided into types of grafting pattern (Table 7.6). Thirty-five hips required structural grafting; particulate or augmentation graft was used in the remainder.

The best way to measure component-construct stability is by obtaining AP X-rays of the pelvis at annual follow-up to measure migration in relation to the tear drop, and determine whether there have been changes. If the cementless acetabular implant migrates, it does so into its own lucency. Looking only at radiolucent zones in a cementless implant is not reliable. The method of Ranawat and Dorr is the most useful radiographic indicator of component failure.

We had 96% stable interface with the porous-coated components. There were five failures with migration of more than 5mm (Table 7.7). The stable, porous implants were all found to have grafts which were united. There was no graft resorption as long as they were stressed. There was some evidence of radiographic union by three months.

Table 7.7. Radiographic results with porous-coated components (n = 134)

Interface	*n*	%
Stable	129	96
Unstable		
Questionably	3	2
Definitely	2	2

The 43 threaded component cups have all shown some degree of migration and six have been exchanged. Threaded components should

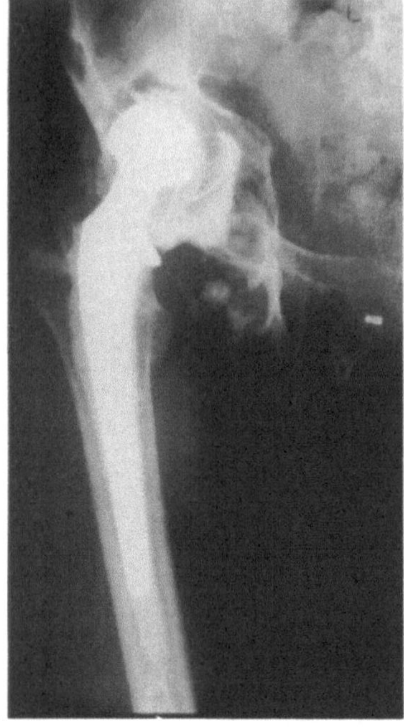

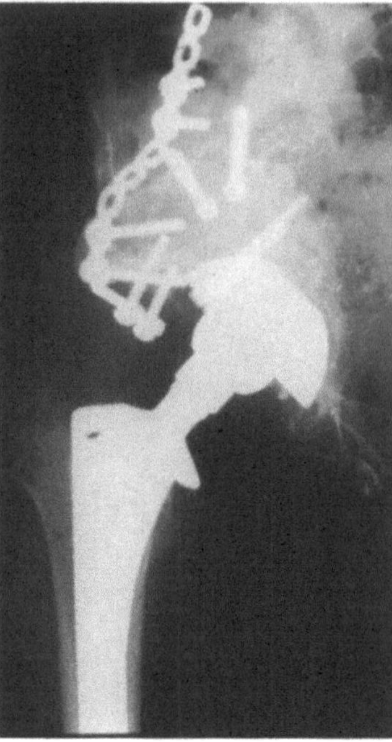

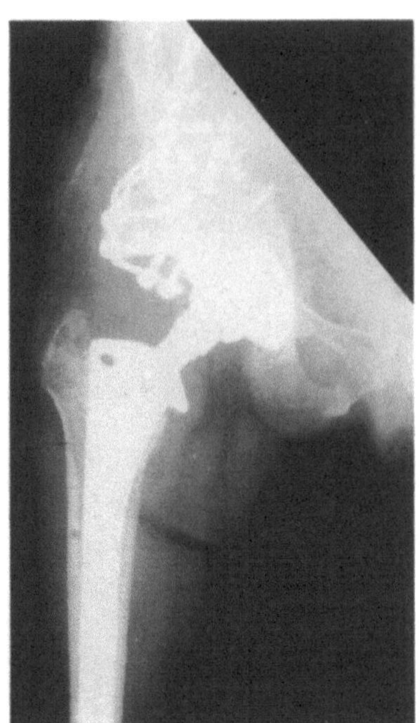

Fig. 7.12a–c. A Type 3B acetabular defect reconstructed with a proximal femur arc graft: **a** Pre-operation; **b** Two years post-operation; **c** Four years post-operation.

not be used in revision surgery, especially when structural allografts are used.

Pain relief was evaluated subjectively pre- and post-operatively with the protocol using the D'Aubigne and Postel Scale for pain and walking.

The average pre-operative score was 4. The average post-operative score was 8.8. The average score for the threaded and bipolar cups was 7.3. The porous-coated cups did significantly better and, other than the five which had migrated and loosened, the patients appeared very satisfied with their results at a mean of five years (Table 7.8).

Table 7.8. Comparison of pain relief with threaded cups (n = 43) and porous cups (n = 134)

	Threaded cups (%)	Porous cups (%)
Start-up pain	48.8	12.7
No start-up pain	44.3	84.9
Constant pain	6.9	2.4

Complications

Reconstructive procedures of this magnitude on multiple operated hips have more complications than do primaries. There were two gram positive streptococcus infections, two sciatic nerve palsies and eight dislocations. There were also three pulmonary emboli and two myocardial infarctions.

Discussion

Reconstruction of the failed acetabulum varies tremendously from case to case. Immediate and long term successful results depend on the ability to recognise pre-operatively and intra-operatively the different levels of difficulty in reconstruction. It is imperative to determine whether or not the acetabular rim which remains will support an uncemented hemispherical acetabular component. In the majority of cases, the rim will support the component and a predictable good to excellent result will occur in almost all cases. However, in those instances where the rim will not support an acetabular component, a structural support allograft should be used. The host–allograft composite which has restored the acetabular rim must provide rigid stability for the acetabular component.

Recognition of such defects and reproducible surgical reconstructive techniques are of paramount importance in order to ensure satisfactory results. We feel that certain mandatory criteria must be followed when using structural allograft in acetabular reconstruction.

Fresh frozen cadaveric proximal tibias and distal femurs must be used. The trabeculae of the allografts must be oriented in an oblique fashion to coincide with the direction of loading of the acetabular component.

The allograft–host construct must be loaded by the acetabular component. The bone graft must be securely fixed to the host independent of the acetabular component. The component must be 1–2mm undersized from the last reamer used. The graft–host composite must provide primary stability to the component, with screws into the component acting only as secondary fixation.

We also recommend that rigid stable fixation with a porous-coated implant be achieved, and that threaded components or bipolar components be avoided since migration is inevitable, resulting in a high degree of failure. In all cases where the initial fixation of the host–graft junction was stable and component migration was minimal, re-operation was easily performed using a larger porous coated component which was inserted into the stable host–allograft structure without the use of any additional bone graft.

All patients must be carefully evaluated annually both clinically and radiographically. Standing AP pelvis X-rays must be done to assess migration so that early revision can be recommended before severe bony destruction occurs. It is our contention that strict adherence to these rules will help to achieve satisfactory and long term results.

8 Acetabular Augmentation in Cemented Arthroplasty: Pre-operative Assessment and Surgical Technique

T.J.J.H. Slooff

In the Netherlands the rate of exchange of one or both components of a total hip arthroplasty ranges from 10% to 20%. In our institution the rate is 10%, of which 30% are combined with autogenous and homogenous grafts. The reasons for failure include aseptic loosening, infection, recurrent dislocation, excessive wear and osteolysis. Aseptic loosening is the major cause.

The patient with a poor result following total hip replacement warrants our full attention. Before the patient is admitted for clinical treatment, it is essential to diagnose the cause of the loosening. It is clear that the surgical approach is quite different between a patient with an infected or an aseptic loosened prosthesis.

Pre-operative Assessment

The assessment of the diagnosis usually becomes clear on the history, medical evaluation, laboratory testing, serial X-rays, nuclear arthrography and the IgG scan. I want to stress the last three investigations in particular.

The careful comparison of serial X-rays in two or more views is essential. Progressive changes are a prerequisite for the radiological and clinical diagnosis of loosening. Migration and subsidence of the components and progressive osteolysis can be assessed when abnormal motion is gross in serial radiographs.

Nuclear arthrography is a more reliable technique than subtraction arthrography since it has a lower false positive and negative rate. It seems to be a more accurate test than the usual technetium scan, and more sensitive than conventional X-ray and arthrography alone. This method consists of a standard technetium scan in order to visualise the prosthesis and adjacent osseous structures. The patient is asked to walk a short distance after the radiographic arthrography in order to increase the intra-capsular pressure. Then scintigraphic images are obtained of the total hip in four views from different angles. The technetium scan and indium images are made with the patient in exactly the same position. Superimposing the region of interest on the indium image on the bone scan image reveals the presence of extra-articular accumulation of the radioactive indium contrast (Fig. 8.1a–d).

The next investigation concerns the human, non-specific, polyclonal immunoglobulin G (IgG) scintigraphy (Fig. 8.2a–c). This technique is especially valuable for detecting low-grade

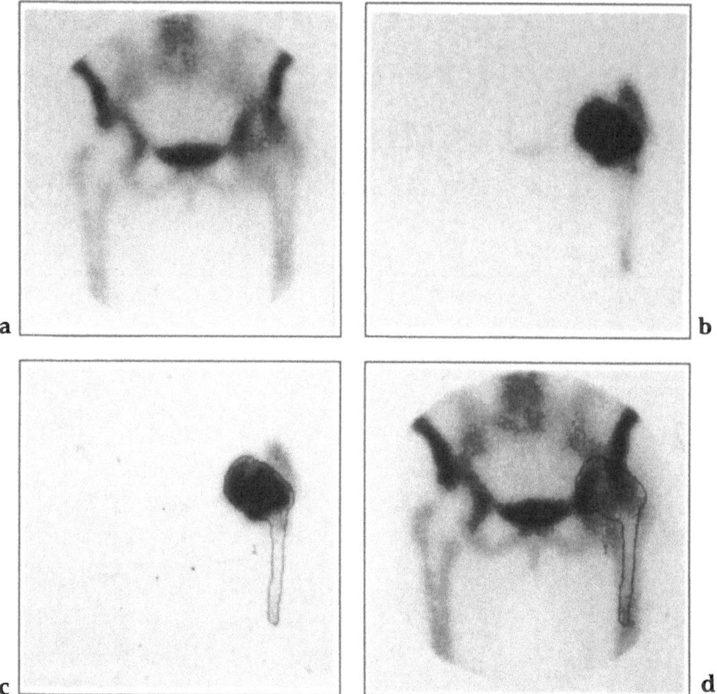

Fig. 8.1a–d. Nuclear arthrography. **a** Technetium-99m-MDP bone scan. **b** Intra-articular indium-111 image. **c** Region of interest drawn on the indium image. **d** Region of interest superimposed on the bone scan image.

infections. IgG is labelled with radioactive indium-111 and injected intravenously. Images of the suspected area of infection are obtained at 4, 24, and 48 hours post-injection. This study is interpreted as positive if focally increasing activity over time is noted. We found IgG scintigraphy to be highly sensitive in detecting infection and/or inflammation.

The indication for revision arthroplasty is primarily the loss of bone, followed by pain and disabled function. Surgery should be considered even before symptoms appear. It is the loss of bone stock which is a very important indication to revise the implant. A periodic yearly follow-up of patients with a total hip arthroplasty is therefore necessary to prevent extensive loss of bone. By this means we control the loosening process very carefully, in order to prevent the use of massive allografts.

Careful pre-operative planning is essential. The evaluation and assessment of hip mobility, the state of the soft tissue, the leg length discrepancy,

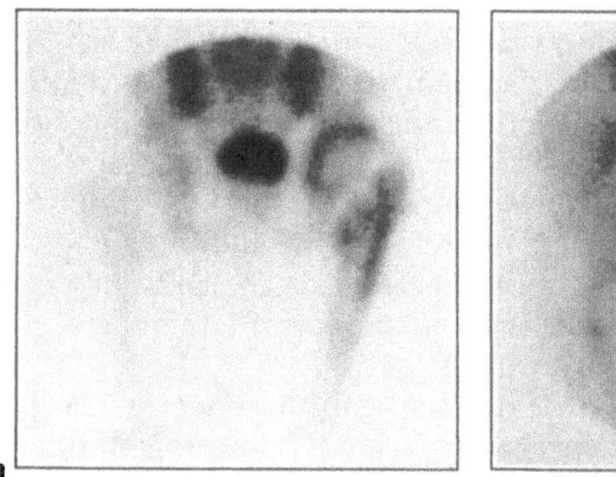

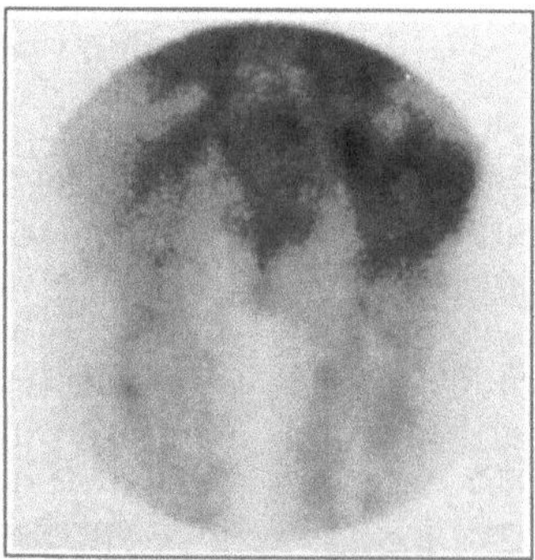

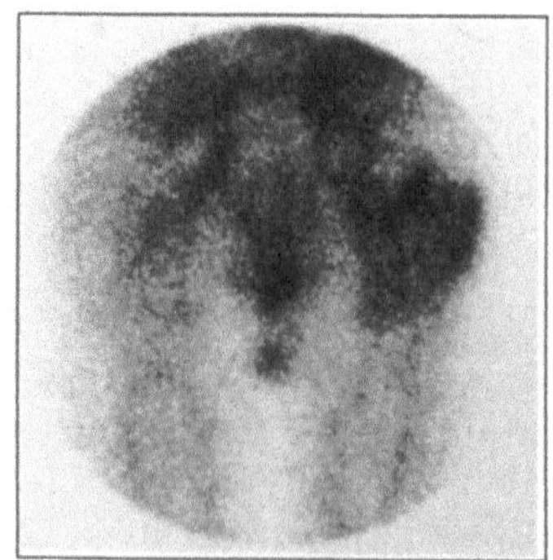

Fig. 8.2a–c. Immunoglobulin G scintigraphy. **a** Technetium-99m-MDP bone scan. **b** Indium-111 labelled IgG 24 hours post-injection. **c** Indium-111 labelled IgG 48 hours post-injection.

age, activity and the general condition of the patient are important. Pre-operatively a decision has to be made between direct or delayed exchange and resection arthroplasty. There is a need for optimal facilities with which to carry out such surgery: availability of time, a skilled team, special instruments and a set of normal sized implants.

A pre-operative work-up also encompasses a careful study of the X-rays to evaluate the severity of the anatomic distortion, the size and the location of bone stock deficiencies and the defects. Attention must also be directed to the femoral cement column and the extension and protrusion of the cement in and out of the acetabulum.

The revision procedure is much more difficult than a primary arthroplasty. Because of excessive scar tissue formed around loose components it takes more time and has a higher infection and complication rate with a less satisfactory overall result. The neurovascular structures can also be endangered by the anatomic distortions.

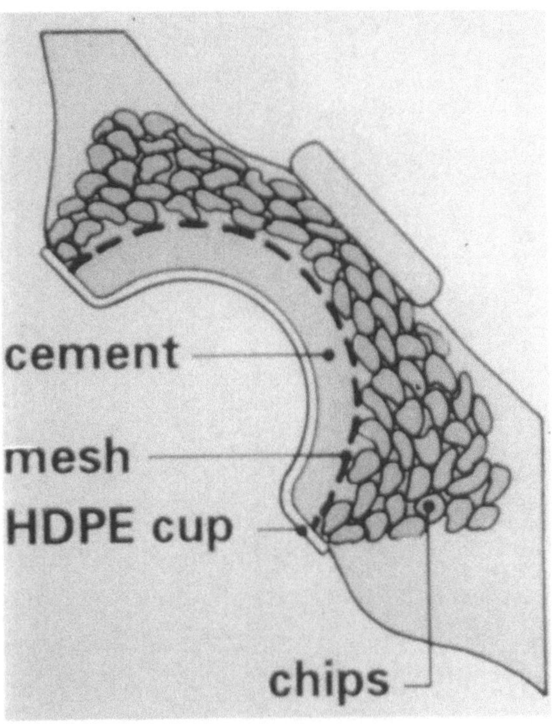

Fig. 8.3. Schematic view of morsellised bone for the socket.

Technique

The surgical repair is directed towards the anatomical location of the centre of rotation, the loss of bone stock and the acetabular containment. These deficiencies must be restored. Technically we use wire mesh to contain the graft, cancellous chips to augment the loss of bone stock, antibiotic-loaded cement and conventional implants.

We use acrylic cement for fixation of the component in the graft. We made this choice based on our experience in more than 400 revisions. Experience with a technique can only be obtained if this technique is not changed or adapted frequently, as is seen with many uncemented systems nowadays. The initial and long-lasting fixation by means of mechanical interlock over a large area is still the best available, especially with the modern cementing techniques. Moreover, using cement is followed immediately by the relief of pain. Based on these considerations, we have felt that there is, to date, no reason to change this fixation concept. We try to refine the cement by using vacuum mixing of the cement and pressurising techniques.

We chose cancellous particulate morsellised chips for the augmentation technique in the acetabulum and femur (Figs. 8.3, 8.4). The defects and perforations can be filled and closed with cancellous chips; the thin sclerotic wall can be augmented and the smooth surface will be replaced by the rough surface of the chips, which provides mechanical and structural interlock for the cement. The chip-like grafts (Fig. 8.5) are rigidly impacted and moulded in such a way with the trial socket that a consistent new layer of bone replaces the defective acetabulum (Fig. 8.6a,b).

The incorporation of this bone graft represents a predictable sequence of events. In particulate cancellous allograft, the process starts with an osteoblastic activity in contrast to the osteoclastic activity in the case of a cortical graft. This means that the incorporation of a cancellous allograft will occur without decrease in the total mechanical strength.

With regard to the incorporation of the graft, there is a difference between cancellous chip-like allograft and solid bone. In the case of chips there is rapid incorporation, complete substitution and retention of mechanical resistance. The ingrowing host bone will be directly loaded and not stress-shielded as with solid graft, provided that the fixation of the allograft is supported by adequate containment.

We must bear in mind that in grafting procedures there is a distinction between

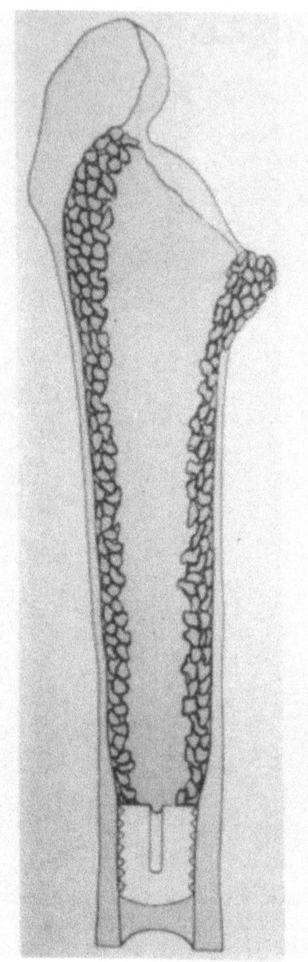

Fig. 8.4. Schematic view of morsellised bone for the femur.

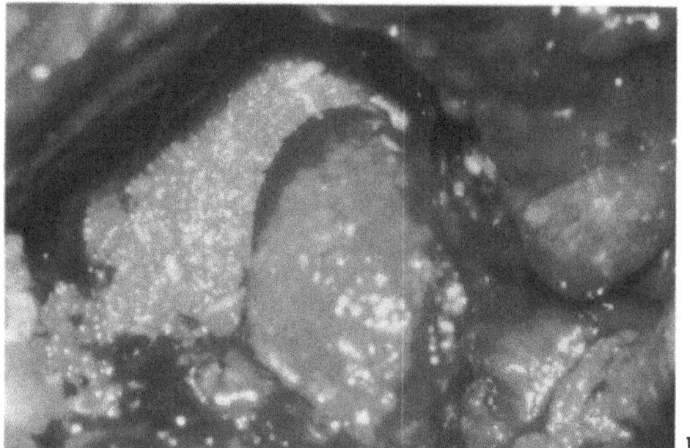

Fig. 8.6. a Impaction of cancellous chips by socket trial prosthesis. b Final result of impacted chips.

consolidation and incorporation of the graft (Fig. 8.7a,b). Consolidation means bony union between the host bone and the graft and this process occurs in weeks, whereas the incorporation process, that is reorganisation into a structural viable bone, will take months. The incorporation of the graft is difficult to assess on X-rays. We use as parameters the homogeneity of the graft and the direction of the bony trabeculae, which must be in accordance with the host bone (Fig. 8.8).

Fig. 8.5. Preparation of particulate morsellised bone chips.

The host bed must be prepared with care for an optimal result of the grafting procedure. Sclerotic bone must be roughened to expose vascular tissue; good contact and secure fixation must be achieved. This will be provided by impaction of the cancellous chips (Fig. 8.5). This impaction in the acetabulum will be provided by using the successive socket trial prostheses (Fig. 8.6a,b). After this impaction the socket is cemented in place using antibiotic loaded cement.

Conclusions

In our opinion, the success of a revision total hip arthroplasty is dependent on the assessment of the cause of the loosening, the periodic follow-up of patients, rigorous pre-operative planning, the use of particulate cancellous chips, a meticulous technique of grafting and cement fixation.

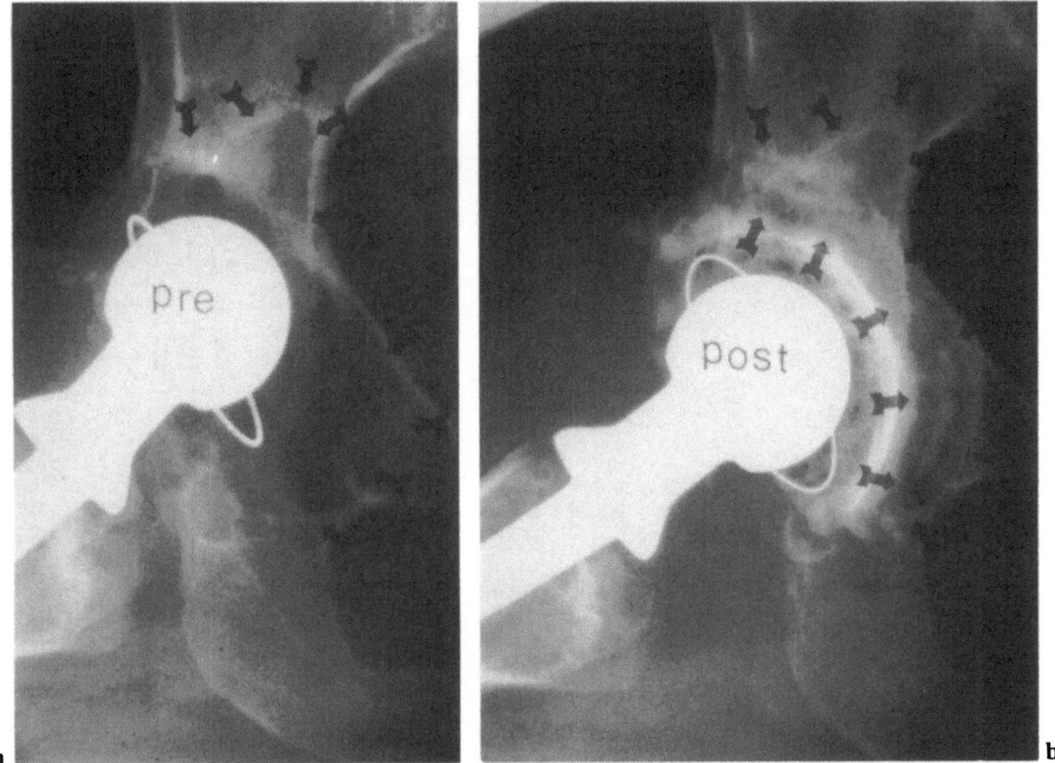

Fig. 8.7. a Pre-operative radiograph. Arrows show extent of bone defect. **b** Post-operation radiograph showing incorporation of bone grafts. Downward arrows refer to the graft–host interface, upward arrows to the graft–cement interface. Between the arrows is the incorporated graft.

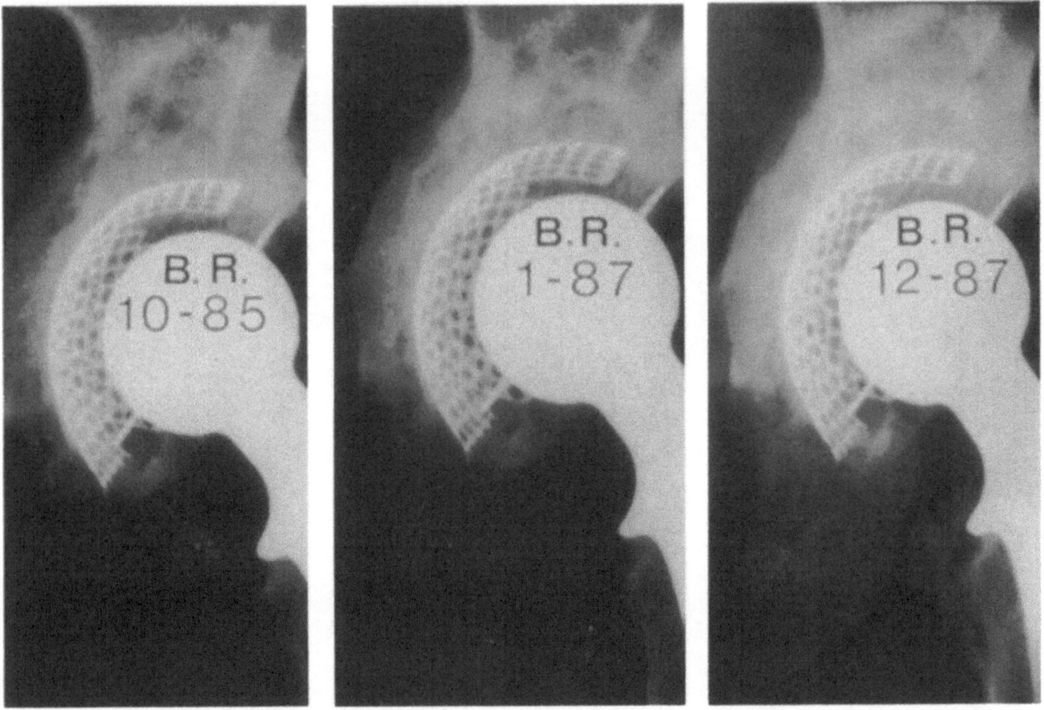

Fig. 8.8. Serial radiographs of acetabulum following augmentation with bone chips.

9 Acetabular Reconstruction with Cancellous Bone Grafts in Revision Hip Arthroplasty: A 10-year Follow-up Study

J.W. Schimmel

The main principles in acetabular reconstruction are restoring the centre of rotation, acetabular continuity and integrity, and replacing the subchondral bone layer with a metal mesh. From a biomechanical point of view it is very important that a stress pattern in the acetabular region is created comparable to the normal situation (Crowninshield et al. 1983). A trend towards a more biological reconstruction with bone grafts has therefore taken place, instead of reinforcing the acetabulum with massive amounts of nonviable materials. We strongly support this biological view, and use an acetabular bone grafting technique developed by Slooff (Slooff et al. 1984) in which we use morsellised bone as much as possible instead of solid blocks.

Technique

Our surgical technique starts with the posterolateral approach, which provides sufficient exposure in most cases. Trochanter osteotomies are seldom necessary. After removing the component and cement, the acetabular bed must be carefully prepared and any residual membrane removed. Sclerotic bone is roughened and multiple perforations are made in the sclerotic surface. Special attention is paid to correct the anatomical location of the socket and to assess where the graft must be applied. A helpful landmark is the transverse acetabular ligament.

Frozen femoral head allografts from the bone bank are used. No matching takes place. Chips are prepared with a rongeur during surgery. Any perforation of the medial wall is closed with a cortico-cancellous shell. The extended acetabulum is filled with a mass of chips and, using the socket trial prosthesis, these are impacted and moulded in such a way that the cup will be correctly positioned anatomically. After reconstructing the acetabulum, the chips are covered with a thin vitallium mesh.

Post-operative care includes anticoagulation therapy, antibiotic prophylaxis, indomethacin therapy to prevent heterotopic ossifications, controlled passive motion and mobilisation after six weeks.

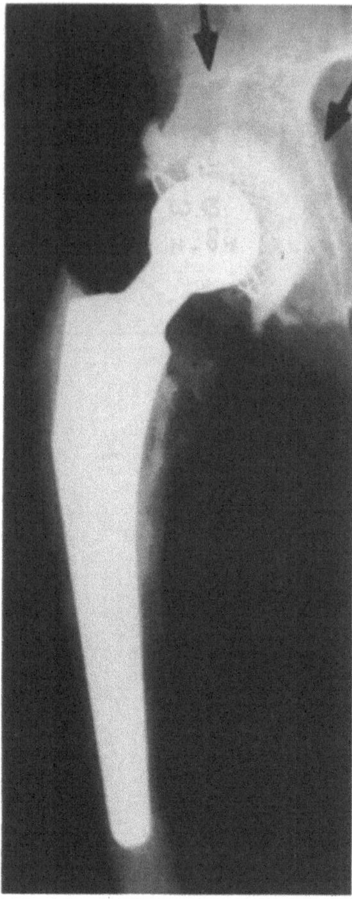

Fig. 9.1. A superior/medial cavity defect after reconstruction with impacted cancellous chip allografts.

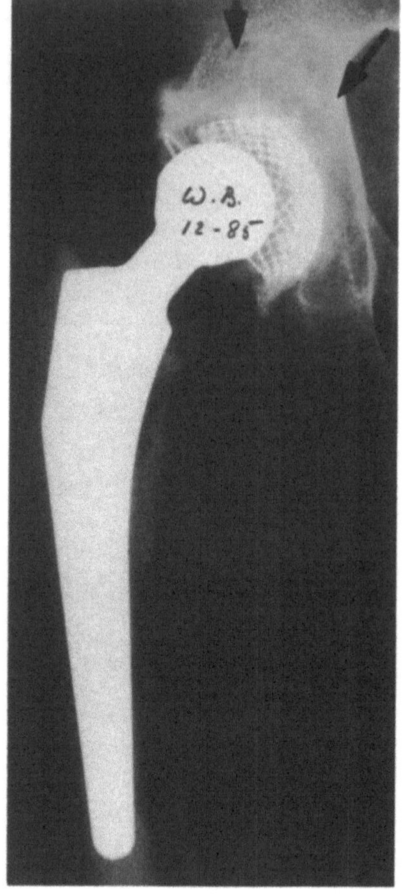

Fig. 9.2. Same hip as in Fig. 9.1 thirteen months after reconstruction showing a continuous trabecular pattern and a homogeneous bone density.

Results

A clinical and radiographic study was made of 97 revision hip arthroplasties in 90 patients between 1979 and 1988. For the follow-up study 84 hips were available in 77 patients with a mean age of 62.4 years. The mean follow-up was 5.3 years, ranging from two and a half to ten years.

We classified acetabular defects according to the American Academy of Orthopaedic Surgeons (AAOS) classification system, which has three basic categories: segmental, cavitary and combined defects (D'Antonio et al. 1989). A segmental defect is any complete loss in the supporting hemisphere of the acetabulum including the medial wall. Cavity defects represent a volumetric loss in bony substance of the acetabular cavity, including the medial wall, but the acetabular cavity and the acetabular rim remain intact.

The most frequent defects we saw were of the superior/medial cavity type in 32% of hips (Figs. 9.1, 9.2). In 21%, defects were of the combined medial segmental/cavity and superior cavity type. There was a continuous pattern in homogeneous bone density. The third group were the medial cavity type of defects in 16%. The last group of defects were the combined superior segmental/cavity and medial cavity type in 13% (Fig. 9.3). The rest of the defects in this series could not be grouped because there were several combined defects.

Morsellised cancellous grafts were used in the majority of hips. Fifteen hips had a combination with cortico-cancellous shell to close the medial wall defects. Most reconstructions were performed with chip allografts.

A modified Harris hip score, including pain, function, activity level and range of motion was used to evaluate our clinical results. The score

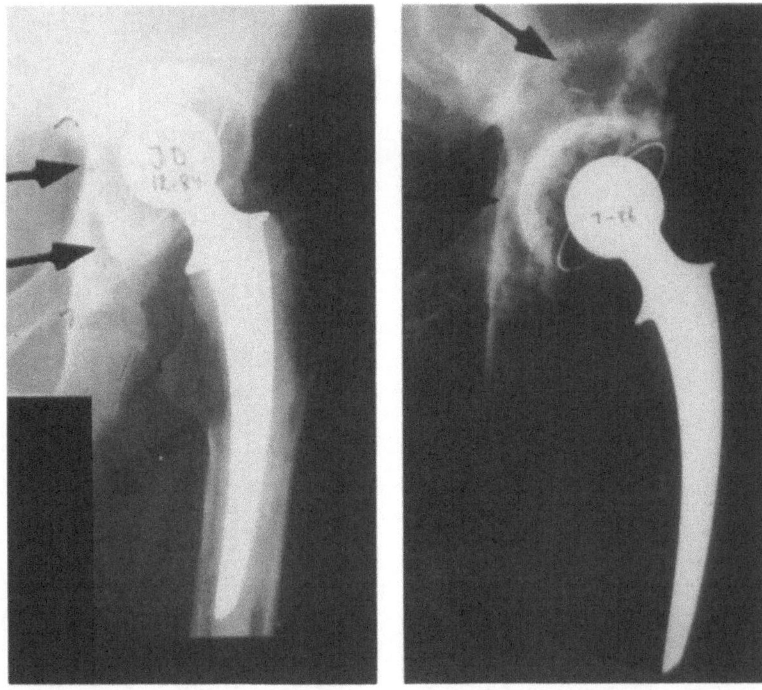

Fig. 9.3. A combined superior segmental/cavity, medial cavity type defect before and after reconstruction with chip allografts.

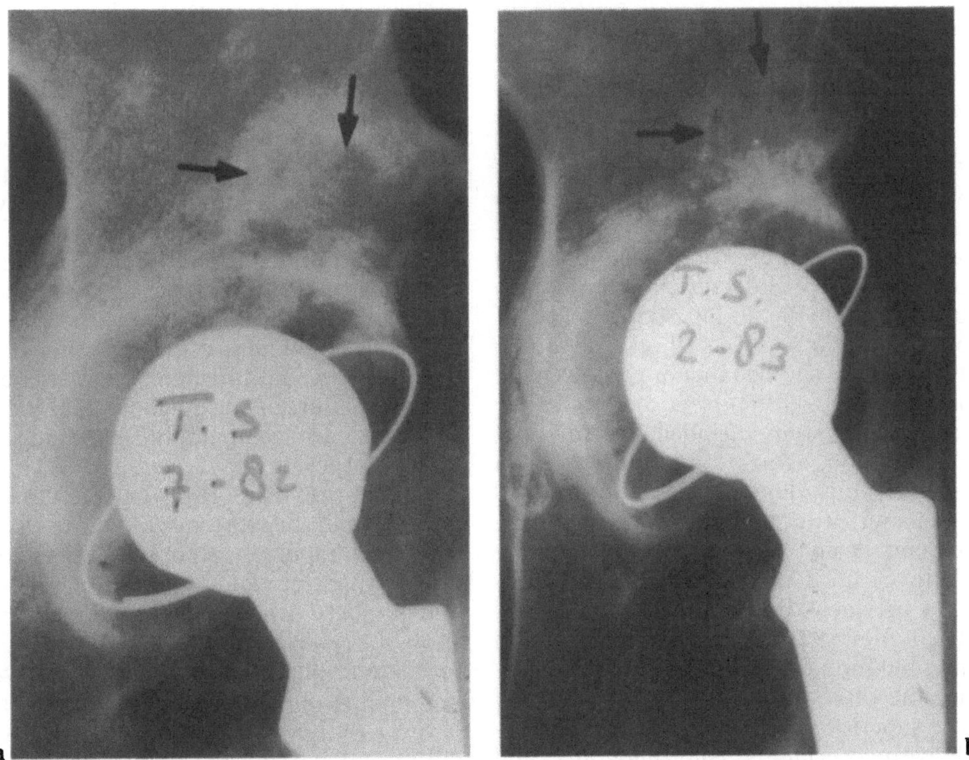

Fig. 9.4. a A reconstructed superior segmental defect. **b** Seven months after reconstruction with chip allografts showing full radiographic incorporation.

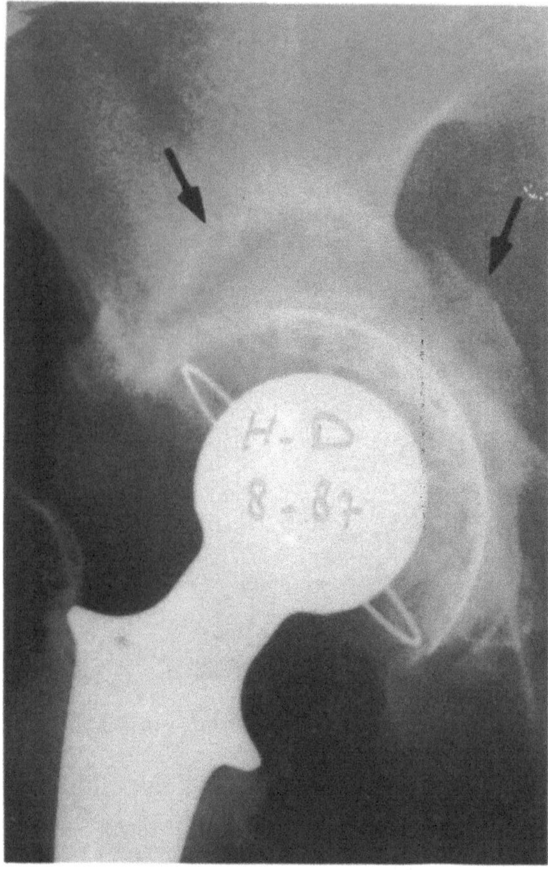

Fig. 9.5. Acetabular reconstruction with chip allografts immediately post-operative.

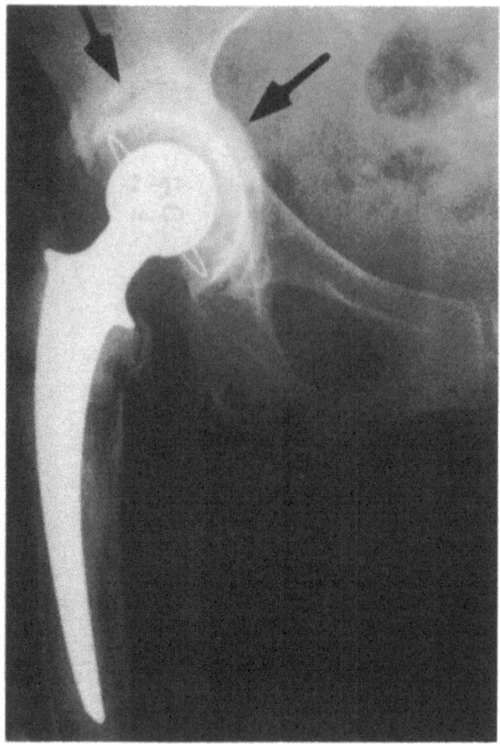

Fig. 9.6. Same hip as in Fig. 9.5 three and a half years after reconstruction showing a graft incorporation failure: there is no obliteration of the graft–host bone interface and no homogeneous bone density.

range was from zero to 100 points. The mean post-operative score was 80 points. In 68 hips the score was 70 points or more. Two implants had to be removed because of infection. In comparison with post-operative Harris hip scores in other series in recent literature (Oakeshott et al. 1987, Marti et al. 1990), it showed that our results were in the same range.

Our radiographic study included assessment of graft incorporation using Conn's criteria of a homogeneous bone density and a continuous trabecular pattern (Conn et al. 1985), migration measurements using Sutherland's technique (Sutherland et al. 1982), and a radiolucency assessment between the bone and cement (De Lee and Charnley 1976). Full graft incorporation was seen in 80 hips (95%) with a variable incorporation time of six months to one and a half years. The superior segmental defect was reconstructed with cancellous chip allografts and showed full incorporation radiographically with a continuous trabecular pattern and homogeneous bone density after seven months (Fig. 9.4). Obvious migration greater than 5mm according to Sutherland's technique, or progressive continuous radiolucency in the bone/cement interface after full graft incorporation, which was defined as radiographic loosening, was assessed in three hips (3.6%).

We had nine failures in this series of 84 hips: two hips became infected; there was an incorporation failure, using the radiographic criteria, in four hips; and radiographic loosening after full graft incorporation was seen in three hips. There seems to be no relationship between the use of an allograft and the existence of failure.

A radiographic incorporation failure of a combined superior and medial segmental/cavity defect closed with a cortico-cancellous shell is shown (Fig. 9.5). Three and a half years after operation, consolidation of the medial wall defect is apparent, but there is no continuous trabecular pattern or homogeneous bone density. Radiographically we consider that this graft is not fully incorporated, although the patient is doing very well (Fig. 9.6).

Conclusion

The technique of acetabular reconstruction presented shows clinical and radiographic incorporation in the majority of hips. The reconstruction provides adequate socket support. Clinically and radiographically there was no apparent difference between the use of auto- and allograft in this series.

References and Further Reading

Conn RA, Peterson LFA, Stauffer RN, Ilstrup D (1985) Management of acetabular deficiency: long term results of bone grafting the acetabulum in total hip arthroplasty. Orthop Trans 9:451

Crowninshield RD, Brand RA, Pederson DR (1983) A stress analysis of acetabular reconstruction in protrusio acetabuli. J Bone Joint Surg (Am) 65:495–499

D'Antonio JA, Capello WN, Borden LS, Bargar WL, Bierbaum BF, Boettcher WG, Steinberg ME, Stulberg SD, Wedge JH (1989) Classification and management of acetabular abnormalities in total hip arthroplasty. Clin Orthop 243:126–137

DeLee JD, Charnley J (1976) Radiological demarcation of cemented sockets in total hip replacement. Clin Orthop 121:20–32

Marti RK, Schuller HM, Besselaar PP, Vanfrank Haasnoot EL (1990) Results of revision of hip arthroplasty with cement. A five to fourteen year follow-up study. J Bone Joint Surg (Am) 72:346–354

Oakeshott RD, Morgan DAF, Zukor DJ, Rudan JF, Brooks PJ, Gross AE (1987) Revision total hip arthroplasty with osseous allograft reconstruction. Clin Orthop 225:37–61

Slooff TJ, Huiskes R, van Horn J, Lemmens AJ (1984) Bone grafting in total hip replacement for acetabular protrusion. Acta Orthop Scan 55(6):593–596

Sutherland CJ, Wilde AH, Borden LS, Marks KE (1982) A ten-year follow-up of one hundred consecutive Muller curved stem total hip replacement arthroplasties. J Bone Joint Surg (Am) 64:970–982

10 Revision of the Acetabular Component

H.P. Chandler

Choice of Acetabular Components

Uncemented hemispherical acetabular components, impacted into an acetabulum that has been under-reamed by 1 to 2mm, can be used in virtually all primary and revision hips. Although the component may be stable with a press fit alone, if screws are necessary, it is important to avoid the anterior superior "quadrant of death" because of potential danger to the iliac vein and artery (Wasielewski et al. 1990, Keating et al. 1990). Cement is never necessary even with massive bone grafts unless contact with living bone is less than 10%.

Bipolar devices, used in conjunction with morsellised bone, have been advocated in revision total hip replacement surgery when acetabular bone stock is deficient (Wilson and Scott 1990). Despite early enthusiasm, the results of this technique have not been as successful as with fixed uncemented hemispherical components. Bipolar components are contra-indicated for use with bulk structural grafts as they will routinely erode through the graft (White and Cook 1989).

Reaming Techniques

When reaming an undistorted acetabulum, the first reamer selected should just fit within its confines. This reamer should then be directed medially until the desired depth of the acetabulum is achieved. Subsequent larger reamers should be used until further reaming would compromise the anterior or posterior walls of the acetabulum. If a small initial reamer is carelessly positioned eccentrically, subsequent enlarging reamers can potentially ream away the anterior or posterior wall of the acetabulum (Fig. 10.1).

When reaming a grafted acetabulum or a dysplastic acetabulum that does not need a graft, the surgeon should start with a small reamer that is carefully centred so that the final reamers will remove equal amounts of bone from the anterior and posterior rim. This first reamer should be directed centrally until the inner table is encountered. The reamer is then orientated in the anatomical position and subsequent larger reamers are

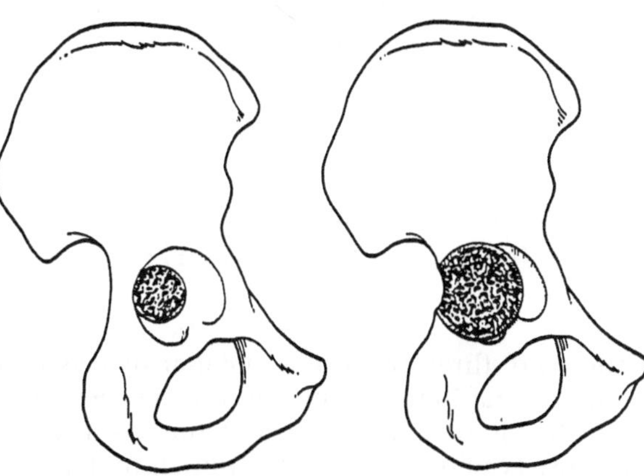

Fig. 10.1. Technique to ensure central reaming.

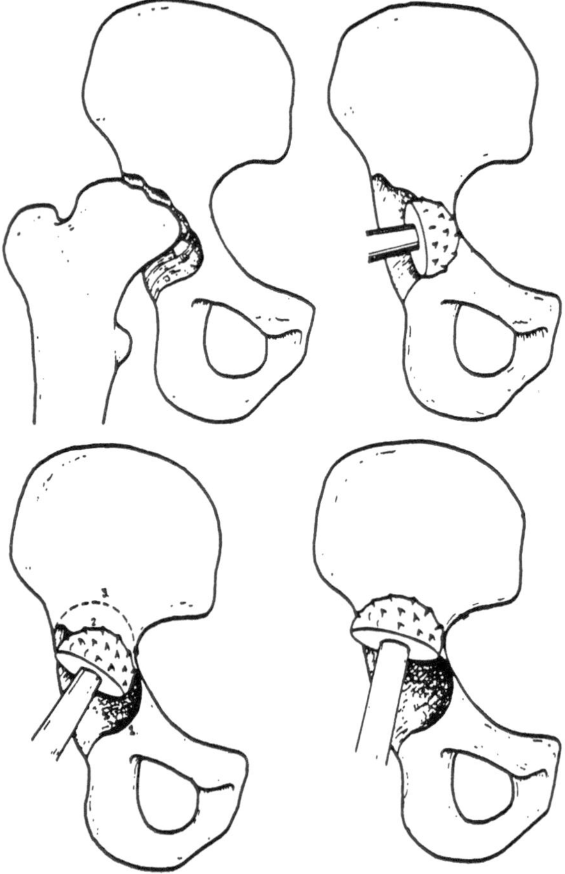

Fig. 10.2. Reaming technique for a grafted acetabulum or a dysplastic acetabulum that does not need a graft.

used until further reaming would compromise rim bone stock (Fig. 10.2). Because minor medial wall perforations are so easily dealt with by using morsellised bone or iliac strip grafts, I do not hesitate to cautiously violate the medial wall if doing so will alleviate the need for a structural weight-bearing peripheral graft (Fig. 10.3).

Types of Acetabular Defects that Require Bone Grafts

There are three types of acetabular defects that require grafting. The first is the deficient rim which requires a structural graft, the second is the enclosed cavity defect and the third is perforation of the medial wall. Combinations of these defects are often present.

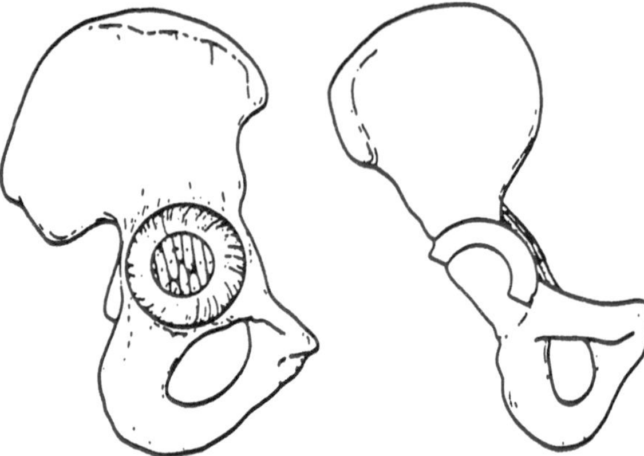

Fig. 10.3. Cautious violation of the medial wall to alleviate the need for a structural weight-bearing peripheral graft.

High Centre of Rotation

A high centre of rotation is not detrimental in itself and can be compensated for by a thicker polyethylene insert, a femoral component with a longer neck, or by distal transplantation of the greater trochanter combined with a shoe lift (Russotti and Harris 1988). However, it should be remembered that the iliac wing narrows dramatically above the true acetabulum and becomes only 6mm thick. Without posterior rim support, the acetabular component can break out when it is subjected to the tremendous forces that are generated by rising from a chair or climbing stairs (Fig. 10.4) (Hodge et al. 1989).

A failed cemented acetabular component that slowly migrates proximally will often have remodelling of the bone about the posterior rim and there may be adequate bone stock to work with at the higher level. However, a dysplastic hip in a false acetabulum does not have this potential and it is more often necessary to use a superior bone graft to bring the hip down to the true acetabulum where there is adequate posterior rim stock with which to work.

Structural Weight-Bearing Grafts

In the past, solid acetabular grafts were more frequently necessary because there was a limited assortment of acetabular components and

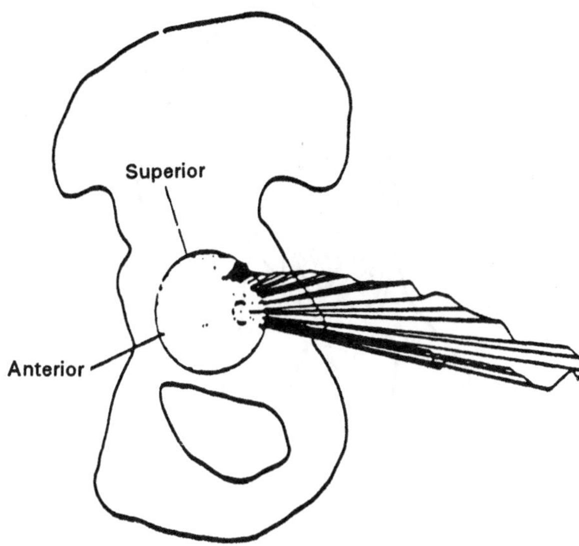

Fig. 10.4. The pressures on the posterior wall of the acetabulum are tremendous when rising from a chair. (From Hodge et al. (1989).)

because it was necessary to use cement. With the variety of sizes of modern uncemented components, the need for solid grafts is less common. If a stable press fit of an uncemented acetabular component can be obtained at the periphery of the rim, this is preferable to a structural graft. However, in those situations where the rim is too deficient to support the acetabular component, structural weight-bearing grafts are still necessary and we have found that they do work, despite reports that they consistently fail (Jasty and Harris 1988, Mulroy and Harris 1990). Of 38 such grafts in our series, followed for a minimum of eight years and an average follow-up of 11 years, the survival of the graft was 97.4%. Failure of structural acetabular grafts is directly related to technique.

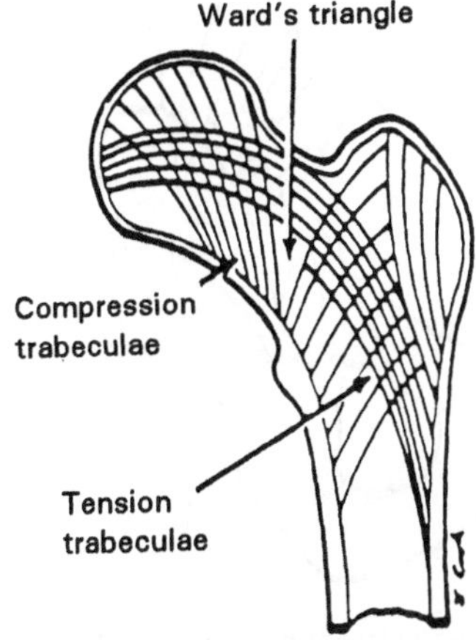

Fig. 10.5. The trabeculae of the donor graft must be realigned in the weight-bearing axis.

Factors that Affect the Outcome of Structural Weight-Bearing Grafts

Choice of Graft Material

Femoral heads can be used if they are harvested from vigorous donors and consist of strong bone. Ideally, these should be donated by younger patients. Osteoporotic heads from elderly sedentary females are acceptable for morsellised bone but are not ideal for structural weight-bearing grafts. Distal femurs or proximal tibias, obtained from formal harvest of young healthy donors, are probably the best graft material especially for large defects.

Trabecular Orientation

In life, the trabeculae of the donor graft were aligned in a complex orientation in response to the forces to which the bone was subjected. Since structural grafts are non-viable when they are used in surgery, it is important to realign the trabeculae once more in the weight-bearing axis (Fig. 10.5). It is difficult to malalign the trabeculae of cortical bone (commonly used with femoral reconstruction) but cancellous bone can potentially be placed with the trabecular pattern transverse to the weight-bearing forces. In such circumstances, the trabeculae can and will fracture and the graft will routinely fill.

Fixation of the Graft

Grafts must be positioned beneath a viable buttress of host bone. This buttress prevents proximal migration of the graft. Care must be taken to ensure the stable host position it assumed when the patient was standing and it is helpful to mark the head and acetabulum in this orientation before the head is excised. Remnants of cartilage on the graft can be removed by an oscillating saw (Fig. 10.6) and from the acetabulum by means of a large curette (Fig. 10.7). With allografts, concentric reamers can be used, but with complex acetabular contours, it is sometimes helpful to

Fig. 10.6. Remnants of cartilage on the graft can be removed with an oscillating saw.

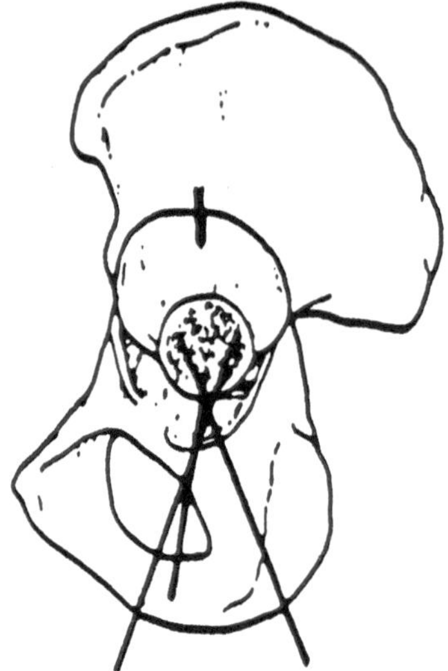

Fig. 10.8. An allograft femoral head usually fits best in the weight-bearing position and it is helpful to make this before the head is excised. The graft is temporarily fixed against the buttress by three closed "X" wires.

make a methacrylate mould of the acetabulum to help in shaping the graft.

Sometimes the graft can be stabilised to the buttress by a press fit, but it is more commonly necessary to use two or three additional fixation screws. Temporary fixation can be achieved by three smooth K-wires, placed in the area of the graft where the new acetabulum will be reamed (Fig. 10.8). This leaves room at the periphery of the graft where final fixation screws can be placed. Cancellous chrome cobalt lag screws (Howmedica, Rutherford, NJ), orientated in the line of weight-bearing, are preferable to fully threaded bolts or screws unless the graft is over-drilled to allow potential impaction against the buttress. It should be re-emphasised that fixation devices do not in themselves bear weight but are used only to hold the graft against the buttress (Fig. 10.9).

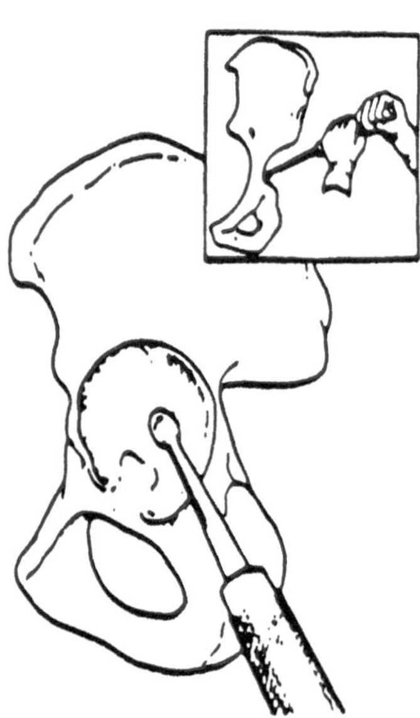

Fig. 10.7. A large curette is helpful in the removal of soft tissues from the acetabulum.

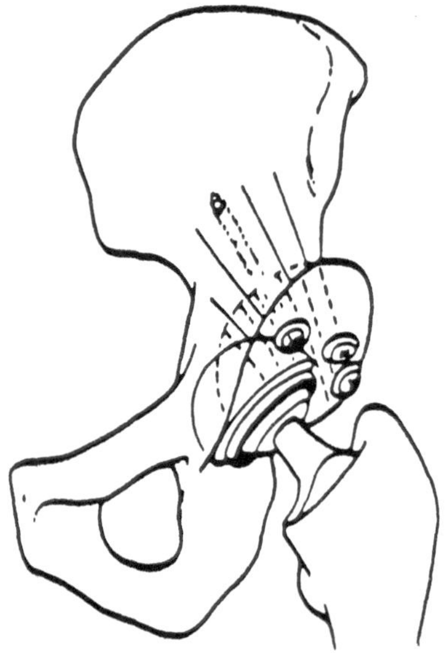

Fig. 10.9. The graft placed beneath the buttress to achieve final fixation and the trabeculae and lag screws aligned with the weight-bearing axis.

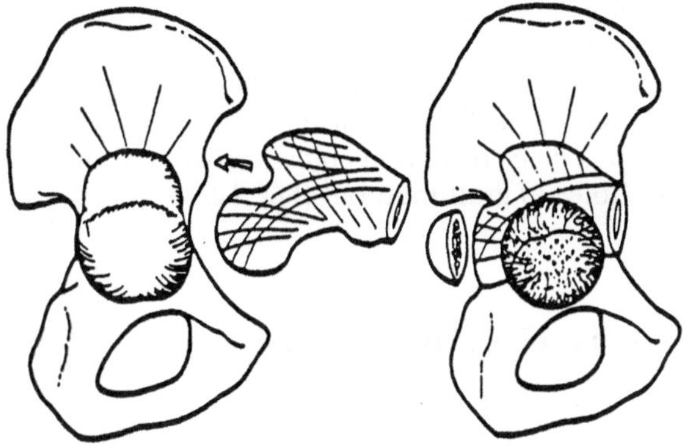

Fig. 10.10. A proximal allograft femur consisting of the head, neck and greater trochanter, used to reconstruct the superior and posterior acetabular rim.

It is much more difficult to reconstruct the posterior rim than it is to graft for superior coverage. However, if the posterior rim is insufficient, the acetabular component will routinely break out with normal stresses such as rising from a chair or climbing stairs. If the superior and posterior portions of the acetabulum are both insufficient, they can be reconstructed together by a single proximal allograft femur consisting of the head, neck and greater trochanter fixed with lag screws. Care should be taken to orientate the compression trabeculae to be in line with the superior compression forces (Fig. 10.10). If there are soft tissue contractures that make it impossible to move the acetabulum distally where there is natural posterior bone stock to work with, an acetabular reconstruction plate must be used in conjunction with a graft and must extend from the ischium to the ilium (Fig. 10.11).

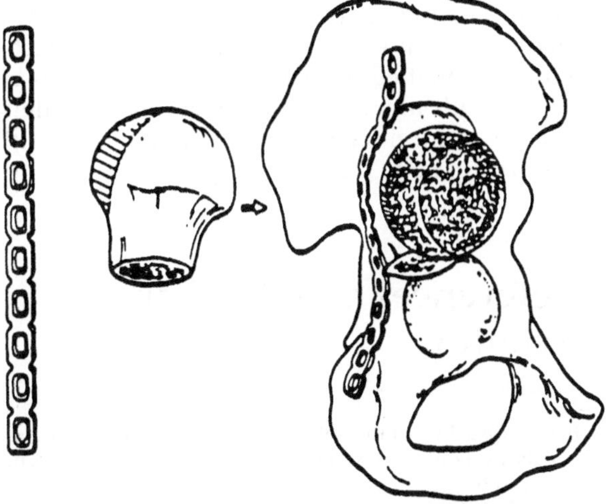

Fig. 10.11. An acetabular reconstruction plate used in conjunction with a graft extending from the ischium to the ilium.

Incorporation of the Graft

Cancellous bone (used exclusively for acetabular reconstruction) incorporates by apposition of new bone on to old trabeculae and therefore becomes stronger if the trabeculae are orientated in the line of weight-bearing forces (Springfield 1987). As with fracture healing, a piezo electric force is probably important in the healing and incorporation of grafts used in total hip replacement (Bassett and Becker 1962). Early weight-bearing is therefore important. A graft that is not stressed will uniformly absorb.

Post-operative Management

Since acetabular grafts are exclusively cancellous bone, they are capable of immediate full weight-bearing if they are placed beneath a firm buttress and if the trabeculae and cancellous screws are orientated in the line of the weight-bearing forces. These grafts will actually get stronger with time and we allow immediate protected weight-bearing of 25 to 35kg for six weeks and then weight-bearing as tolerated, governed only by the patient's comfort and by the status of the soft tissues – particularly the abductors. The status of union or incorporation of the graft does not influence the decision as to when to allow weight-bearing.

Cavity Defects (Cavity Type II)

If the surrounding bone is strong enough to support the acetabular component, cavity defects should be packed with morsellised bone or reamings obtained from either the femur or the acetabulum (Fig. 10.12). It is probably better to mix cancellous autograft with allograft if the latter is used. Small and large cement hole defects appear to consistently fill in if packed with morsellised bone. Even large contained defects such as those seen with protrusio acetabuli can be packed with cancellous bone if the rim is over-reamed and supports the acetabular component. Solid grafts, used to fill large protrusio defects, will also incorporate but are technically more demanding and not necessary.

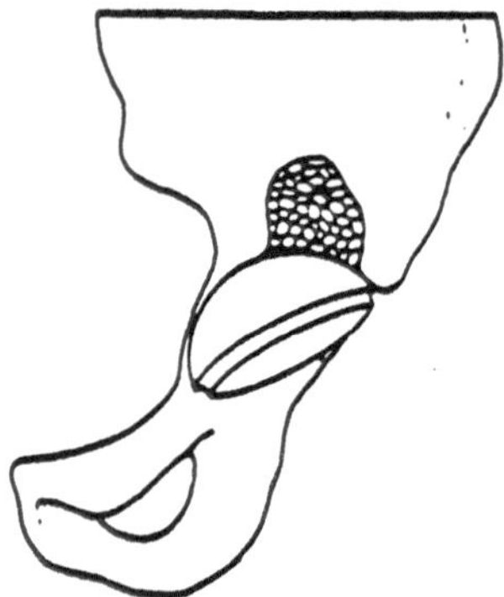

Fig. 10.12. Even significant intra-acetabular defects can be managed by morsellised bone if it is contained and is not required for structural support.

Medial Wall Perforations (Central Segmental Type 1B)

The acetabulum with a central defect should be cautiously reamed so that a hemispherical acetabular component larger than the defect can be supported by the remaining peripheral bone. The defect itself can be grafted with any type of bone including autogenous iliac strips, morsellised bone, or solid allograft bone. The acetabular component acts as a mould and the graft is pushed against it by intra-abdominal pressures. All types of grafts unite quickly because of the rich blood supply of the iliacus and perhaps because of the piezo electric effect related to breathing and the valsalva manoeuvre. All grafts will quickly remodel to form a medial wall that is close to the normal thickness of an uninvolved acetabulum. Because cortical and cancellous strips from the patient's iliac crest are usually available, they are my choice of graft material (Fig. 10.13).

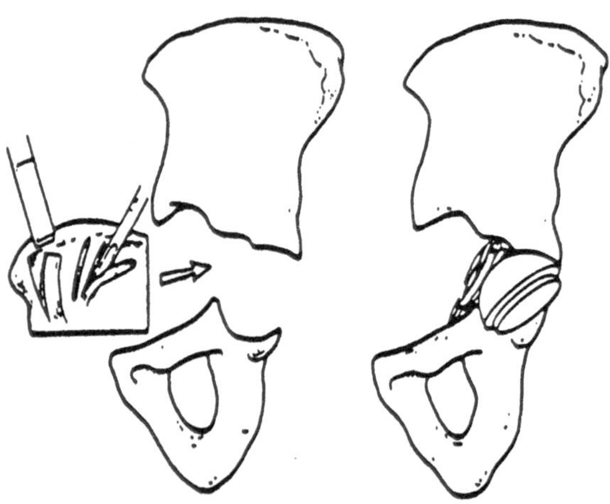

Fig. 10.13. Cortical and cancellous strips when available from the ipsilateral iliac crest used for graft material with perforation of the medial wall.

Pelvic Discontinuity

If both columns are incompetent, at least one column should be reconstructed by an acetabular reconstruction plate, and it is my preference to reconstruct the posterior column in this situation because it experiences much more force than the anterior one. If the posterior column is intact, it is not necessary to reconstruct the anterior column just because it is deficient. However, if the posterior column is deficient, even if the anterior column is intact, it is more likely that a reconstruction plate will be necessary for the posterior column. Grafts may be independently necessary to reconstruct defects so that an uncemented acetabular component can be made stable.

Conclusions

Uncemented hemispherical acetabular components can be used in virtually all primary or revision acetabular reconstructions. Major rim (peripheral segmental type 1A) or intra-acetabular defects (cavity type II) may need structural weight-bearing autografts or allografts. The success of such grafts is directly related to technique. Contained defects (cavity type II) can be treated with morsellised bone. Medial wall deficiencies (central segmental type 1B) do well with any type of bone grafts.

References and Further Reading

Bassett CAL, Becker RO (1962) Generation of electrical potentials by bone in response to mechanical stress. Science 137:1063

Hodge WA, Carlson KL et al. (1989) Contact pressures from an instrumented hip endoprosthesis. J Bone Joint Surg (Am) 71:1378–1386

Jasty M, Harris WH (1988) Results of total hip surgery with structural femoral head allografting for major acetabular defects. Orthop Trans 12:716

Keating EM, Ritter MA, Faris PM (1990) Structures at risk from medially placed acetabular screws. J Bone Joint Surg (Am) 72:509–511

Mulroy RD, Harris WH (1990) Failure of acetabular autogenous grafts in total hip arthroplasty. J Bone Joint Surg (Am) 72:1536–1540

Russotti GM, Harris W (1988) High placement of the acetabular cup: A long-term follow-up study. Orthop Trans 12:90–91

Springfield DS (1987) Massive autogenous bone grafts. Orthop Clin North Am 18(4):249–256

Wasielewski RC, Cooperstein LA, Kruger MP, Rubash HE (1990) Acetabular anatomy and the transacetabular fixation of screws in total hip arthroplasty. J Bone Joint Surg (Am) 72:501–508

White RE, Cook J (1989) Resorption of large intra-pelvic bone grafts for medial acetabular defects in cementless revision total hip replacement. AAOS 56th Annual Meeting, p.46

Wilson MG, Scott RD (1990) Reconstruction of the deficient acetabulum using the bipolar socket. Clin Orthop 251:126-133

Part IIB

Hip: The Femur

11 Banked Allograft Bone for Proximal Femoral Deficiency

A.E. Gross

This is a presentation of proximal femoral grafts, but intra-medullary grafting is not described. If most of the femoral cortices are intact, it is best to insert morsellised bone inside the cavity and use a press fit implant. This chapter concerns cortical grafts that are necessary when host bone is inadequate for intra-medullary grafts (Fig. 11.1a,b).

The cortical grafts are non-circumferential strut grafts and circumferential. Either a proximal femur or a proximal tibia made to look like a proximal femur is used for large fragment femoral grafts longer than 3cm. A long stem femoral component is used with cement against the allograft but no cement in the host. Revision surgery is very difficult, although not impossible, if cement is used with a long stem femoral component. We do not cement the host femur therefore, but do not hesitate to cement the allograft.

Grafts

Grafts are fixed with a step-cut and cerclage wire and the patient's residual bone is wrapped around the allograft. The autograft bone has a soft tissue envelope, therefore it has a blood supply, so there is viable bone graft around. The greater trochanter is rewired to the allograft. A special allograft prosthesis is used with flutes at the distal end. The donors of these femurs and tibias have much smaller medullary canals than the average patient undergoing revision. The prosthesis has been designed so that it is narrow at the top and it is unnecessary to ream away the allograft to insert a reasonable sized stem. Union occurs and there will be stable reconstruction if there is a stable step-cut with a good bone graft and sound cerclage technique. It is impossible to get a press fit in long grafts into metaphyseal bone. The stability will depend on the step-cut and cement to the allograft.

Case Reports

Case 1

The patient had massive loss of proximal femoral bone stock after three total hip replacements. A larger proximal femoral allograft was necessary (Fig. 11.1a,b).

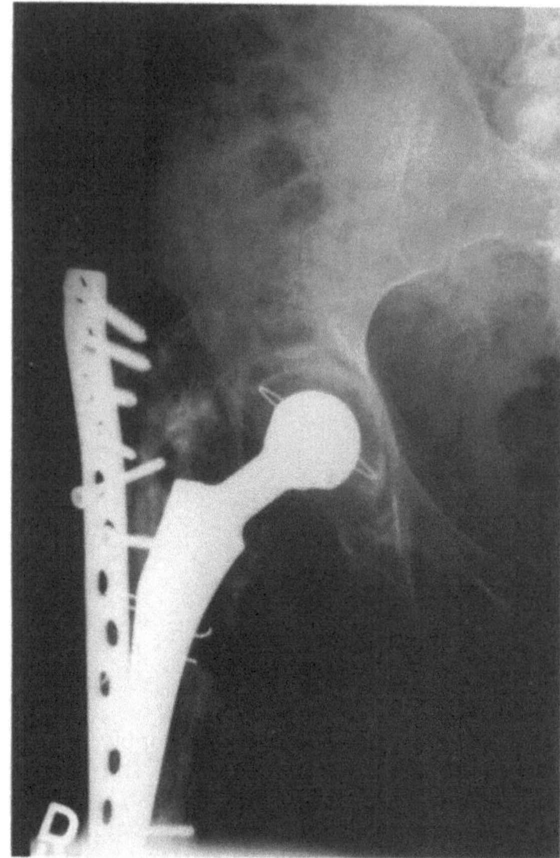

a

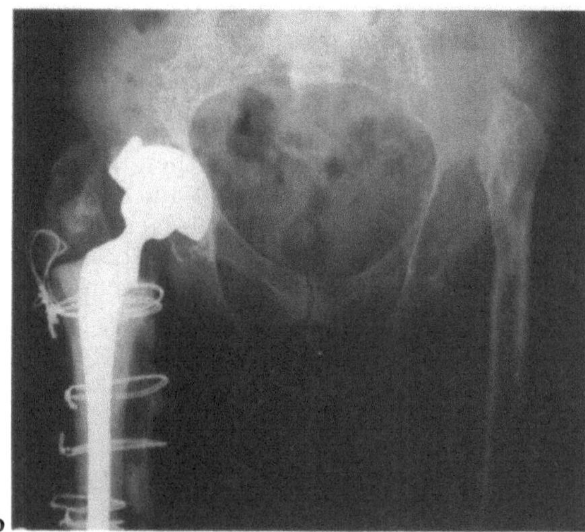

b

Fig. 11.1. a Radiograph of right hip illustrating massive loss of proximal femoral bone stock. **b** Five years after proximal femoral allograft.

Case 2

If there is non-union at the junction between allograft and host bone it is necessary to plate and bone graft them but we try not to plate these grafts initially. It is very important to recognise the weaknesses of allografts; the less violation of their cortical integrity the better. Holes are not put through them except to re-attach the greater trochanter, and cement is used to make the allografts even stronger.

We prefer proximal femoral allografts over tumour prostheses because bone and soft tissue cannot be re-attached effectively to a tumour prosthesis. There is also a much better gradation of forces down the shaft than is possible with a tumour prosthesis where all the forces would be focused at the junction to host bone. Also, a proximal femoral allograft creates a much better situation for the patient should a further operation be needed.

The calcar grafts are less than 3cm and the results are not as good. In most patients grafts this short are not necessary because the length can be made up in the prosthesis.

Cortical strut grafts make excellent grafts especially for bypassing windows and stress fractures. The window is replaced and then a cortical strut graft applied. This should be longer than the window and will remodel extremely well. Copious autografts from the iliac crest placed around the junctions are always useful.

Results

A prospective study of 40 patients with proximal femoral allografts had a follow-up average of 36 months. A trochanteric escape of more than 1cm was present in only 21% of patients. There were three non-unions necessitating plating and bone grafting. There has been no significant resorption or fragmentation of these large grafts. In these difficult situations the success rate was 85% using our strict scoring protocol.

Complications of Revision Surgery Using Allograft Bone: Pelvic and Femoral

Where allograft has been used in revision surgery the infection rate is only 3.02%. That is acceptable for any revision series and even more so with allografts. It is probably low because if there is any suspicion of infection where an allograft is going to be used, surgery is done in two stages with three months between the operations.

The dislocation rate of 7% is acceptable for a revision series. Six excision arthroplasties were done, one for recurrent dislocation and the others for infection. There was one amputation for pain. Five deaths occurred; three were unrelated to revision surgery and one was due to haemorrhage.

Overall Results: Femoral and Pelvic

The success rate with morsellised allografts which contained cavitary defects was 94%, cortical struts 86% and large proximal femoral allografts 85%. The minor column bulk acetabular allografts were 85% successful but the larger ones only 62%. These were brought up to 81%, however, with an additional revision. Despite a high revision rate, it was possible to achieve the correct level because of the restored bone stock.

Discussion

In my opinion, cement may be used in allograft bone but not in host bone unless the patient is low demand. Cemented, long stemmed femoral components should not be used unless further surgery is highly unlikely.

The primary goal in revision arthroplasty is to restore the anatomy using uncemented components supported by host bone without allografts at all if possible. Bulk allografts are best avoided and the use of morsellised allograft bone preferable, but the problems presented to the surgeon may require a major decision. Should the anatomy be sacrificed in order to place uncemented components primarily against host bone? If this would raise the centre of rotation 2cm and support the cup against good host bone, that is the right choice, but if good host bone is not available at an acceptable level then bulk grafting is necessary.

Major pelvic allografts restore bone stock for the present but, more importantly, for the future. This is an advantage if they have to be revised since it can be done at the right level. Other problems such as resorption and fracture lead to a higher revision rate. This can be improved using better bone, cortical bone, true acetabular allografts, male femoral heads and distal femur which do not expose cancellous bone surfaces to the host granulation soft tissue. Better internal fixation is needed. Plates or rings from host bone to host bone should be used and when screws are necessary they should be in an oblique to vertical orientation.

12 Allograft Bone in Major Revision Hip Replacement Surgery

B. Loty and M. Postel

Major bone defects are frequently encountered in revision total hip arthroplasties. We report here the Cochin Hospital experience in the use of bone allografts for reconstructing the acetabular and femoral sides, to allow successful implantation of a new prosthesis. Our experience in revision prosthesis allografting began in 1975 and increased rapidly as revisions became more and more frequent.

Our bone bank provides two types of allografts to support this activity. Frozen femoral heads are used for all kinds of acetabular defects. The small femoral defects are also reconstructed with femoral heads, but irradiated massive cortical bones are used for major deficiencies of the femur.

Femoral Head Allografts

We began femoral head bone banking in 1975. Procurement is performed during total hip arthroplasty (Tomeno et al. 1988). Exhaustive examination and systematic serology allow strict donor selection to give ideal procurement conditions. Femoral bone samples are sent for bacteriological controls and cultured for a month. After harvesting, grafts are stored at −30 °C. Though we procure more than 200 femoral heads a year (Fig. 12.1) that number only meets our own need and all are implanted at Cochin Hospital.

Acetabular Reconstruction

Materials and Methods

Acetabula are reconstructed with one or several bone blocks shaped to fit the defect and receive the prosthesis (Fig. 12.2). Cartilage is removed and the grafts are impacted and screwed to obtain primary stability. Additional cruciate acetabular plate is used each time for large grafts or when the acetabulum is fractured (Fig. 12.3). We consider secure fixation mandatory for good graft healing and to prevent collapse during the remodelling period. A Charnley cemented cup is finally implanted.

Our first ten years' experience concerned 122 major acetabular reconstructions, excluding the

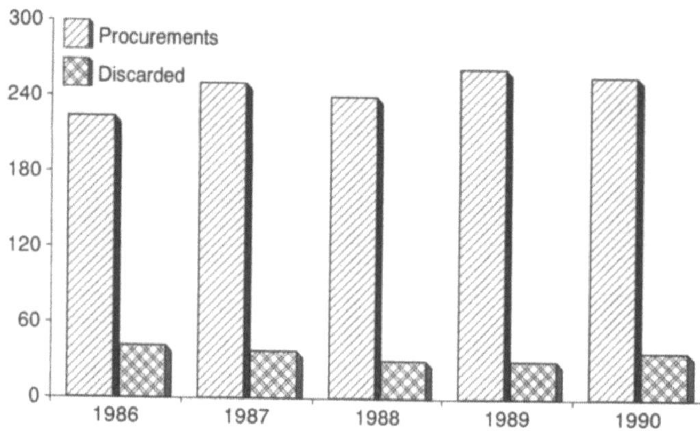

Fig. 12.1. Procurement of femoral heads at Cochin Hospital, 1986–1990.

small grafts. The follow-up period was one to ten years (Hedde et al. 1986).

Complications

Two infections occurred, but the graft did not appear to be the cause. Only one graft failure occurred, in a patient with a large acetabular defect operated on in 1984 (Fig. 12.4). The bone deficiency was reconstructed with a large complete femoral head secured only by screws. The graft progressively collapsed from the sixth month and the cup finally loosened completely after two years. At re-operation, the superior part of the initial graft was perfectly healed. The collapsed inferior part of the graft was reinforced with a new femoral head allograft and the fixation was then secured with a cruciate acetabular plate. The final result is good with a three year follow-up after re-operation.

Radiological Incorporation

Space between graft and host acetabulum disappeared between the sixth and the twelfth month, but incorporation still progressed during two or three years. At three year follow-up, all grafts showed union and consolidation and no secondary modification was observed (Fig. 12.5).

Small Femoral Defect Reconstruction

Banked femoral heads were also used for femoral grafting. The small grafts used for repairing a cortical hole fused and, in spite of superficial resorption, allowed a good cortical reparation (Fig. 12.6). Good results were also obtained from

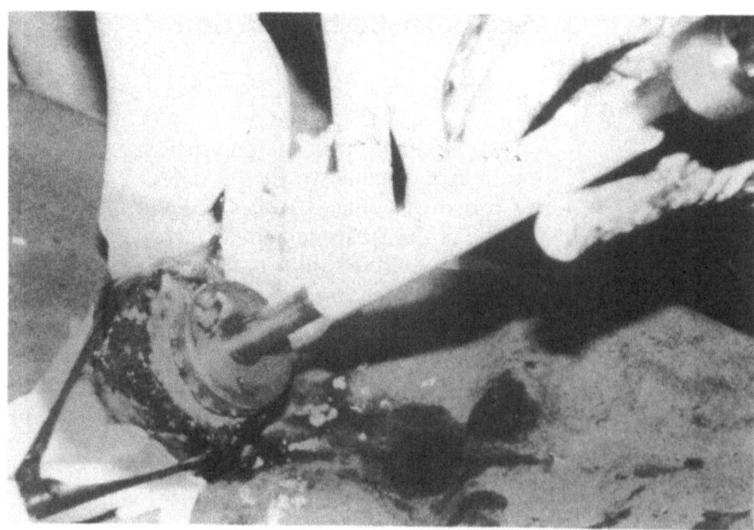

Fig. 12.2. Acetabulum reconstruction: femoral head preparation.

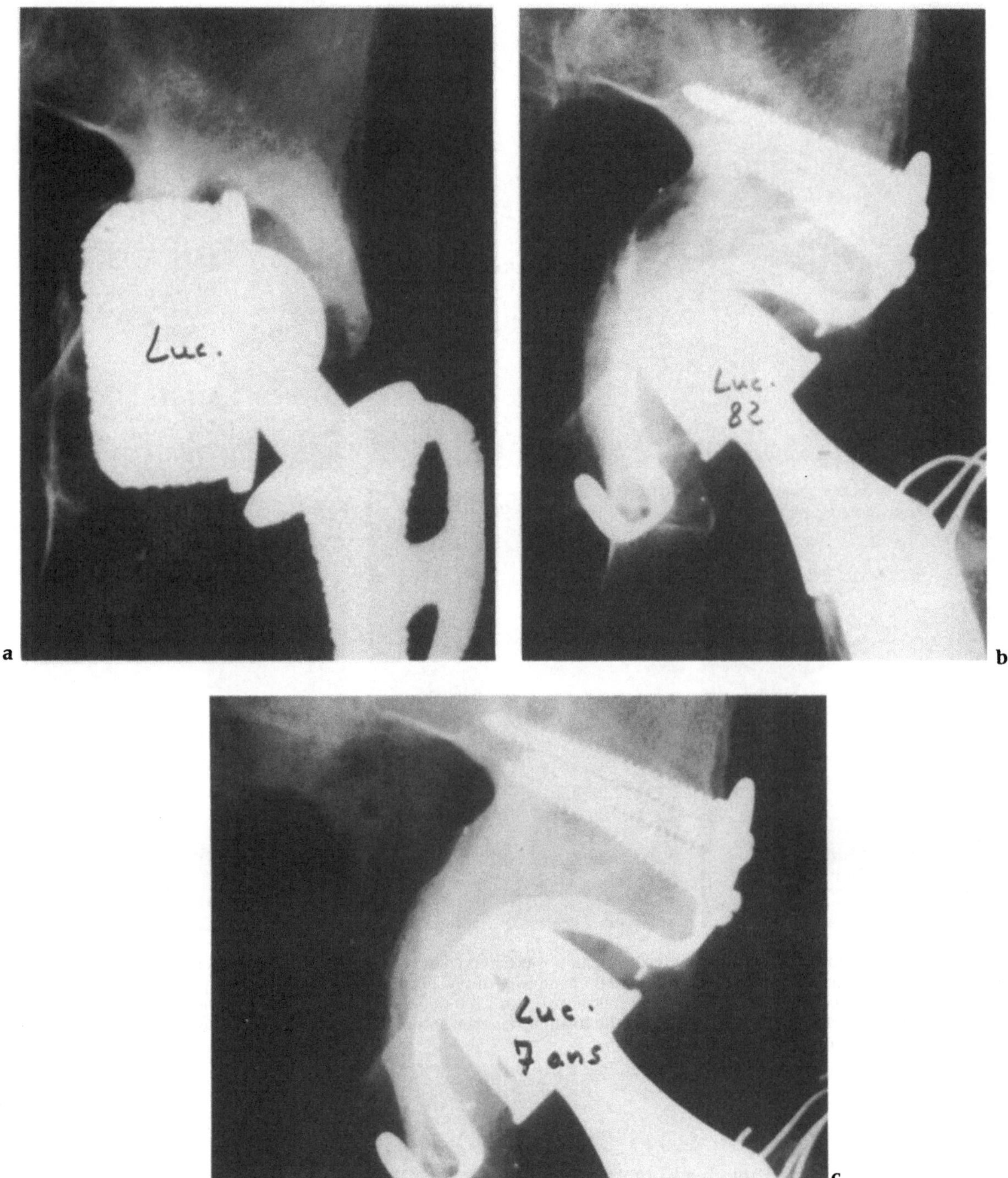

Fig. 12.3a–c. Acetabulum reconstruction reinforced with a cruciate acetabular plate. **a** Pre-operation. **b** Post-operation. **c** Good result after seven years' follow-up.

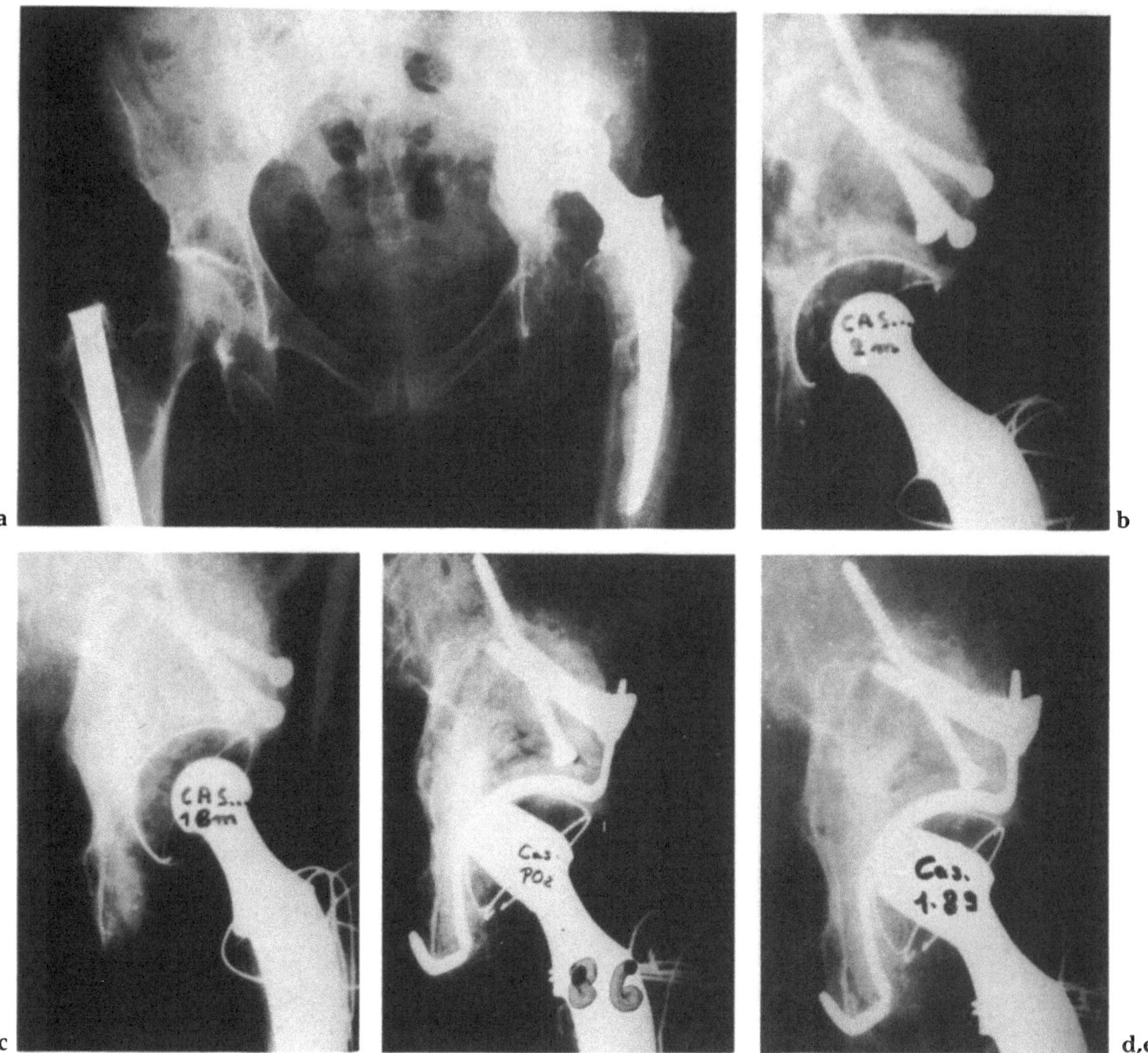

Fig. 12.4a–e. One failure after acetabulum reconstruction was observed. **a** Large acetabular defect operated on in 1984: pre-operative aspect (left hip). **b** Reconstruction with a large complete femoral head secured only by screws, post-operative aspect. **c** The graft progressively collapsed and the cup loosened at two years. **d** Re-operation: the collapsed inferior part of the graft was reinforced with a new femoral head allograft and the fixation was then secured with a cruciate acetabular plate. **e** Good final result with a three year follow-up after re-operation.

calcar reconstructions after neck grafts from femoral heads.

Tough, major femoral bone loss is not easily reconstructed with femoral heads. Primary stability is hard to obtain and implies the use of long stem prostheses. Even with these long stem reconstructions, we observed resorption of the grafts (Fig. 12.7). Since 1985, we have therefore used massive cortical allografts in large femoral defect reconstructions.

Massive Cortical Allografts

Cadaver procurement has a higher risk of graft contamination and led us, after routine donor selection, to sterilise massive grafts by gamma rays which have excellent tissue penetration. A 25 kGy dose ensures reliable bacteriological sterilisation, and increases viral security. There is no risk of induced tissue radioactivity (Loty et al.

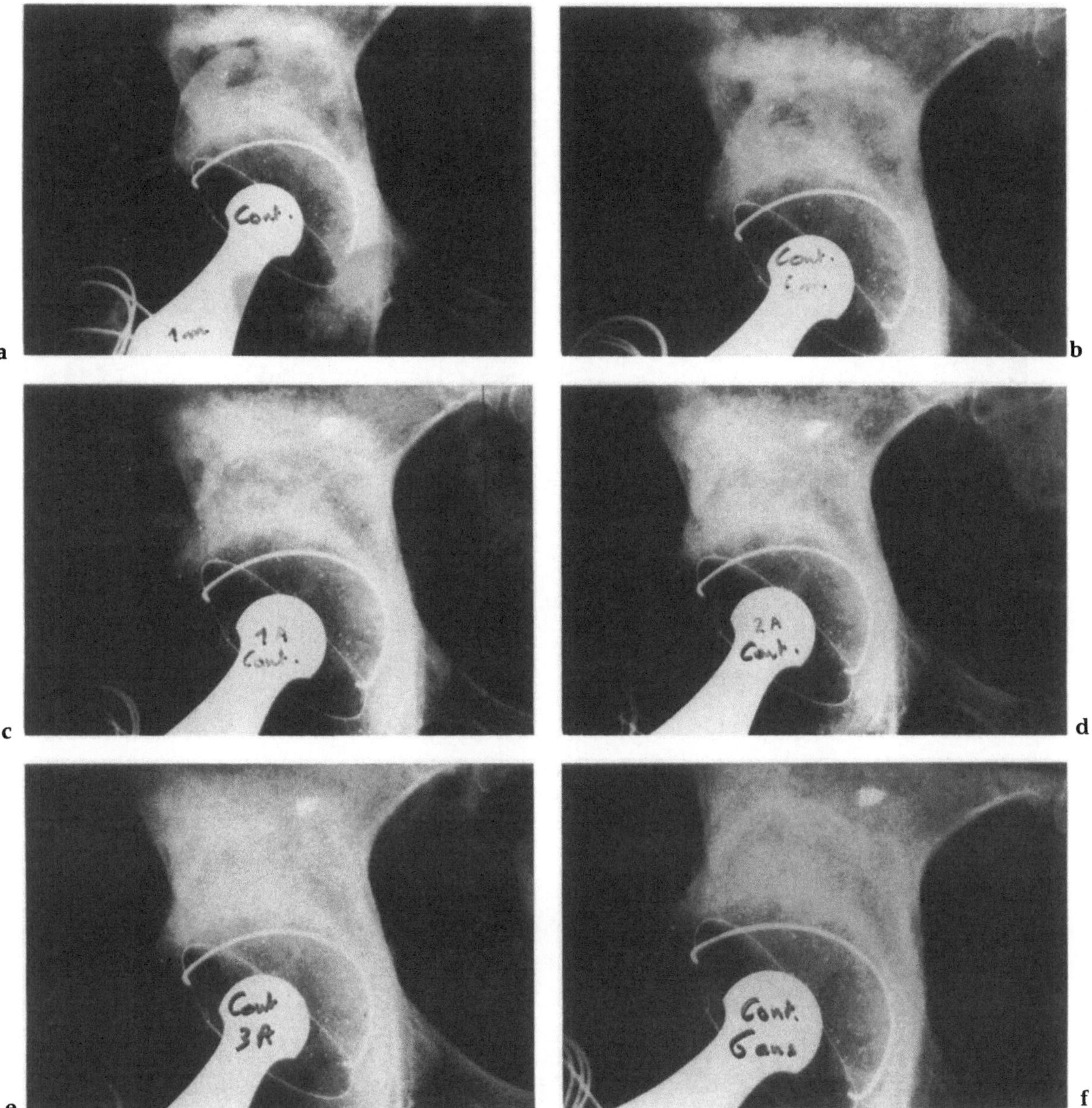

Fig.12.5a–f. Incorporation of acetabulum grafts. **a** Post-operation. **b** Six months post-operation: progressive union of the graft. **c** One year post-operation: the gap between graft and host bone has disappeared. **d** Two years post-operation: union still progressing. **e** At three years the definitive aspect of the graft was obtained. **f** No more changes occurred at six years.

1990, Tomeno et al. 1988). Bending tests performed on human bone samples showed an acceptable 80% strength after 27,000 Gy, but the deleterious effect of overdoses must be avoided by accurate dosimetry (Loty 1988).

Bone banking procedure follows strict guidelines. After checking the removal authorisations, donor selection is founded on medical history, routine serology, and complete autopsy. Sterile conditions are not required for harvesting but any massive contamination is avoided. Irradiation is performed with a precisely controlled dosimetry.

A very important aspect of safe bone banking is record-keeping. For each graft, information is recorded concerning the donor's characteristics and serology, procurement conditions,

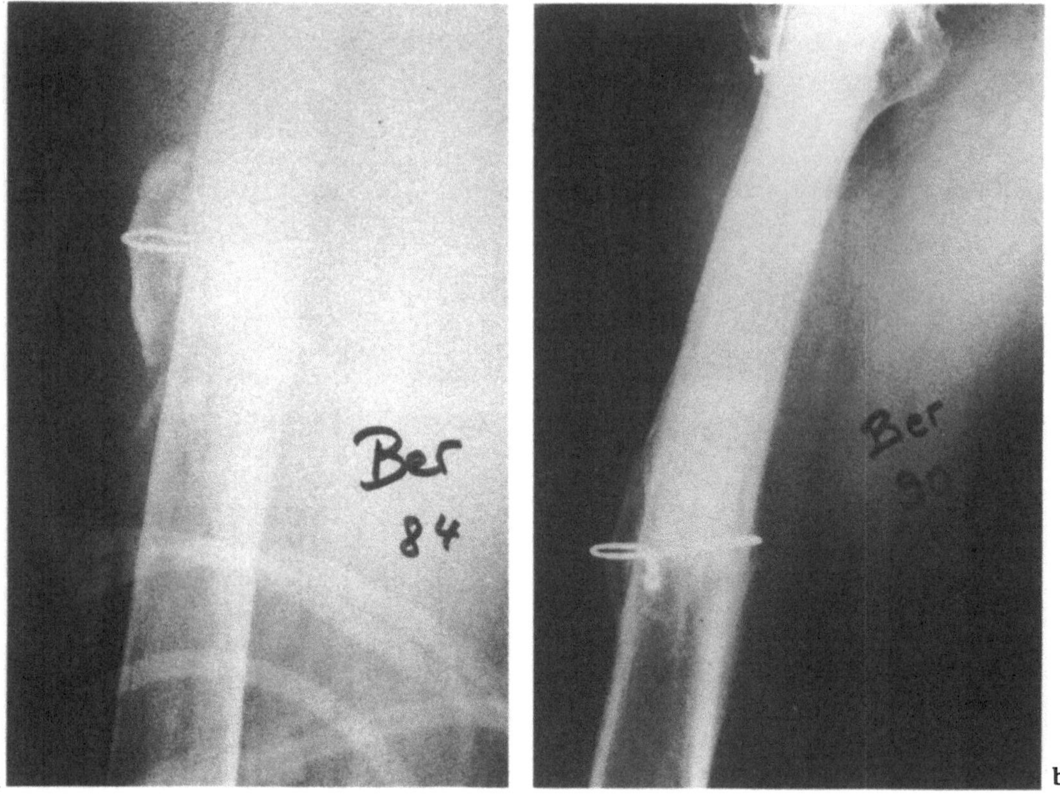

Fig. 12.6a,b. Small femoral cortical defect repaired with a femoral head fragment. **a** Post-operative aspect of the graft secured by cerclage wiring. **b** Follow-up at six years: good cortical reparation in spite of superficial resorption of the graft.

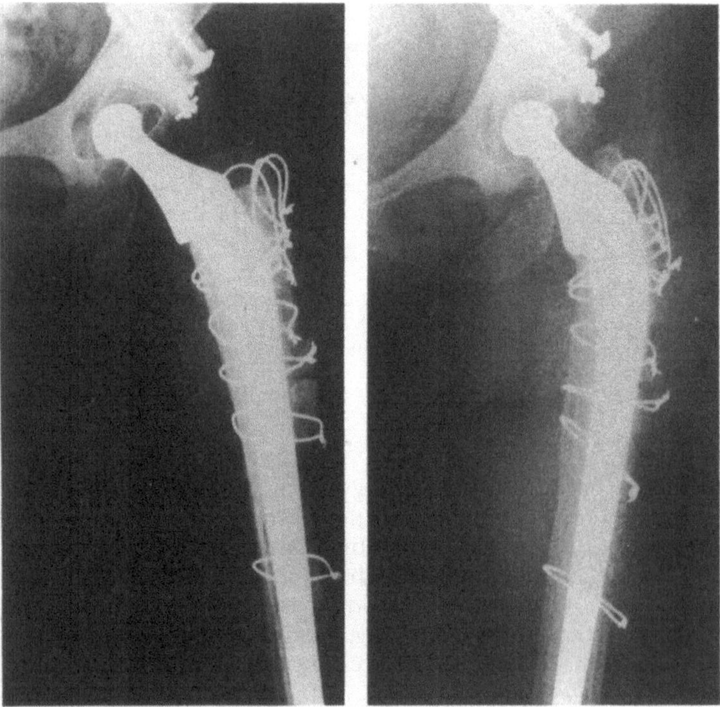

Fig. 12.7a,b. Resorption after femoral head grafting in large femoral defect. **a** Large lateral defect in the femoral shaft, repaired with femoral head fragments secured by cerclage wiring: post-operative aspect. **b** Follow-up at seven years: no loosening of the long stem prosthesis, but resorption of the grafts.

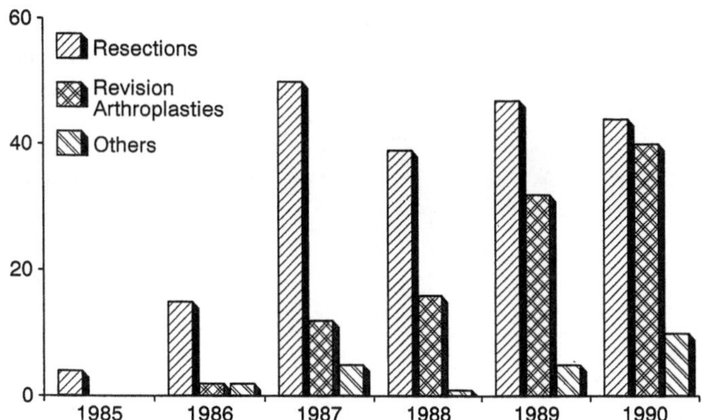

Fig. 12.8 Massive allografts used at Cochin Hospital, 1985–1990.

irradiation doses, and the graft's anatomy. Complete documentation about the recipient, surgical procedure and graft outcome is also always required. This computerised documentation allows stock management, adapted choice for each reconstruction and regular analysis of procedural efficiency.

Large Femoral Defect Reconstruction

Materials and Methods

Initially, we mostly used massive allografts for bone tumours, but the need for revision arthroplasties has increased during recent years (Loty et al. 1991). One hundred massive allografts are now prepared and inserted each year, an equal number being used for tumours and revision arthroplasty reconstructions (Fig. 12.8).

Among 101 upper femur reconstructions performed at Cochin Hospital from 1985 to 1990, 73 concerned major revision arthroplasties, excluding tumour reconstructions (Postel et al. 1991). The average age of the patients was 63 years. The average follow-up was two years (Fig. 12.9). The approach was always transtrochanteric. Cemented total hip prostheses were combined with the allografts (Fig. 12.10). Antibiotics were always added to bone cement and patients had systemic treatment for a month. Partial weight-bearing with two crutches was allowed immediately, and full weight-bearing began at three months.

Different surgical procedures using circular or semi-circular allografts were used (Table 12.1)

Table 12.1. Massive allograft femur reconstructions: procedures

73 massive femoral allograft reconstructions in revision total hip arthroplasties (Cochin 1985–1990)
35 sequential allografts 6–28cm (average 14cm)
18 covered with remnant deficient cortex
17 shaped and impacted in enlarged deficient femur
38 semicircular allografts 7–20cm (average 14cm)
9 cortical replacements
29 cortical reinforcements

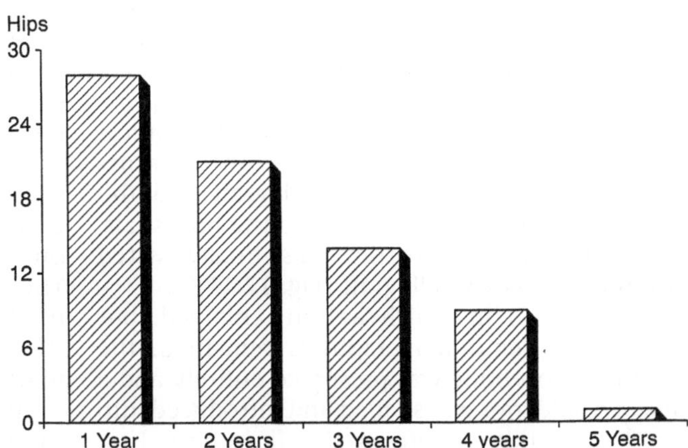

Fig. 12.9 Follow-up of 101 upper femur reconstructions performed at Cochin Hospital, 1985–1990.

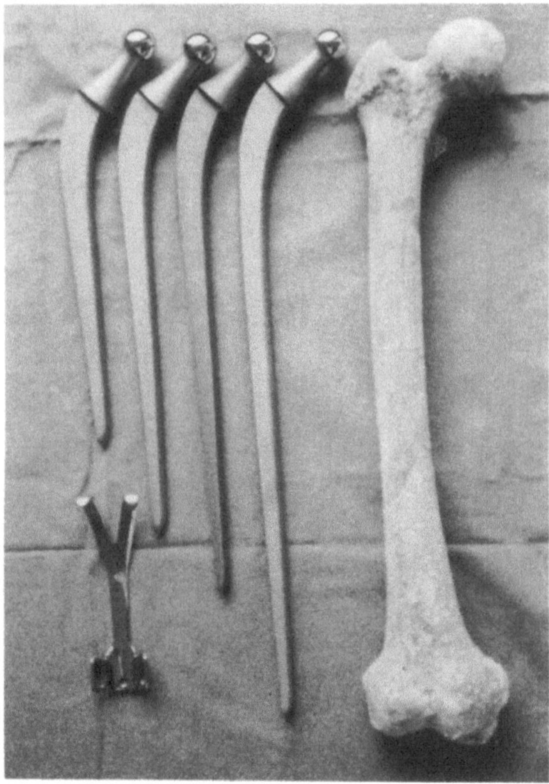

Fig. 12.10. Revision Charnley–Kerboull prostheses we used for composite reconstructions.

according to the type of bone deficiency encountered, the goal being to preserve as much host bone as possible for implantation of the new prostheses.

Segmental reconstructions were performed in 35 cases. They consisted of massive allografts replacing the proximal femur or impacted into it (6–28cm, average: 14cm). In 18 cases in which there was insufficient cortex, the upper femur was replaced by segmental allograft as in tumour reconstruction, but the remnant host femoral cortex was carefully preserved with its soft tissue attachments. The allograft was cut to the exact size and reamed to receive the long stem prosthesis. Care was taken to obtain the anatomical femoral length, alignment and rotation. The prosthesis was cemented in the allograft and the composite implant cemented into the host shaft. The remnant host cortex was finally wired around the massive allograft (Fig. 12.11).

In 17 cases in which a thin, enlarged, but continuous femoral metaphysis remained, the remnant host bone was preserved entirely and a massive allograft was shaped and introduced into it. The allograft was prepared to fit into the host bone as far as possible, and impacted. The prosthesis was then introduced (Fig. 12.12). In six of these grafts a long stem prosthesis was cemented, both into the host femur and the allograft. In the other eleven cases after the allograft had been introduced, it appeared so secure that a standard short stem prosthesis only was used, cemented only into the allograft. In 38 cases, the loss of bone concerned only one side of the femoral shaft and we replaced it by a semi-circular allograft (7–20cm, average: 14cm).

Nine cortical defects were replaced. The allograft was shaped to fit the defect and wired with the remnant cortex before cementing the prosthesis (Fig. 12.13). Attention was paid to the removal of the cement from the graft-femur junctions. Twenty-nine weak or fractured cortices were reinforced with long semi-circular allografts laid on the host femur and secured with cerclage wiring (Fig. 12.14).

Forty-two greater trochanters were directly reattached to the allograft; they were strongly wired or fixed by a trochanteric plate in eight cases (Fig. 12.11).

Complications

Apart from one hemiplegia, no serious medical complication occurred. There were no infections among the 55 aseptic revision operations we performed. One infection recurred after a one-step exchange arthroplasty for sepsis, but did not concern the allograft strut which was conserved when the prosthesis was rewired.

Dislocations occurred without recurrence in two hips and required no further treatment. Two others due to trochanteric non-union were repaired, and another one due to mal-implantation of the prosthesis had to be modified.

Among the 42 patients where the greater trochanter was fixed only on the allograft, there were 11 non-unions. The majority had only a radiolucent line without displacement or clinical consequences, but three required a further operation for plating.

The frequency of trochanteric complications suggests a need for strong initial fixation, apparently better assumed by plates. We observed only one failure after trochanteric plating (Fig. 12.15). The greater trochanter united but a partial fracture of the allograft occurred at this site, though the initially implanted trochanteric plate avoided major displacement.

Some millimetres of subsidence of the prosthesis through the graft occurred in two cases. In

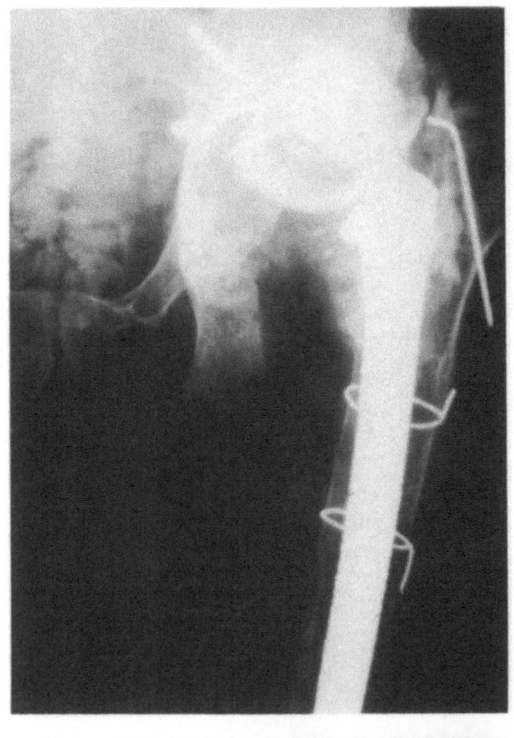

a

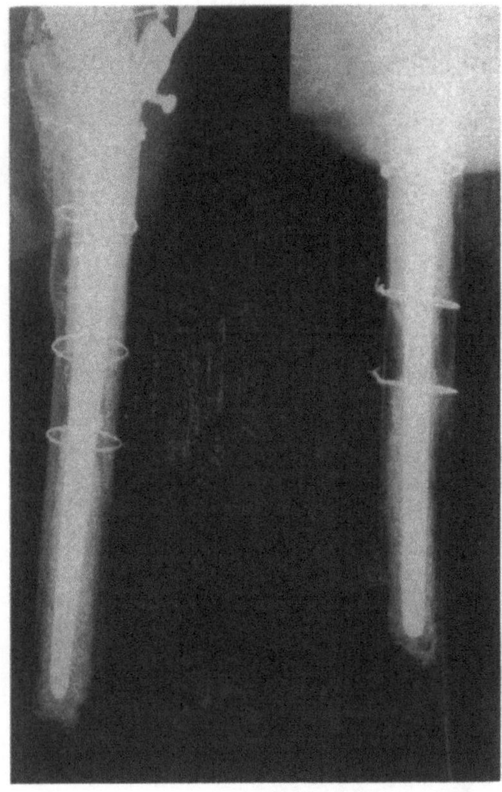

c

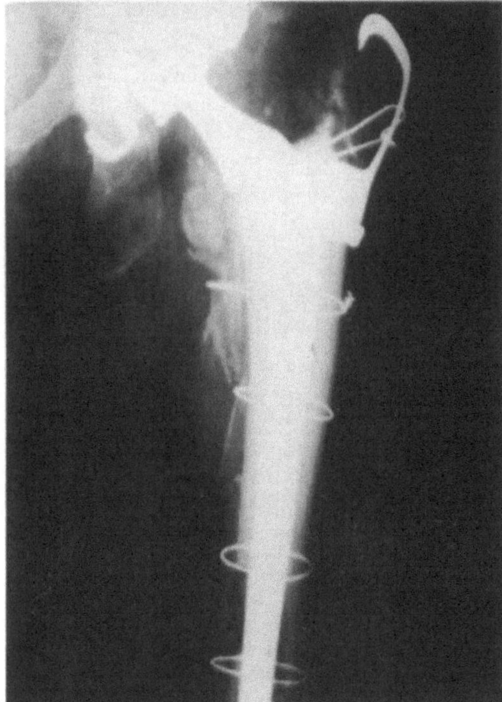

b

Fig. 12.11a–c. Segmental allograft covered with remnant deficient cortex. **a** Pre-operation. **b** Post-operation. **c** Follow-up at two years: good fusion at the junction and with the wired bone.

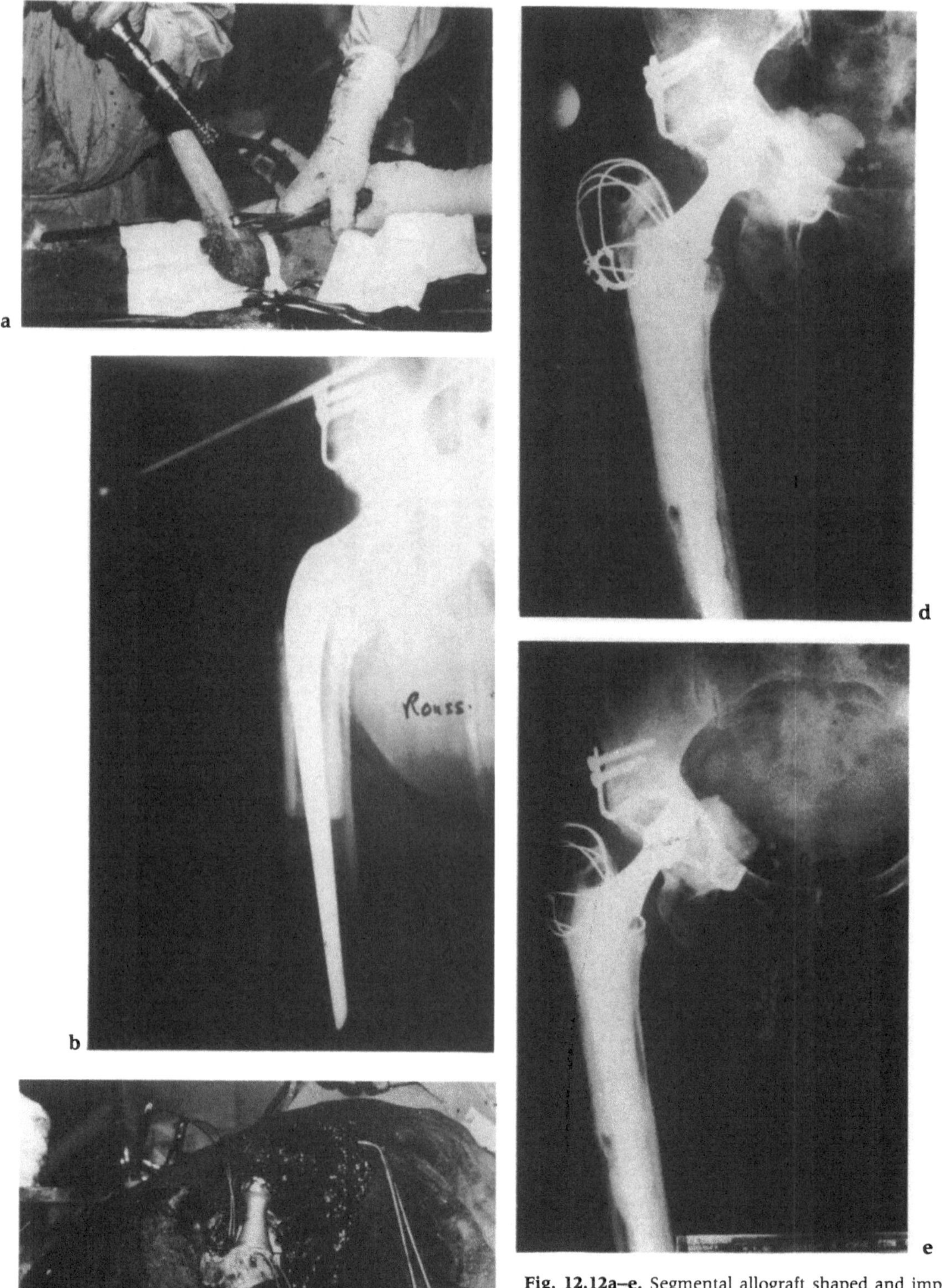

Fig. 12.12a–e. Segmental allograft shaped and impacted in enlarged deficient femur. **a** Preparation of the allograft. **b** Allograft introduced into the host shaft: pre-operative X-ray. **c** Prosthesis is cemented, both into the host femur and the allograft: pre-operative view. **d** Post-operative aspect. **e** Follow-up at three years: fusion appears to be favourable though it was difficult to appreciate with embedded allografts.

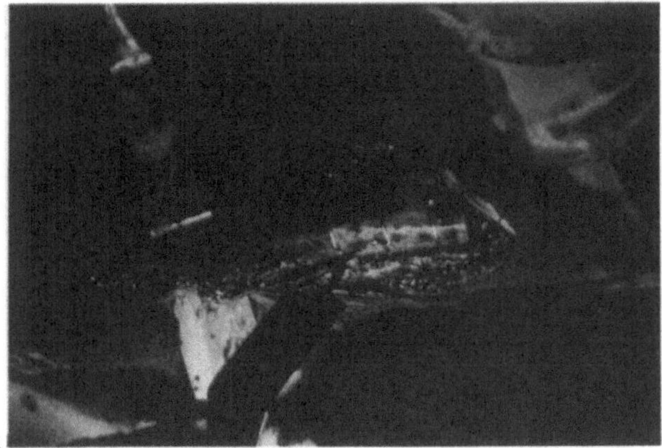

a

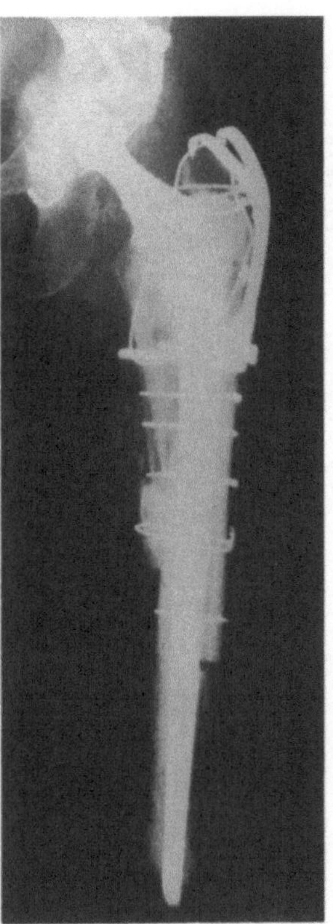

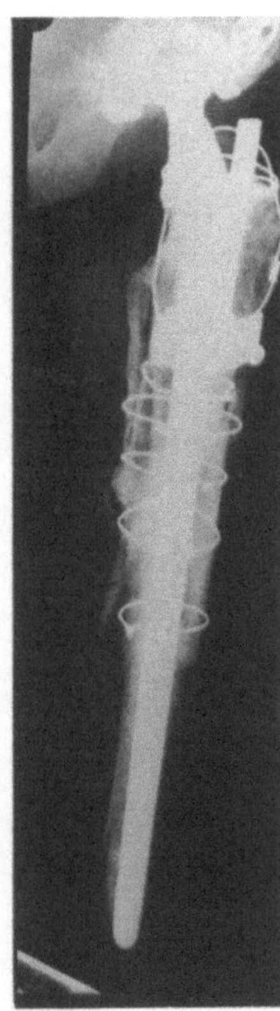

b

▲ **Fig. 12.13a,b.** Semi-circular allograft: lateral cortical replacement. **a** Pre-operative view. **b** Post-operative aspect (AP and lateral view). ▶

Fig. 12.14a–c. Semi-circular allograft: cortical reinforcement. **a** Pre-operation: lateral cortex defect. **b** Post-operation. **c** Follow-up at two years: good fusion of the lateral semi-circular allograft. ▼

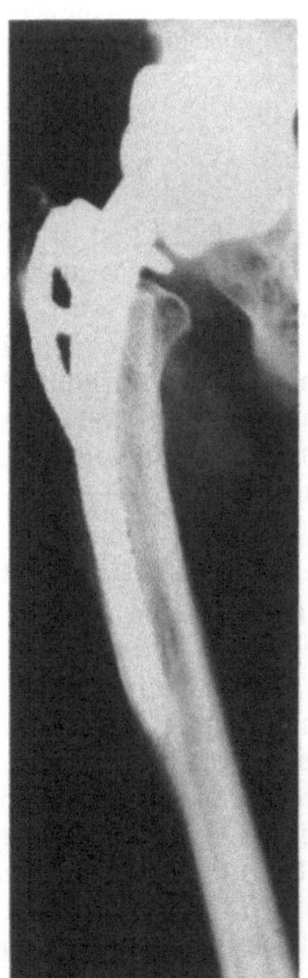

a

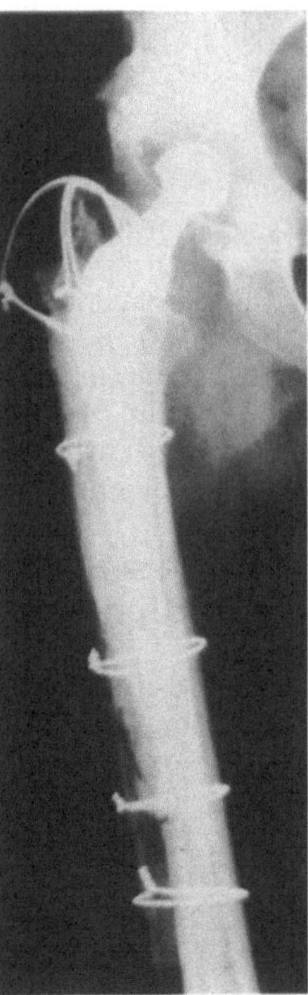

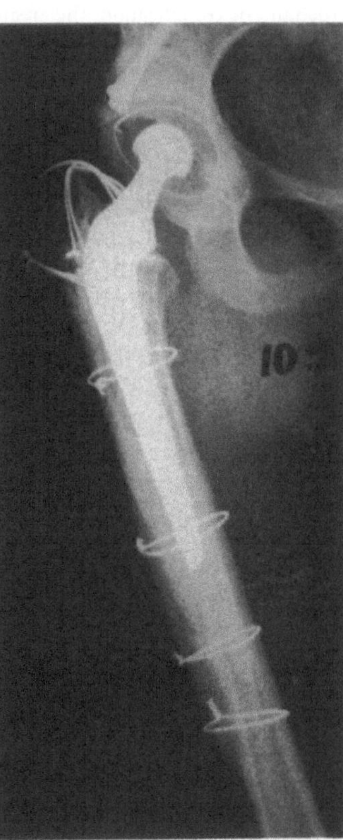

b,c

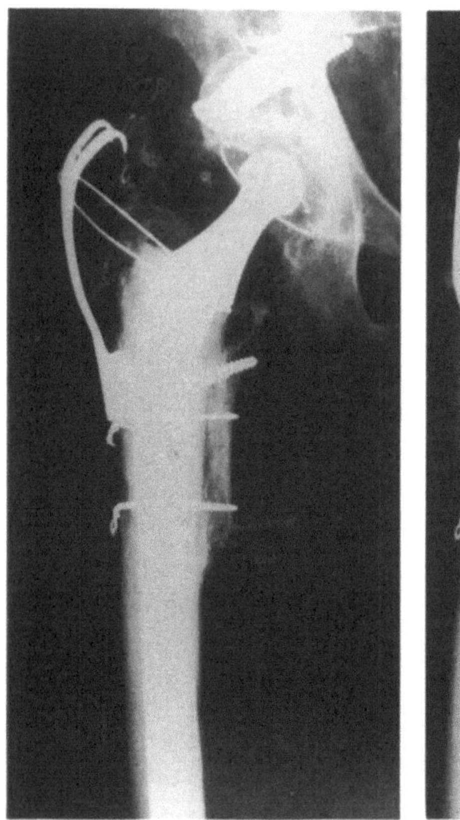

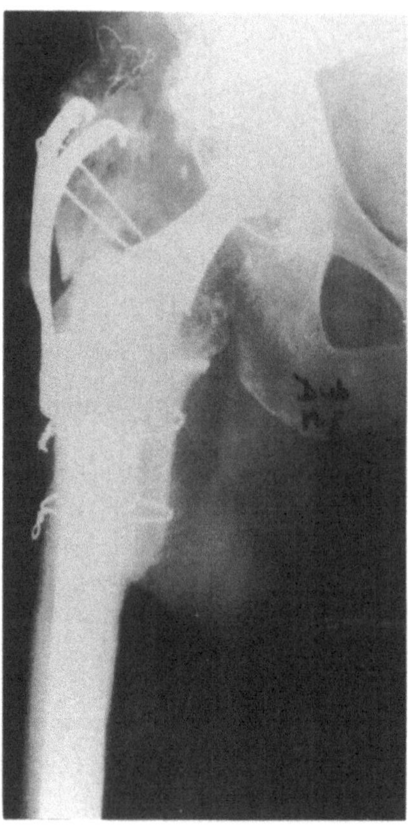

Fig. 12.15a,b. Trochanteric complication: the greater trochanter united but a trochanteric fracture of the allograft occurred. **a** Post-operative aspect. **b** The plate avoided important trochanteric displacement. Note the good fusion of the segmental allograft at the junction, and with remnant wired cortex.

both cases loosening was revealed by a distal fracture of the cement (Fig. 12.16a,b). There was very little increase in migration of the prosthesis during the study and the clinical function remains reasonably good after four to five years. In another hip, the allograft and the prosthesis migrated together distally during the first months and osteolysis developed at the interface. Although function is perfect at one year follow-up, the future is uncertain.

Functional Results

The Merle d'Aubigne score was used for functional evaluation. Pain and stiffness were not a problem. Flexion was always over 90° although three hips had poor stability.

The radiological evolution differed according to the type of reconstruction performed: (1) Circular allografts covered with remnant cortex showed fusion at the junction and with the wired bone within the first year (Fig. 12.6a,b). (2) Fusion was harder to appreciate in segmental allografts embedded into the host femur, but always appeared to be favourable except in one early migration already reported (Fig. 12.12a–e). (3) Semi-circular allografts always showed obvious and early union, even when located on the lateral side of the femur (Fig. 12.14a–c).

The Use of Allografts in Infected Total Hip Replacements

Eight hips with a past history of infection received massive femoral allografts. Ten were operated on for septic loosening of the previous arthroplasty including two reconstructions after a Girdlestone arthroplasty (Fig. 12.17). Debridement appeared satisfactory in the other and the reconstruction was performed in one procedure. All these patients had antibiotic-loaded cement and systemic treatment for a minimum of six months. Only

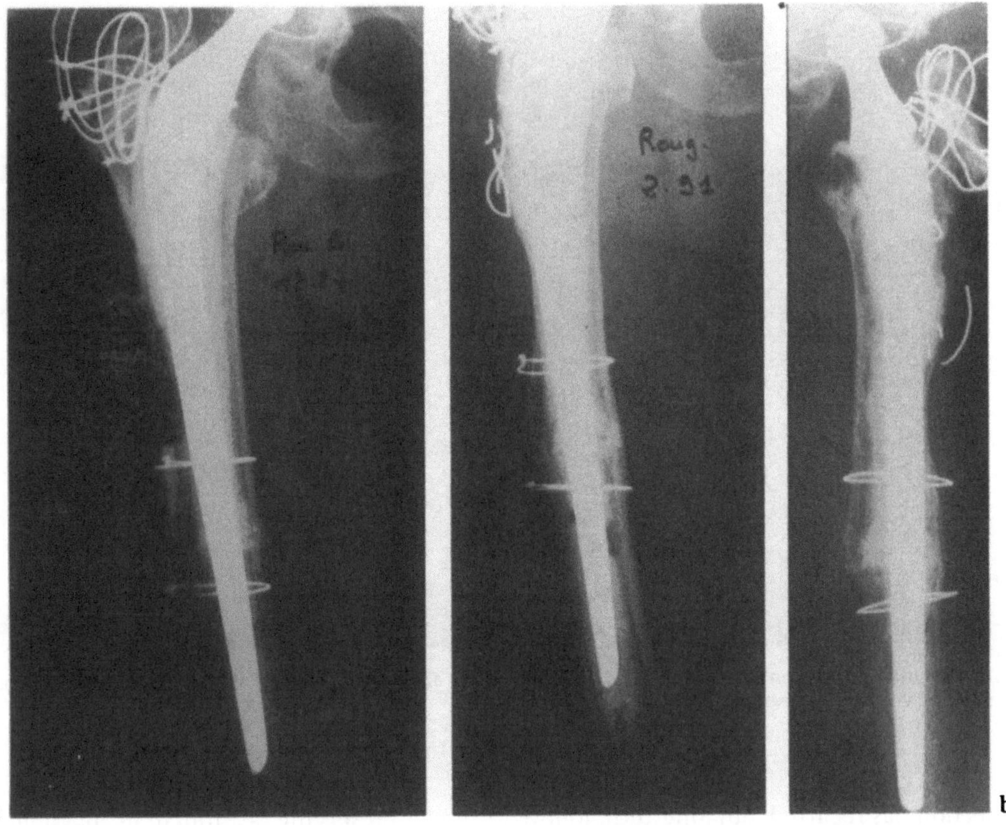

Fig. 12.16a,b. Subsidence of the prosthesis through the graft, distal fracture of the cement. **a** Post-operative aspect. **b** Follow-up at four years: migration of the prosthesis appears to be stabilised, the clinical function remains quite good.

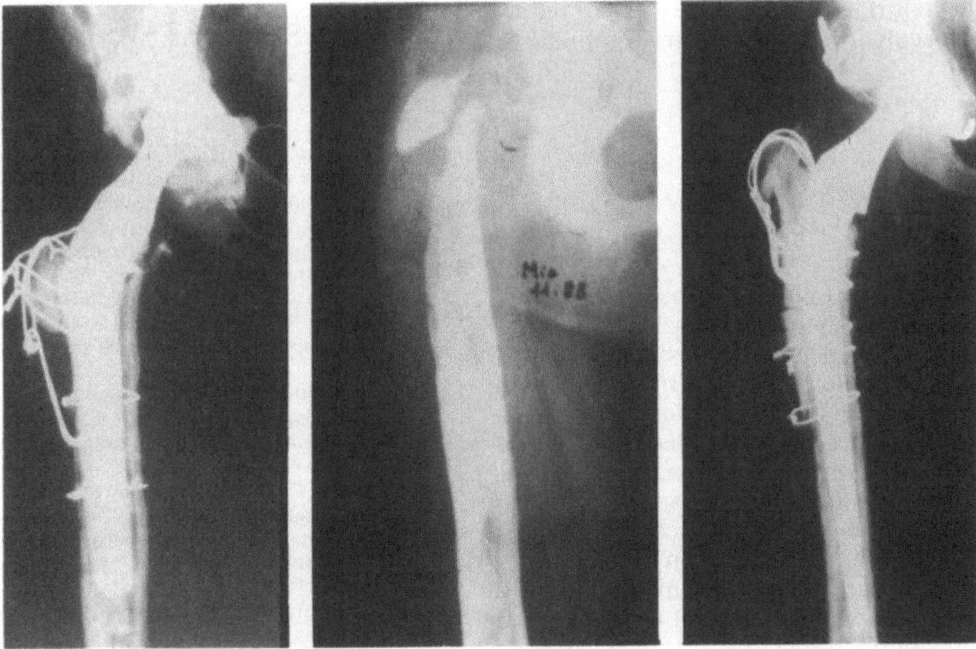

Fig. 12.17. Two-step revision of a septic arthroplasty with a massive femur allograft.

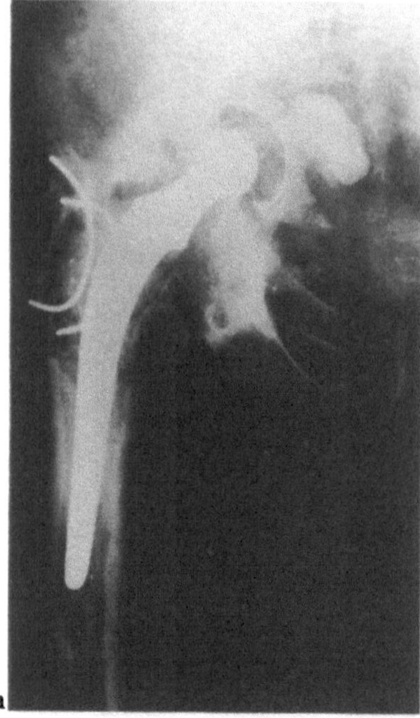
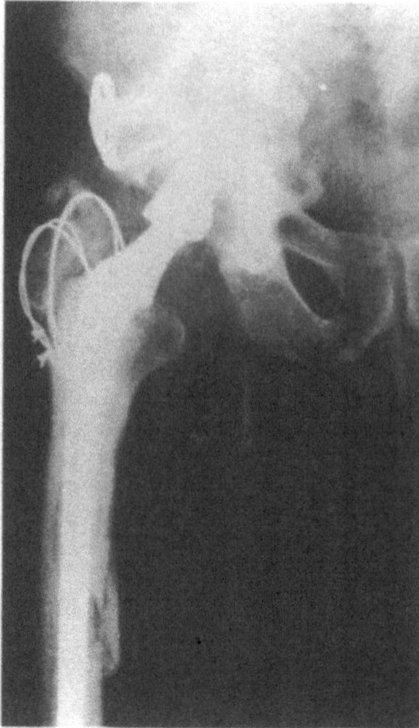
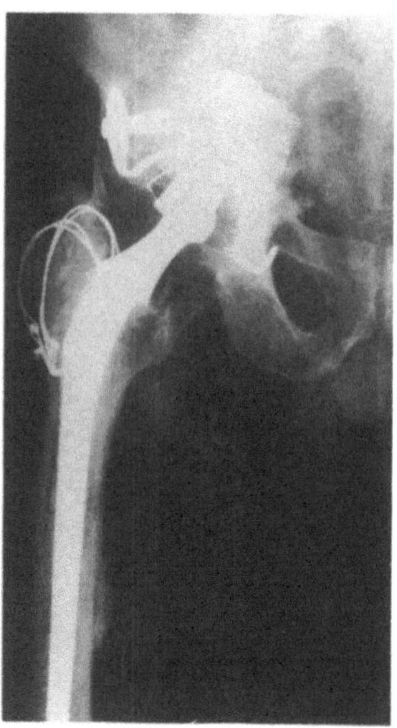

a b,c

Fig. 12.18a–c. One-step revision of a septic arthroplasty with a large acetabulum allograft. **a** Pre-operation. **b** Post-operation. **c** Follow-up at six years: good result.

one infection has recurred, but long-term follow-up is required for conclusive evidence.

In another comparative study we recently reviewed a series of one-step septic revision arthroplasties in which 46 of the 90 received femoral head allograft (Fig. 12.18). There was no statistical evidence that the use of allografts influenced infection.

Conclusion

After more than 15 years' experience, frozen femoral heads have proved to be reliable in reconstructing the acetabulum. Massive irradiated allografts appear to be a promising method in major femoral reconstructions but their follow-up is still short. Irradiation avoided the infection risk due to cadaver transmission. These results rely on safe bone banking procedures which require predefined and strictly controlled methods. Good results also depend on adapted surgical guidelines, particularly efficient graft fixation. The aim of the different surgical techniques we used in revision arthroplasties was to obtain good anatomical situation, alignment and length, while preserving as much as possible of the remnant host bone.

References and Further Reading

Hedde C, Postel M, Kerboul M, Courpied JP (1986) La réparation du cotyle par homogreffe osseuse conserve(a1)e au cours des revisions de prothèse totale de hanche. Rev Chir Orthop 72:267–276

Loty B (1988) Irradiation des allogreffes osseuses. In: Gerard Y (ed) Symposium sur les Banques d'Os, 62e Réunion Annuelle de la Sofcot. Rev Chir Orthop 74:109–159

Loty B, Courpied JP, Tomeno B, Postel M, Forest M, Abelanet R (1990) Radiation sterilised bone allografts. Intern Orthop 14:237–242

Loty B, Courpied JP, Tomeno B, Kerboull M, Postel M, Forest M (1991) Allogreffes osseuses massives sterilisées par irradiation: bilan apreazs 5 ans d'utilisation. Acta Orthop Belg 57(II):35–43

Postel M, Loty B, Courpied JP, Tomeno B, Kerboull M (1991) Upper femur massive composite prostheses after tumour surgery. In: Limb Salvage: Major Reconstructions in Oncologic and Nontumoral Conditions. Springer-Verlag, Berlin Heidelberg, 415–421

Tomeno B, Courpied JP, Loty B (1988) Techniques et indications des greffes et transplantations osseuses et osteo-cartilagineuses. Encycl. Med. Chir (Paris, France). Techniques Chirurgicales – Orthopédie, 44030, 11–16p

13 The Use of Femoral Strut Grafts in Cementless Revision Arthroplasty

W.G. Paprosky

We use implants without cement, try to avoid massive structural grafts and currently use augmented strut grafts. We believe the polyethylene agents are more responsible for the massive destruction than the cement and metal, certainly on the femoral side. No matter how good the initial fixation, a failure will ultimately occur. In revision surgery of the femur, once the cement has been removed, there is significant osteolysis and the bone is of such poor quality that revision with cement may be impossible.

There are numerous series of recorded failed cement revisions. One out of four had failed by three years (Table 13.1). There is also significant evidence of radiographic mechanical loosening in up to 50% of these patients. Revision with cement and not adjuvant femoral graft replacement will not restore bone and leads to early failure.

Table 13.1. Percentage failure rate of cemented revision hip arthroplasties

Series	%	Years
Hunter et al.	22	2
Hoaglund et al.	23	3
Stanford Sherman	29	3
Mayo Clinic	24	3

There are several treatment options: to use cement only, cement with segmental allografts, or cementless implants with massive allografts. I caution you with regard to using the latter because of the inability to obtain good fixation distally to control rotation. There are large distal canals and the implant distally will be larger than the massive proximal allograft will be able to accept proximally. You want a narrow proximal stem, but it is then difficult to get control distally. We prefer to use a cementless implant with strut grafts. This relies on distal fit and extensive porous surfaces as opposed to custom implants, modular devices and proximal fitting implants with porous surfaces. Long-term ingrowth cannot be achieved with proximal coating because the poor proximal bone stock is unable to support the implant long term and provide rotational stability.

Femoral Defect Classification

Femoral defect classification, based upon preoperative assessment and intra-operative pattern recognition, is of paramount importance in

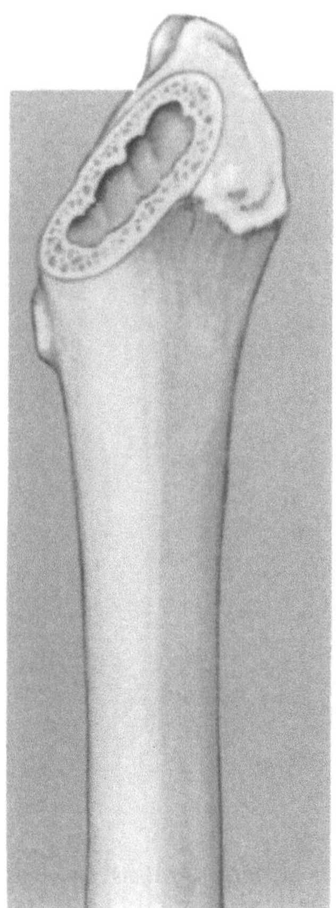

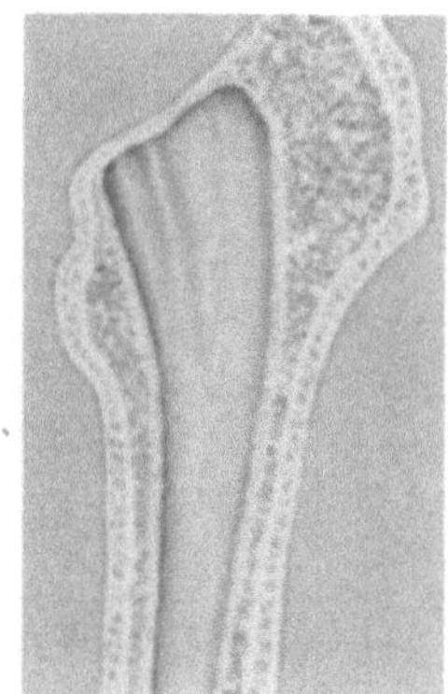

Fig. 13.1. Femoral defect: Type 1.

femoral reconstruction to determine the appropriate reconstruction pattern and type of femoral component to be used. This classification is based upon the extent of femoral bone loss as well as the amount of diaphyseal support available to the implant, independent of the condition of the metaphysis.

Type 1

The Type 1 defect is similar to that of a femur in a primary total hip arthroplasty. (Fig. 13.1). The metaphysis and isthmus are intact with minimal bone loss. There is only partial loss of the calcar and minor anterior/posterior bone loss. The diaphysis is intact with minimal osteolysis.

Minor intermedullary voids are filled with cancellous bone. Sequential reaming is used to maximise live bone regeneration and rigid fixation in the diaphysis. Optimum fixation can be achieved using an extensively coated standard length implant. Good proximal bone stock enables one to achieve proximal fixation as well. Thick diaphyseal cortices provide stability against rotational and axial forces to give rigid distal fixation.

Type 2

The Type 2 femoral defects are characterised by more extensive metaphyseal bone loss. The metaphysis is not intact but ballooned and funnel shaped. There is complete absence of the calcar and sclerotic, minimally supportive bone. The diaphysis is intact.

There are three subtypes of bone loss:

Type 2A (Fig. 13.2a)

Bone loss does not extend to the subtrochanteric region where the bone is fairly supportive. However, there is less proximal rotational stability afforded by the subtrochanteric bone because of lysis of the existing bone and loss of anterior/posterior bone stock.

Grafting techniques will include treatment with cancellous slurry and matchstick grafts inside the metaphysis and a napkin ring calcar graft from a femoral head allograft.

The implant which comes to rest on the subtrochanteric bone will be further advanced in the diaphysis than the Type 1 because of the absence of calcar bone stock. Therefore the diaphyseal portion of the implant will approach the anterior cortex in the bowed portion of the femur. The loss of calcar (and the further distal seating of the implant) will require a longer neck to give post-reduction stability. At the same time, distal advancement of the implant could impinge against the anterior cortex.

Implant selection, therefore, will be a distally advanced, extensively coated standard length stem with extra long neck; a more proximal seating of an 8 inch extensively coated device supported by calcar allograft; or a calcar replacement prosthesis with extensive coating and a length of 7 inches.

Type 2B (Fig. 13.2b)

The antero-lateral portion of the subtrochanteric bone of the metaphysis is absent. The metaphysis no longer provides support or any resistance against rotational forces and there is severe anterior/posterior bone loss. The metaphysis of the femur now appears funnel shaped. A standard length stem will no longer be adequate since complete stability will come from distal diaphyseal fixation to prevent rotation and axial migration of the component.

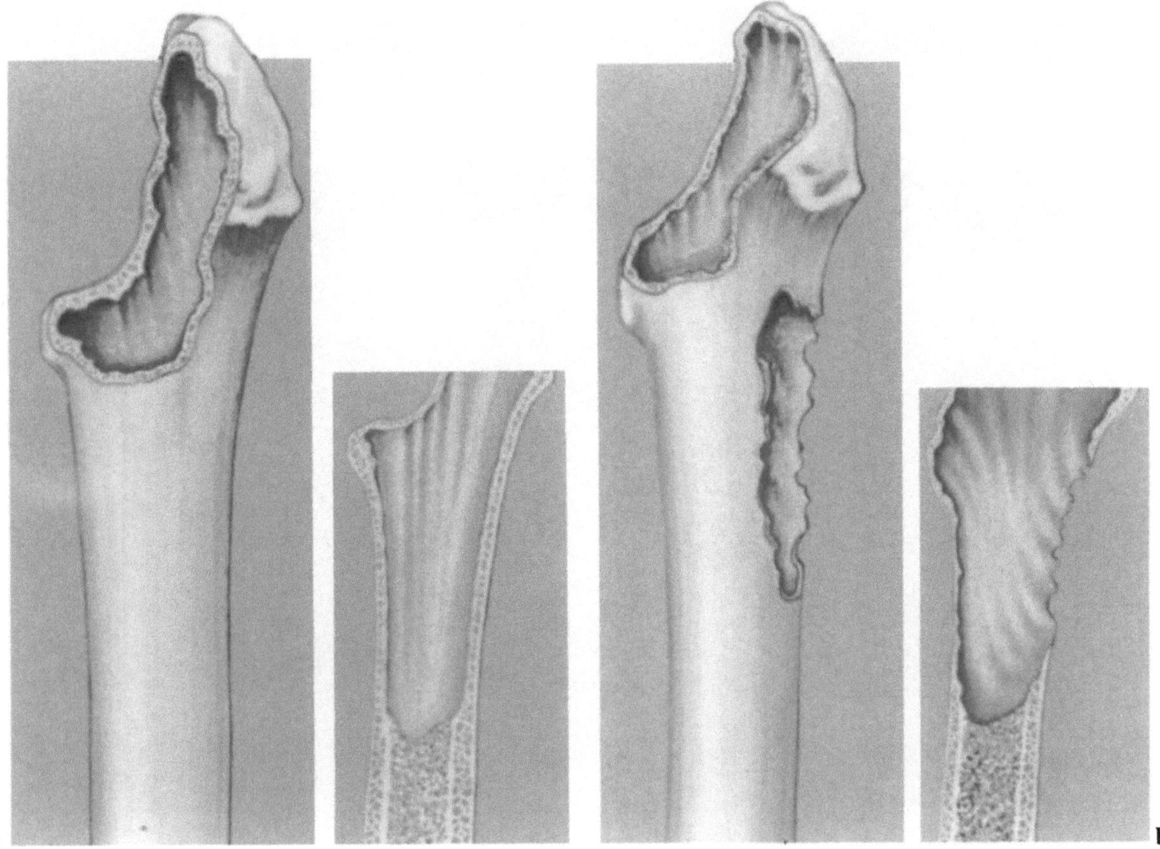

Fig. 13.2. Femoral defects: **a** Type 2A; **b** Type 2B.

The bone defects are replaced by cancellous slurry or matchstick grafts inside the metaphysis. A sagittally cut femoral head onlay strut may be cerclaged over the lateral metaphyseal defect. A cortical strut graft cut from a proximal tibia or distal femur may also act as an onlay graft secured with cerclage wires (Fig. 13.3). Component selection includes an 8 inch extensively coated straight or curved stem. If added neck length is required for post-reduction stability, a calcar replacement extensively coated stem is used.

Type 2C (Fig. 13.4)

The femur is characterised by a metaphysis that is non-supportive, but in this type the posterior-medial metaphyseal wall is absent or non-functional. There is absolutely no resistance to rotational forces or axial migration provided by the metaphysis; therefore more distal fixation will be needed. Medial wall and calcar restoration will be required. A proximal tibia allograft is step cut using the tibial plateau portion as the calcar and the tibial metaphyseal extension as the medial strut portion (Fig. 13.5a,b). Component selection includes an 8 inch straight or curved stem or, on some occasions, a 10 inch curved stem.

Case 1

A 36 year old male patient had a revised total hip prosthesis which had loosened following an infection with *Staphylococcus aureus*. This presented as a Type 2C defect (Fig. 13.6a). In a two-stage revision procedure, a femoral component with a 6 inch stem and a proximal medial bone graft was used (Fig. 13.6b). Radiographs at seven years show incorporation of the bone graft (Fig. 13.6c).

Type 3 (Fig. 13.7)

These are characterised by extensive metaphyseal and diaphyseal bone loss. The metaphysis and isthmus are osteolytic with sclerotic tissue, and non-supportive. The calcar is completely absent and the diaphysis is not intact above the isthmus.

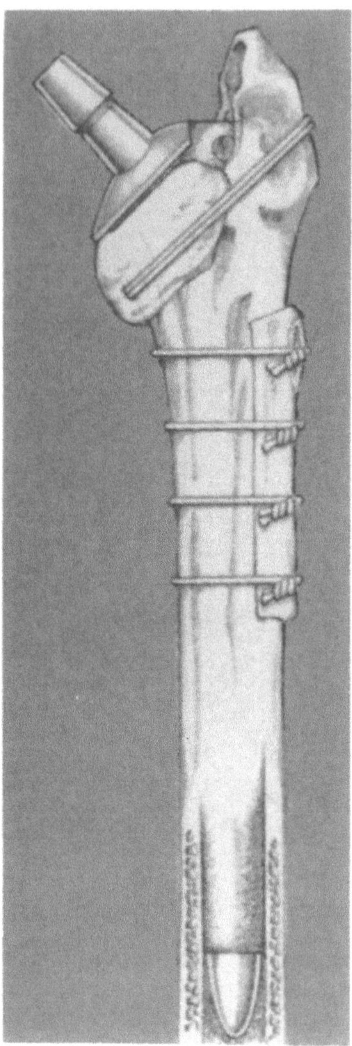

Fig. 13.3. Onlay graft from proximal tibia or distal femur secured with cerclage wires.

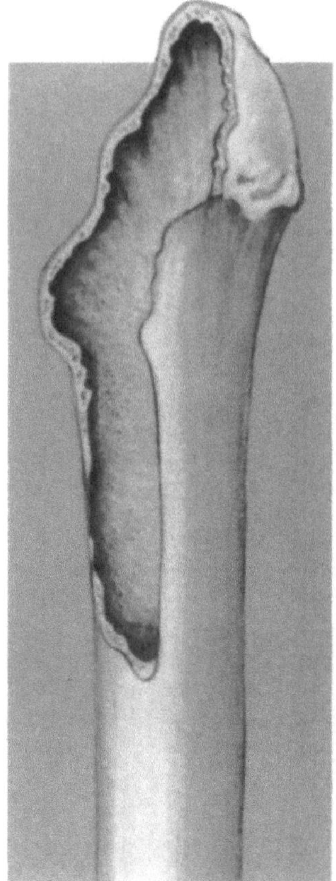

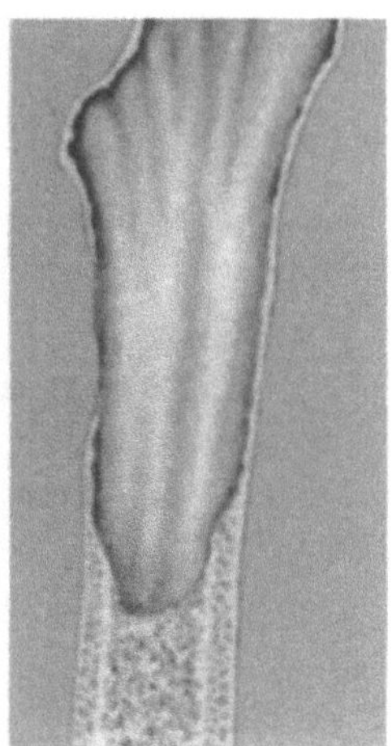

Fig. 13.4. Femoral defect: Type 2C.

Moreover, there is an intact diaphysis below the isthmus.

The extensive bone loss is treated by cancellous slurry and matchstick grafts inside the femur. Large cortical strut allografts with cerclage wires are securely fastened over the entire defect. If a calcar allograft is required, then the same technique used in Type 2C defects is employed. A longer section of proximal tibia is left on the allograft to cover the diaphyseal defect. The diaphysis will not support or give any rotational stability to a standard length stem and, for the most part, does not offer rotational stability to an 8 inch stem. The only significant stability against rotation comes from the use of an extensively coated 10 inch stem. However, because of the anterior bow of the femur, a 10 inch stem must be curved to avoid fracturing the femur anteriorly.

Case 2

A 66 year old female patient with three previous revision hip operations presented with a Type 3 defect (Fig. 13.8a,b). The hip was reconstructed using a femoral implant with a straight extensively coated 8 inch stem and an inverted tibial augmentation graft (Fig. 13.8c,d).

Case 3

A 43 year old female patient with two previous revision hip operations is another example of a

Fig. 13.6a–c. A Type 2C femoral defect reconstructed in a two-stage procedure: **a** Pre-operation; **b** Post-operation; **c** Seven years post-operation.

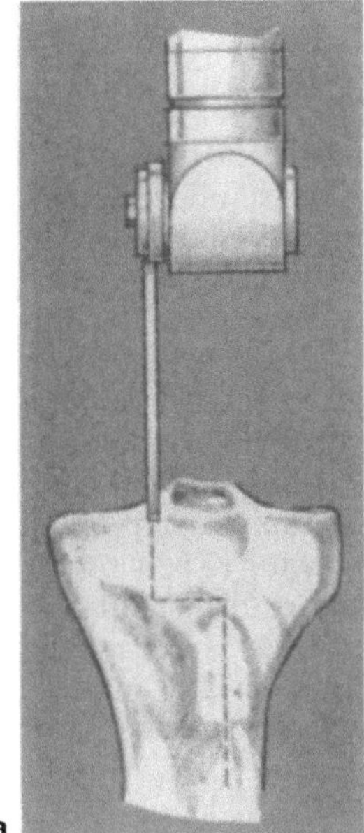

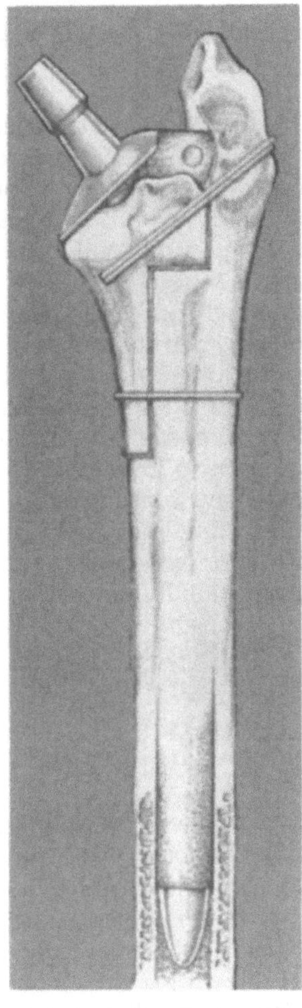

Fig. 13.5a,b. Restoration of medial wall and calcar: **a** Allograft step cut from tibial plateau; **b** Cerclage wire secures tibial buttress graft.

Type 3 defect (Fig. 13.9a,b). An inverted L-shaped tibial augmentation graft was used to restore lost calcar and antero-lateral bone. A 10 inch curved extensively coated stem was inserted (Fig. 13.9c,d,e).

Strut Grafts

Fixation in the diaphysis is essential for the implant to achieve stability against rotation. An onlay strut is used to cover the large lytic areas. Rather than sacrifice all the metaphyseal and proximal diaphyseal bone, we use the allograft cortical onlay segments to form a host-allograft construct securely fixed with cerclage wires. When whole segmental or intercalcary grafts are used, we would prefer not to use allografts as primary support. Once you do that, you have an implant that is relying primarily on the graft for support. We prefer the grafts to be used as secondary support, augmentation, to promote bone regeneration. Rather than a massive graft in the upper part of the femur, it is better to use struts to

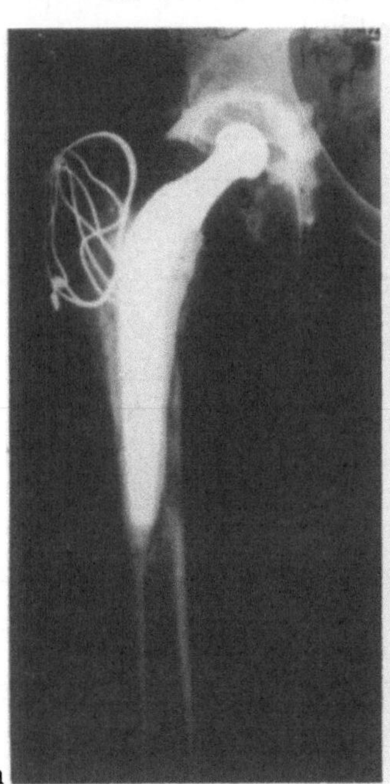

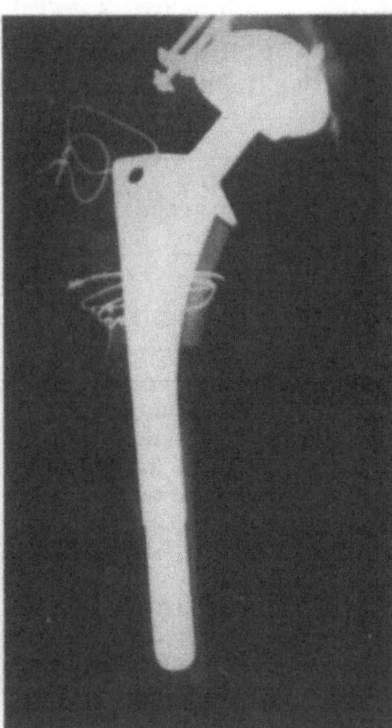

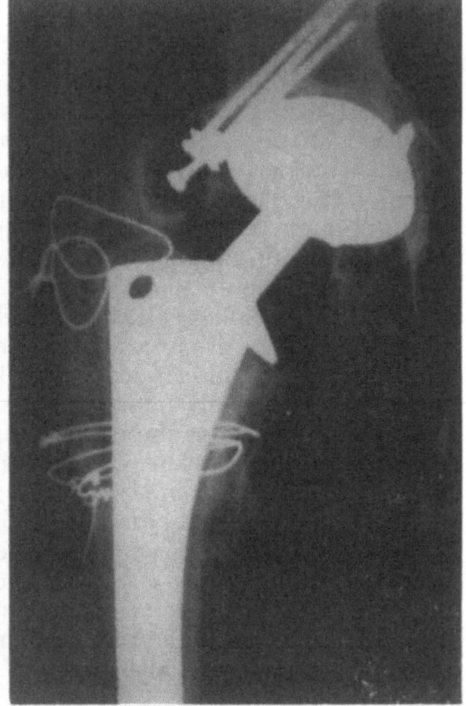

a b,c

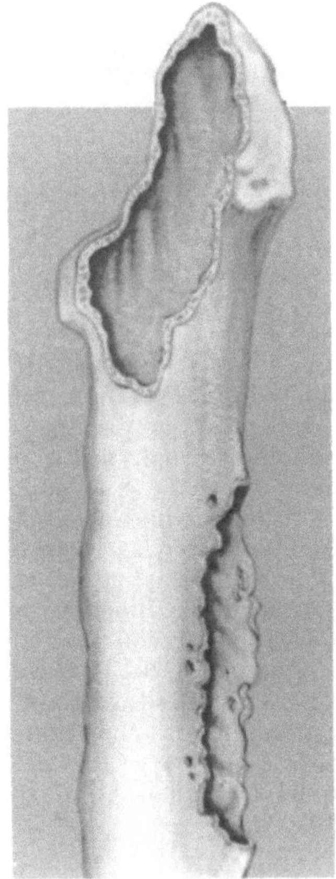
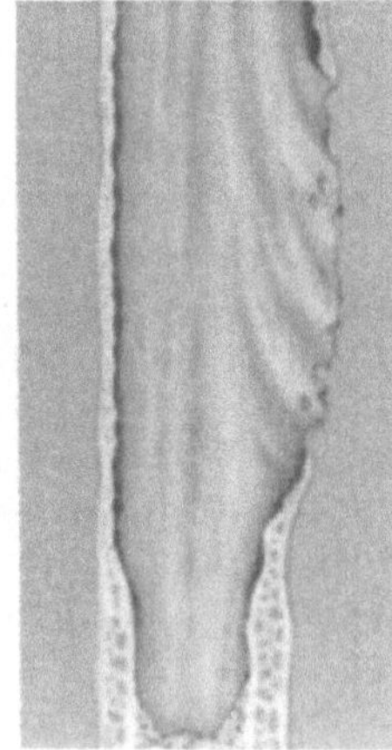

Fig. 13.7. Femoral defect: Type 3.

conserve host bone. If, however, not enough diaphysis is left between the isthmus and distal metaphysis to provide primary implant stability a massive graft may be needed. By achieving a more favourable modulus, dense cortical struts may protect against stress shielding. Adjusting the height of the implant by using calcar grafts will restore leg length. Although an attractive postulate, the promotion of bone ingrowth into the porous surface is probably highly unlikely.

What conditions are necessary to use cortical struts successfully? Success depends on the ability of the femur primarily to support the femoral implant. The implant is the primary support, as opposed to structural graft where the graft supports. Once you have bone loss extending to the distal metaphysis where you are unable to get primary implant support, you must revert to the method described by Gross which is to cement a stem proximally into a segmental graft and fix the graft-stem construct to the remaining host bone with long stem intermedullary fixation or extramedullary plate fixation.

Implant Selection

A defect in which the metaphysis and diaphysis is supportive allows you to use a standard length implant with proximal or extensive porous coating only when there is minimal metaphyseal damage and good diaphyseal support. In situations where 50% of the circumferential bone is missing from the metaphysis with an intact diaphysis, you must obtain control against rotation using longer implants with extensive coating.

It is essential to have control against rotation for long term stability. If you are going to rely on the implant to give primary support, distal flutes or extensive coating must be used. Many authorities now state that a smooth stem distally without flutes or extensive coating will always make long term stability difficult.

Results

We reviewed 245 femoral revisions that we had done between 1982 and 1989. Of these, 17 were lost to follow-up leaving 228 revisions of which 141 were females (Table 13.2). The most common cause of femoral failure was aseptic loosening (Table 13.3). The mean age was almost 60 years. Our mean follow-up was almost five years in all patients. The mean number of previous operations was 3.4 (Table 13.4).

Table 13.2. Number and sex of patients

	n
Patients	228
Females	141
Males	87

Table 13.3. Cause of femoral failures

	n
Aseptic loosening	189
Septic loosening	15
Femoral fractures	14
Femoral component fractures	8
Femoral malpositions	2

Table 13.4. Age, follow-up and previous surgery

Age at operation (years)	59.6 (range 24–86)
Follow-up (years)	4.75 (range 2–8)
Previous surgery (no. of operations)	3.4 (range 1–7)

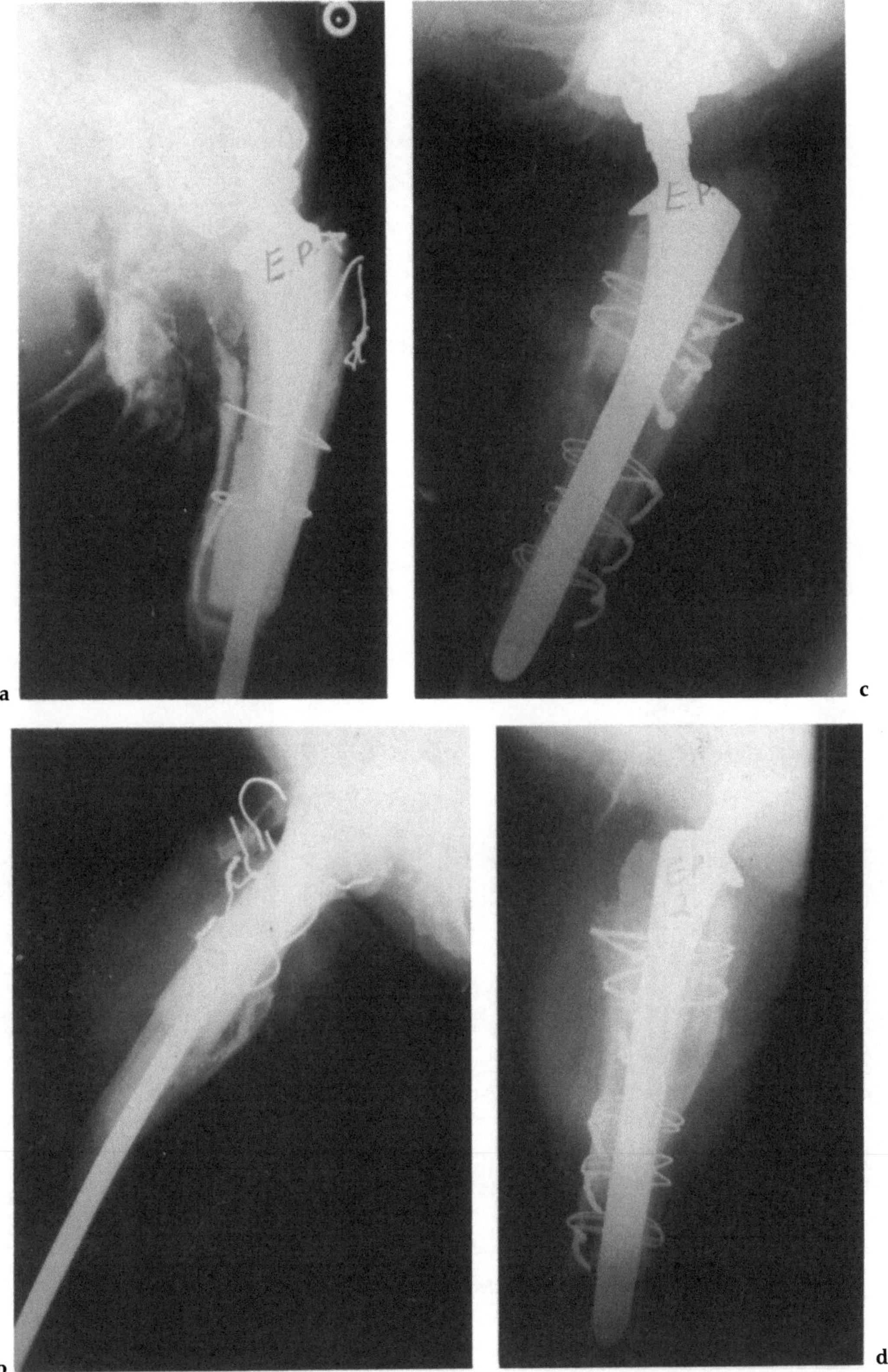

Fig. 13.8a–d. A Type 3 femoral defect: **a,b** Pre-operation; **c,d** Post-operation.

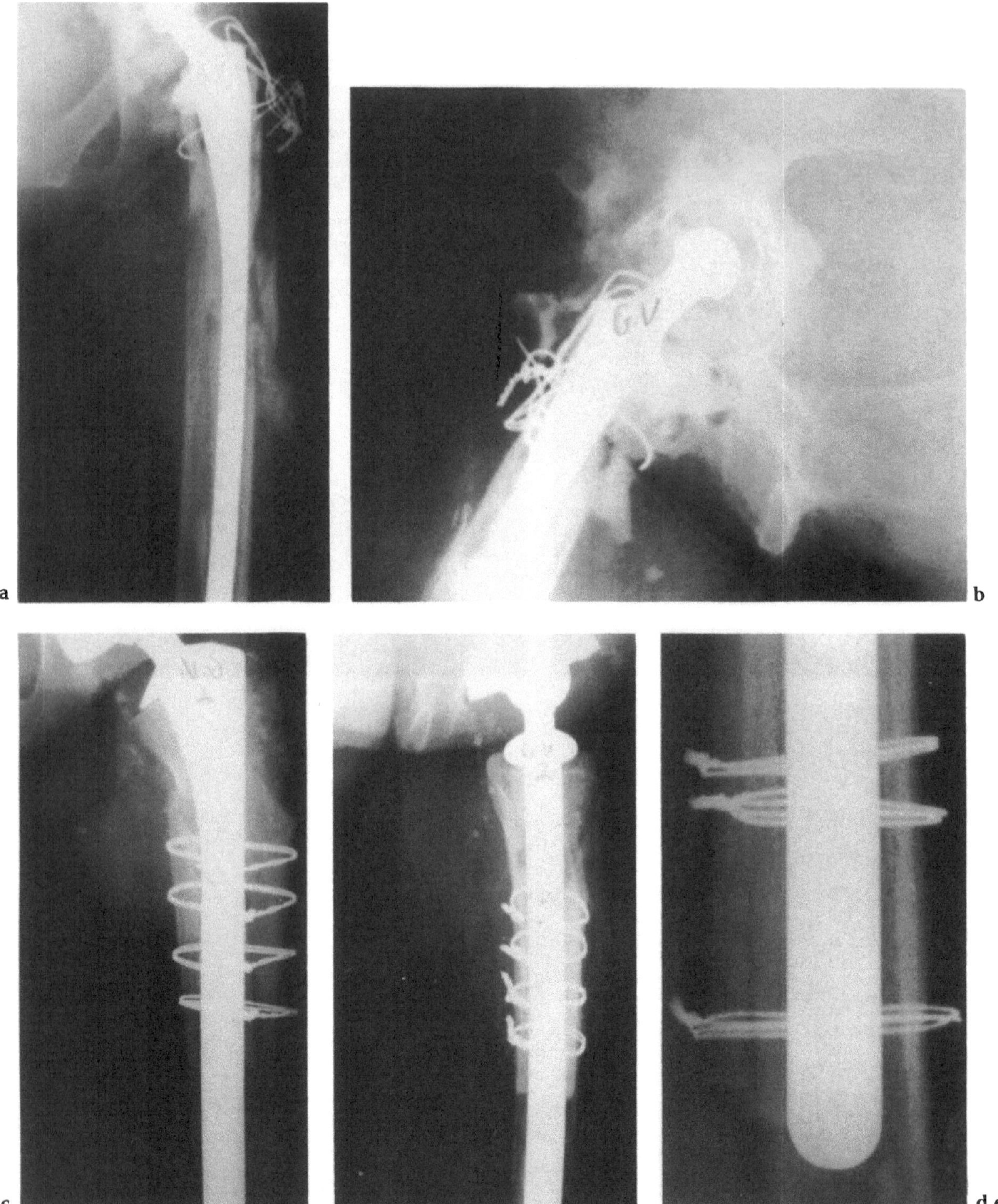

Fig. 13.9a–e. A Type 3 femoral defect. **a,b** Pre-operation. **c,d,e** Post-operation.

With femoral defects, the Type 1 and 2A groups were approximately 15%. Half of them were Type 2B and 2C and one out of three were Type 3 (Table 13.5).

Table 13.5. Incidence of the types of femoral defects

Type	*n*	%
1, 2A	36	15.8
2B, 2C	120	52.6
3	72	31.6

Of the cementless revisions performed, 11 were lost to follow-up and fresh-frozen cortical or calcar strut grafts were used in 124 revisions. We do not irradiate or treat chemically as it reduces the strength of the graft. In the Type 3 category, there were 57 diaphyseal struts and 18 calcar type grafts, 17 antero-lateral strut grafts and 21 postero-medial strut grafts (Table 13.6).

Table 13.6. Types of strut or calcar graft used

Type	*n*
2: Calcar	18
2B: Antero-lateral strut	17
2C: Postero-medial strut	21
3: Diaphyseal struts	57

Of the cortical struts, 87 of the 94 incorporated; seven were resorbed, because the femoral components were mechanically unstable and undersized. Five of those were revised; two are mechanically unstable but they are not painful enough to revise. Eleven of the calcar grafts resorbed but these were only in the ones that were less than 2cm in height. It is absolutely critical for the femoral component to be stable or it will loosen resulting in failure of the entire construct (Table 13.7).

Table 13.7. Results

	n
Cortical struts	94
Incorporation of cortical struts	87
Non-union and resorption	7
Revised femoral components	5
Painful unrevised components	2

We evaluated all cases with respect to the stability of the interfaces, according to Engh's criteria. Of the Type 3 group, seven of the eight were unstable but proximally coated devices had been used (Table 13.8). Our overall mechanical failure rate (a combination of the number of revisions divided into the total number of revisions) gave a rate of 3.5%.

Table 13.8. Radiographic evaluation: Engh's classification

Type	Stable interface	Unstable interface
1, 2A	36	0
2, 2C	119	1
3	65	7

We used the D'Aubigne Postel pain and walking scores. Types 1 and 2 had significant improvement in their post-operative pain scores. Type 3 were not quite as good but this was due in part to seven failures in this group where undersized stems had been used which became loose (Table 13.9).

Table 13.9. D'Aubigne and Postel scores

Type	*n*	Pre-operative score	Post-operative score
1, 2A	36	4.5	8.89
2B, 2C	120	4.4	8.71
3	72	4.1	7.80

When we evaluated the incorporation of the strut grafts, the post-operative scores were 8.73. However, when those cortical grafts resorbed, the post-operative scores were 5.6. This again was attributed to loosening of the femoral component. Whether the calcar graft incorporated or resorbed, the result was still the same provided the femoral component was stable (Table 13.10). The calcar grafts were mostly decorative and provided no overall function.

Table 13.10. Comparison of D'Aubigne and Postel scores when femoral component is stable but calcar graft is incorporated or resorbed

Calcar graft	*n*	Pre-operative score	Post-operative score
Incorporated	7	4.35	8.19
Resorbed	11	4.25	8.65

Complications

Three infections occurred. Only one implant had to be exchanged. The other two required debridement only. There have been two sciatic palsies with one recovering and the other remaining symptomatic.

There have been eight dislocations. We have eliminated dislocation in past months by the use

Table 13.11. Complications

Complication	*n*
Gram positive *Streptococcus* infections	2
Sciatic nerve palsy	2
Dislocations	8
Pulmonary emboli	3
Myocardial infarction	2
Cerebral vascular accident	1

of a post-operative orthosis (a lightweight custom device) which is worn by the patient post-operatively (Table 13.11).

Conclusions

In order to produce consistent results, it is mandatory to strictly adhere to important principles. It is critical that stability of the femoral component comes from the host. Engh demonstrates that distal fixation of the extensively porous coated femoral component is the most reliable means of obtaining stable femoral fixation. The seven cases of femoral loosening in our study showed less than ideal distal fixation with the femoral stem being undersized.

The defect must be appropriately identified and treated in a reproducible fashion. Our classification system was able to accommodate all femoral defects and this allowed us to allocate the appropriate graft type to the identified defects in a consistent manner. The cortical strut grafts are added to the host-prosthesis composite as an augmentation to the femoral fixation rather than a structure support. As much host bone as possible is preserved, which in turn preserves the vascularity.

A calcar graft may not be necessary in the presence of a tight distal fitted stem which must be extensively porous-coated or have distal flutes to provide rotational stability.

As the cemented femoral revision arthroplasties continue to show an unacceptably high failure rate, we have demonstrated one method that effectively addresses the host bone loss and provides stable femoral prosthesis fixation in a reproducible manner. Our results showed incorporation of the cortical strut allograft and restoration of the host bone in the presence of a stable fitted femoral prosthesis. This may enhance the long-term outcome of revision arthroplasty.

14 Allografts in Major Revision Total Hip Surgery

A.K. Hedley

This presentation concentrates primarily on the correction of deficiencies such as ectasias where the femur has "ballooned out" or segmental defects. The ectasia can be treated either by filler bone graft or larger prostheses. Ideally, templates of every prosthesis currently available should be at hand instead of being restricted to one particular implant. Basic principles should be followed throughout treatment of the defect.

Case Reports

Case 1

The patient had a six year hip prosthesis associated with massive bone loss. The polyethylene liner had rotated inside the metal shell and generated a tremendous amount of debris. It was handled quite successfully with a press fit prosthesis and bone grafting from the inside. If the component is stable we anticipate good healing (Fig. 14.1a,b).

In my opinion, there is a very definite indication for a collar on the femoral prosthesis in this type of surgery. It is not possible to generate hoop stresses in bone that is ectatic and a collar serves a very good purpose by providing axial stability. When no standard component fits we are obliged to use custom components to achieve stability. There may be definite indications for long stems, but my own preference is to use as short a stem as possible.

Case 2

This patient had four previous revisions with a very ectatic femur and absolutely no possibility of getting any "off the shelf" prosthesis to fit except a custom component. In this patient a mid-length stem was used and at three years his cortex had reformed very well (Fig. 14.2a,b).

Case 3

I approach some problems by slitting the femur from below the stem all the way up, including the greater trochanter, opening it like a door, taking the prosthesis out and then replacing the piece with cables at closure. This technique allows easy prosthesis removal as dislocation is not required. It also allows excellent access to the femur for cement removal and reconstruction. I have used Dall cables with sleeves and cable grips, which I recommend as outstanding devices solving many problems.

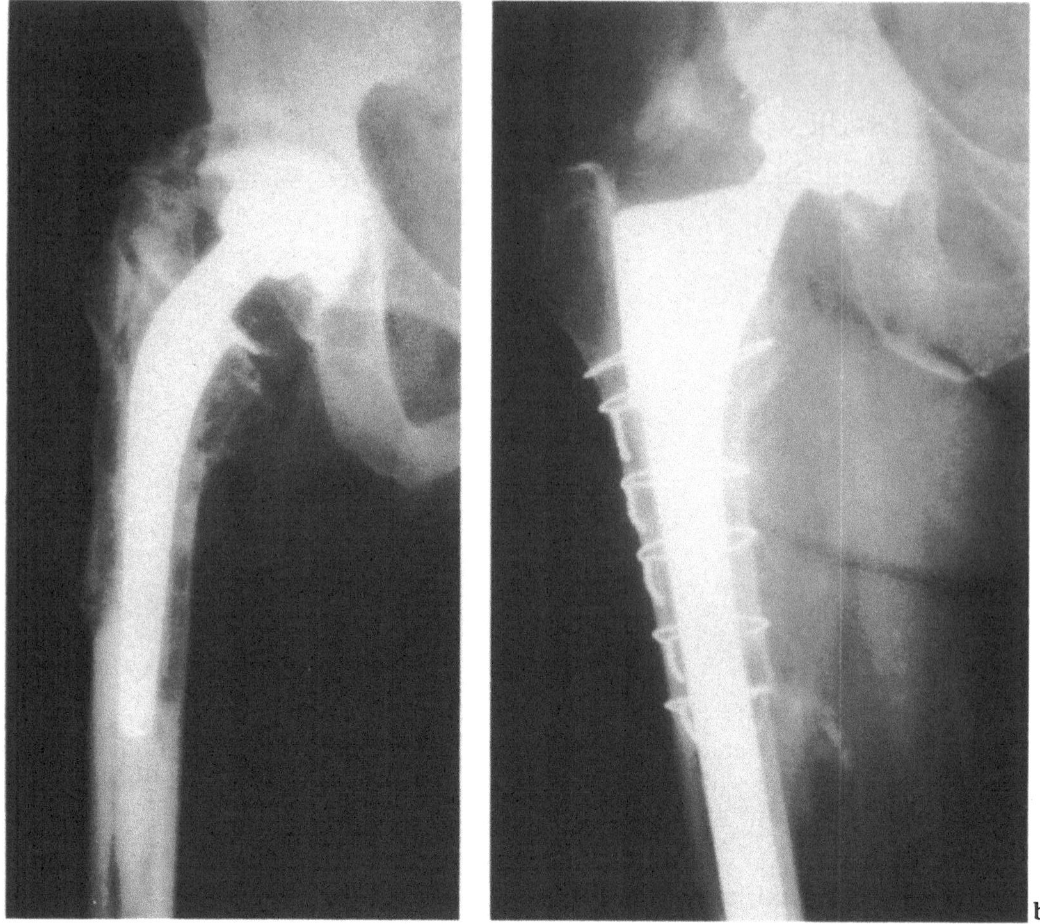

Fig. 14.1. a AP radiograph showing massive osteolysis of both femur and acetabulum. **b** Revision with press fit long stem and cerclage wires protecting a very tenuous proximal femur. Bone paste was placed inside the femur.

Dealing with a large number of ectatic femurs I conclude that the prosthesis should be stable without filler graft. If you rely on matchstick grafts jammed down inside the femoral canal for stability of your prosthesis, it will fail. The prosthesis needs to be stable and the graft acts only as a filler.

Case 4

Regarding segmental grafts, the defects can be handled either by proximal allografts or special prostheses. Tumour prostheses should not be used. If such a prosthesis is cemented distally and fails a very difficult revision circumstance is created. The proximal femur is already lost and the distal end of the femur will follow. It becomes increasingly difficult to solve.

The other problem with this type of prosthesis is their notorious instability. With abductor attachment to an allograft, there is a very different result. Once the proximal allograft has been "matched" to the host bone with a step we work across the interface with a very fine burr until the contact areas are absolutely perfect and then add a cerclage wire for added stability. After one year the junction cannot be seen. If the component is cemented into the allograft and not cemented distally, the distal femur will recover well, the "cement disease" will disappear and there will be a stable reconstruction.

The proximal allografts can either be circumferential or partial. Acetabular reconstructions are done with bulk allografts and press fit components, using the screws through bone and not the metal shell. I believe metal-to-metal contact

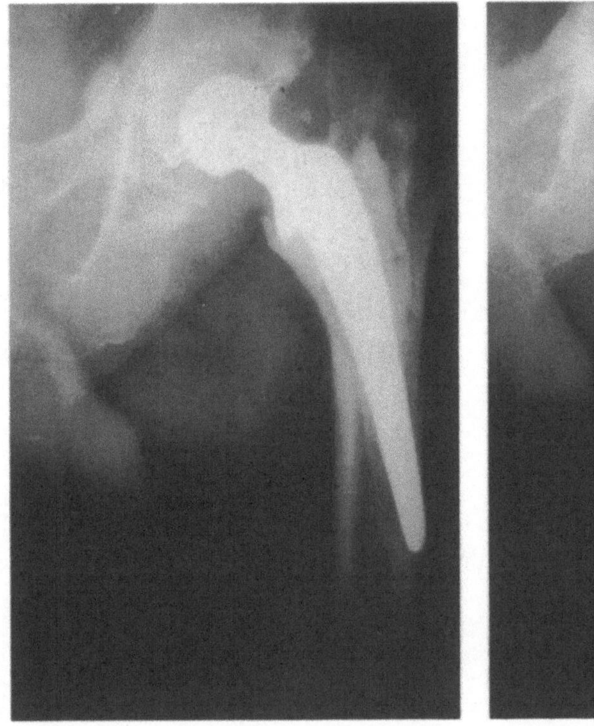

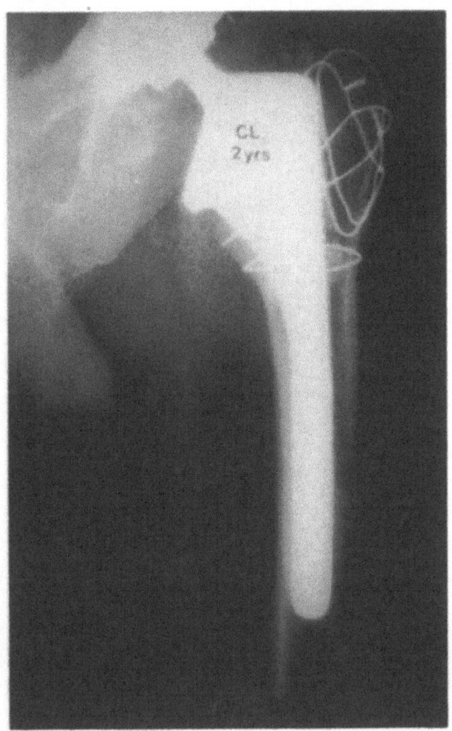

Fig. 14.2. a The femur is very ectatic proximally resulting in a varus position of the femoral component. The proximal femur is very wide while the isthmus is relatively normal. **b** A custom component was used with a large metaphyseal portion and a normal diaphyseal portion. Note the collar on the femoral component.

causes excessive metallosis the results of which will cause trouble for years to come.

Case 5

The patient had a press-fit prosthesis inserted on to old cement (Fig. 14.3a). He also had a fractured distal femur and needed a proximal femur allograft. A long stem was used so as to fix his distal femoral fracture (Fig. 14.3b,c). He united not only the proximal femur allograft but using the Dall cable grip has progressed to successful stability. I recommend that device very highly.

Case 6

The methods of approximation are important. Oblique cuts tend to slide. The host bone is inferior so a step-cut is far better because it does not telescope, and an accurate leg length adjustment can be obtained (Fig. 14.4).

Case 7

The patient presented with knee pain. He had had a previous femoral allograft that had been put down inside the femur and which subsided. The knee pain was due to the stem impinging on the knee prosthesis. We removed the old allograft and used an allograft with a step-cut, augmented with cerclage wires which went on to unite (Fig. 14.5 a–d).

Case 8

This was an interesting lady with an absolute "hardware store" inside her hip. Her affected leg was three inches too short (Fig. 14.6a). She required a proximal femur allograft with a step-cut and cerclage wire (Fig. 14.6b,c). Later the socket moved a little and was touching metal, so at four years was revised. Much of this allograft was still dead, but the junction was perfect. There was bleeding at the junction although the remainder of the graft still appeared fairly avascular. The long term success of these allografts may well depend on the bulk of the allograft remaining avascular.

Case 9

Two attempts were made at grafting this patient and twice she rejected the grafted bone.

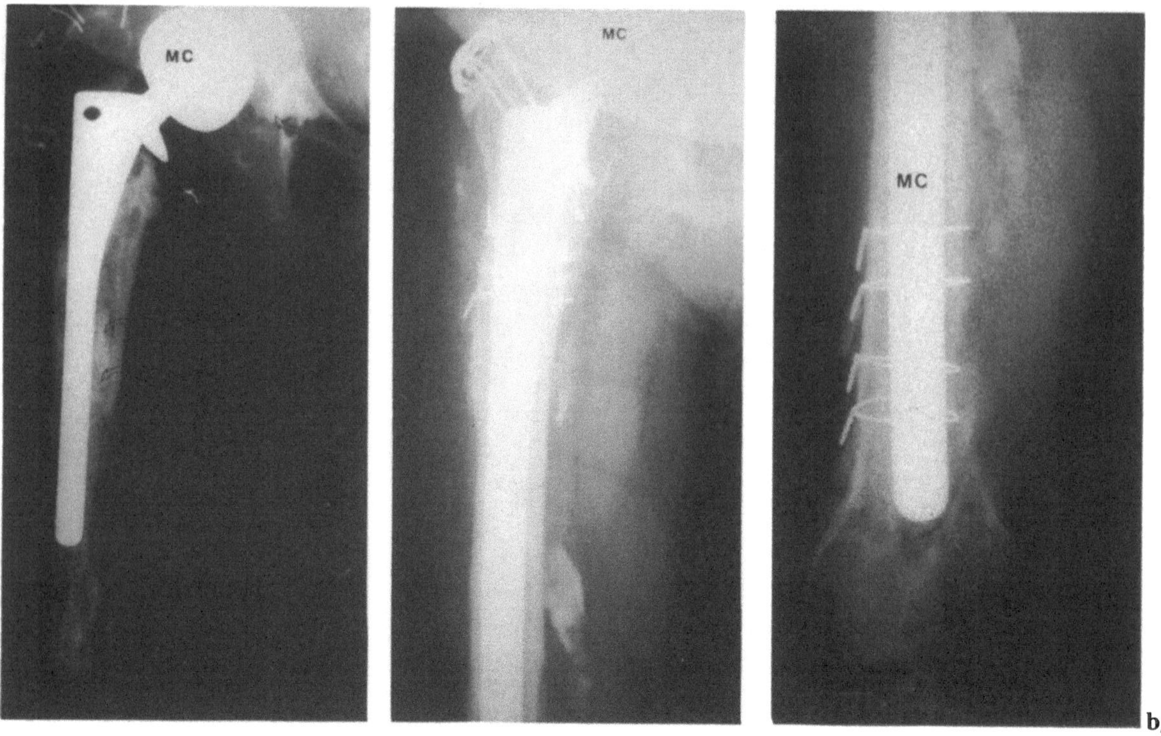

Fig. 14.3. a Prosthesis sealed on cement not removed at the time of revision. **b** Post-operative long stem with proximal femoral allograft. A Dall cable grip was used to secure the trochanter to the allograft. **c** One year post-operation the distal femur fracture has united.

Antibodies were demonstrated and she was the first patient in whom we actually type-matched for bone. She had a massive antibody profile and we typed specific bone. Two years after surgery none of the matched graft had resorbed. Case 7 (Fig. 14.7a,b).

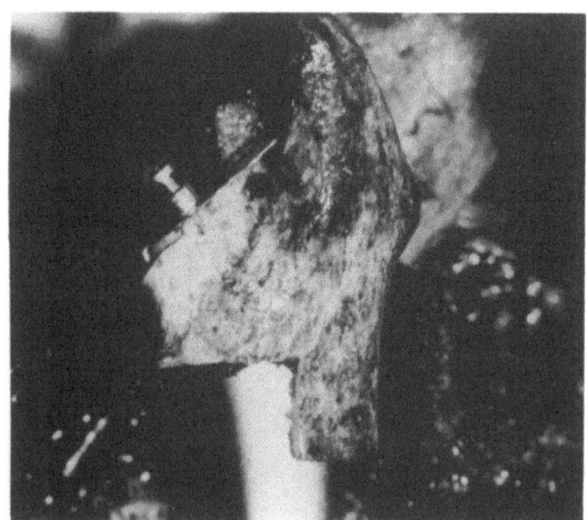

Fig. 14.4. A step-cut allograft in contact with host bone.

Case 10

The patient was a young man who had an arthrodesis of the hip with 1cm of the pin protruding into his socket. He had had a pseudo-arthrosis for 20 years. After plate removal we fashioned a cortical "plate" to fit exactly into the lateral osseous defect with good results (Fig. 14.8a,b).

Case 11

If you use a calcar graft, a collar is important to load the graft. The patient had a calcar graft that healed very well in this way. He had a lateral cortical strut and a successfully united trochanter (Fig. 14.9).

Case 12

This patient was reconstructed by recreating his fracture and then reassembling the femur. A lateral allograft strut with cerclage was used with a mid stem (Fig. 14.10a,b).

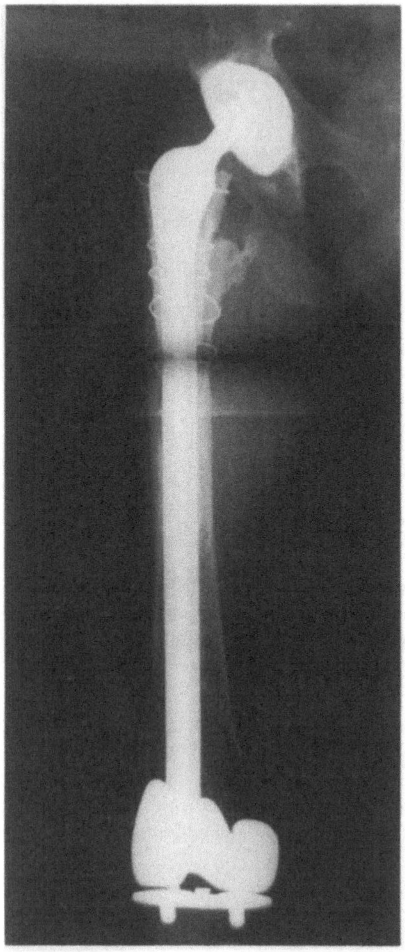

a

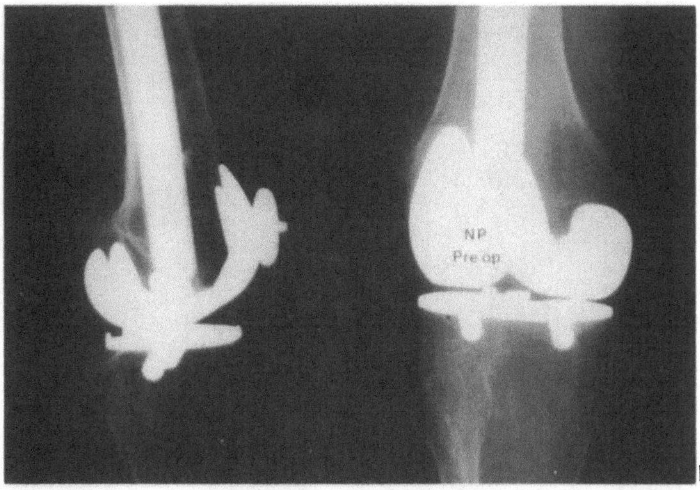

b

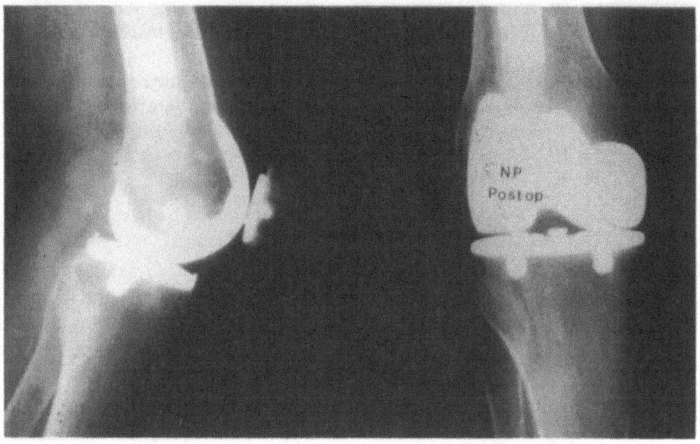

d

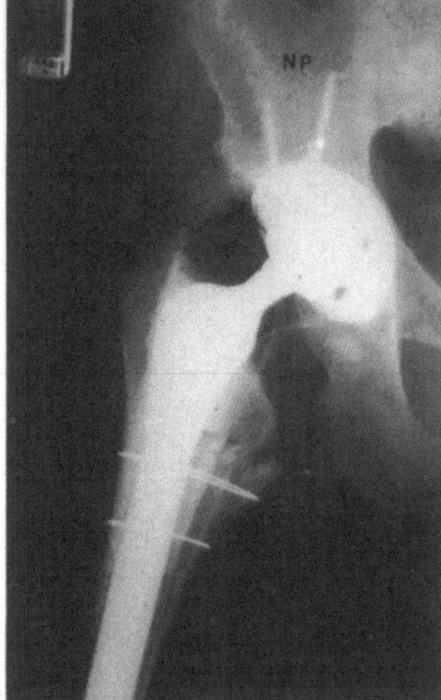

c

Fig. 14.5. a,b The allograft has subsided inside the femur allowing the femoral stem to make contact with the knee prosthesis. **c** A new step-cut allograft was used providing a more stable reconstruction. **d** The femoral component appreciably higher than the knee after step-cut allografting.

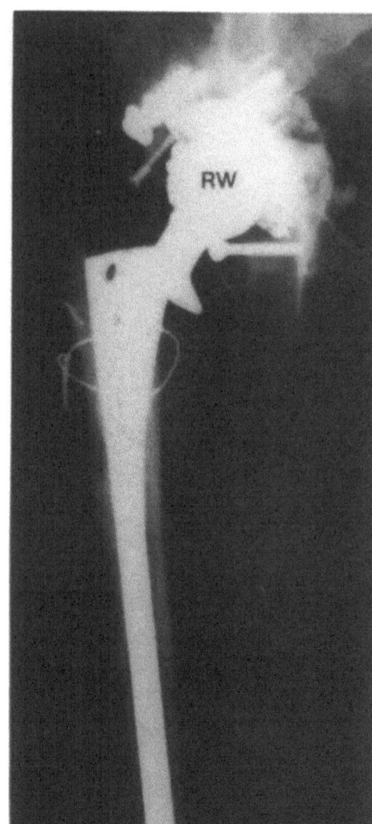

a

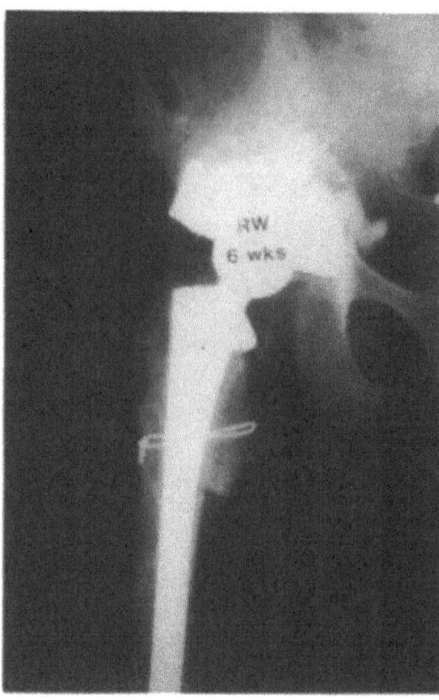

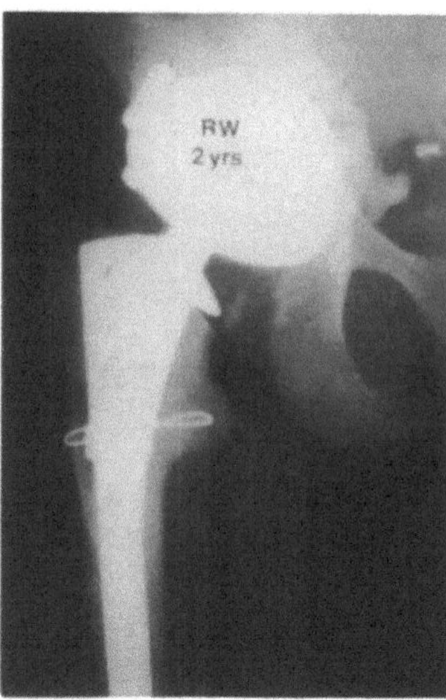

b,c

Fig. 14.6. a A pre-revision X-ray showing loose acetabular component and short femur. **b** Proximal step-cut allograft clearly seen at six weeks post-operation. **c** Two years later, there is union of the allograft with successful restoration of leg length.

Fig. 14.7. a Pre-revision X-ray showing absence of bone and subsidence of revision femoral component. **b** Post-reconstruction with type matched bone.
▼

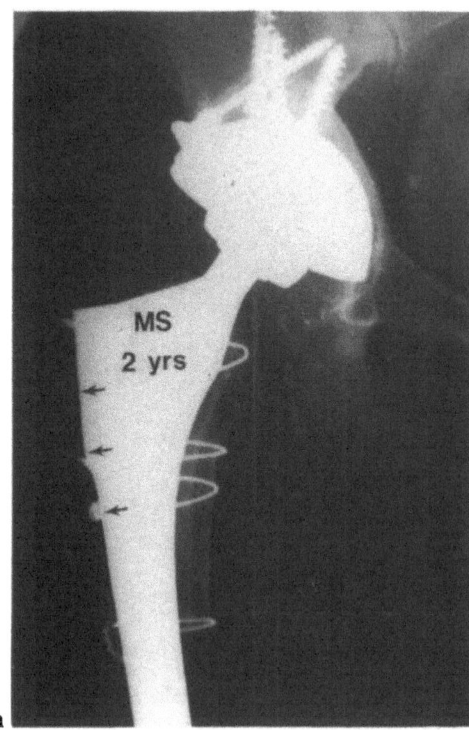

a

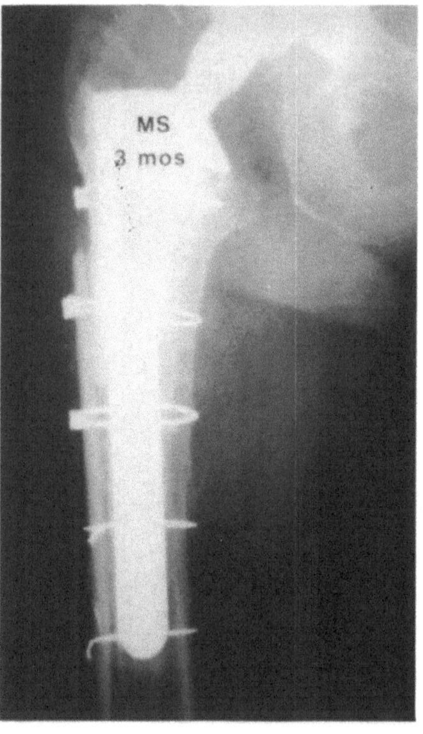

b

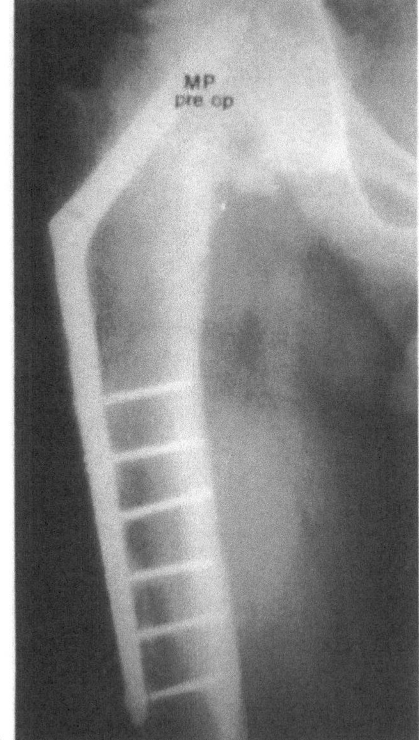

a

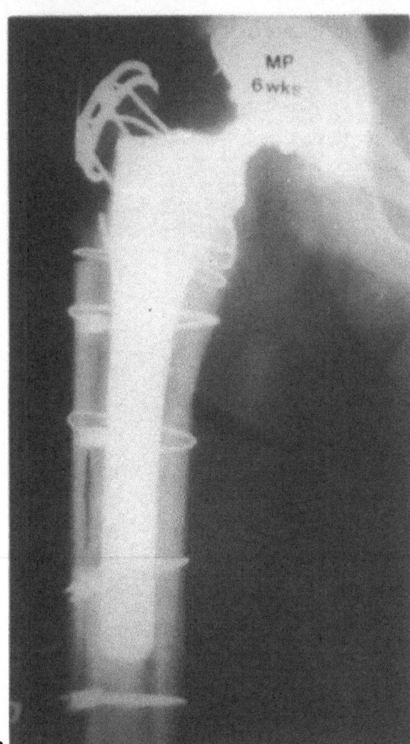

b

Fig. 14.8. a "Arthrodesis" of the hip with the Jewett nail protruding 1cm into the dome of the acetabulum. Note the plate is "intra-osseous". **b** Post-conversion to total hip replacement. The lateral slot was fixed with an allograft strut.

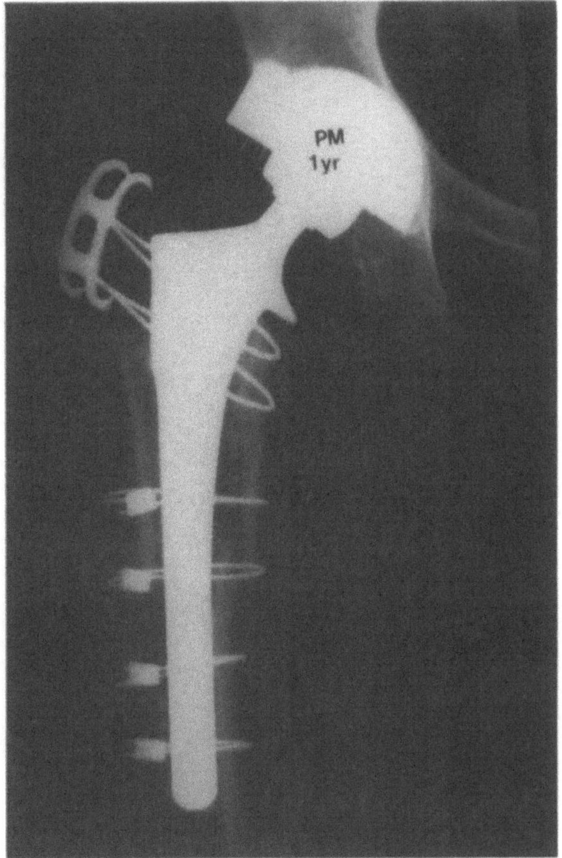

Fig. 14.9. Post-revision. The calcar is reconstructed with allograft loaded through a collar. Note the Dall cable grip on the trochanter and the cable and sleeve used to secure the allograft.

Case 13

Fractures can be treated well with the same technique. The patient had a fractured femur at the tip of a long stem. An Ogden plate, augmented with an allograft strut, was used. I now would dispense with the plate and simply use allograft struts. (Fig. 14.11a–c).

Case 14

The patient fractured at the tip of a fully coated stem and was plated with a single plate which failed as did a double plate. An electric stimulator was inserted and she was placed in traction for six months. That did not work either.

A long stem prosthesis was inserted augmented with an allograft and an autograft. At three years the allograft strut has remodelled, the femur has

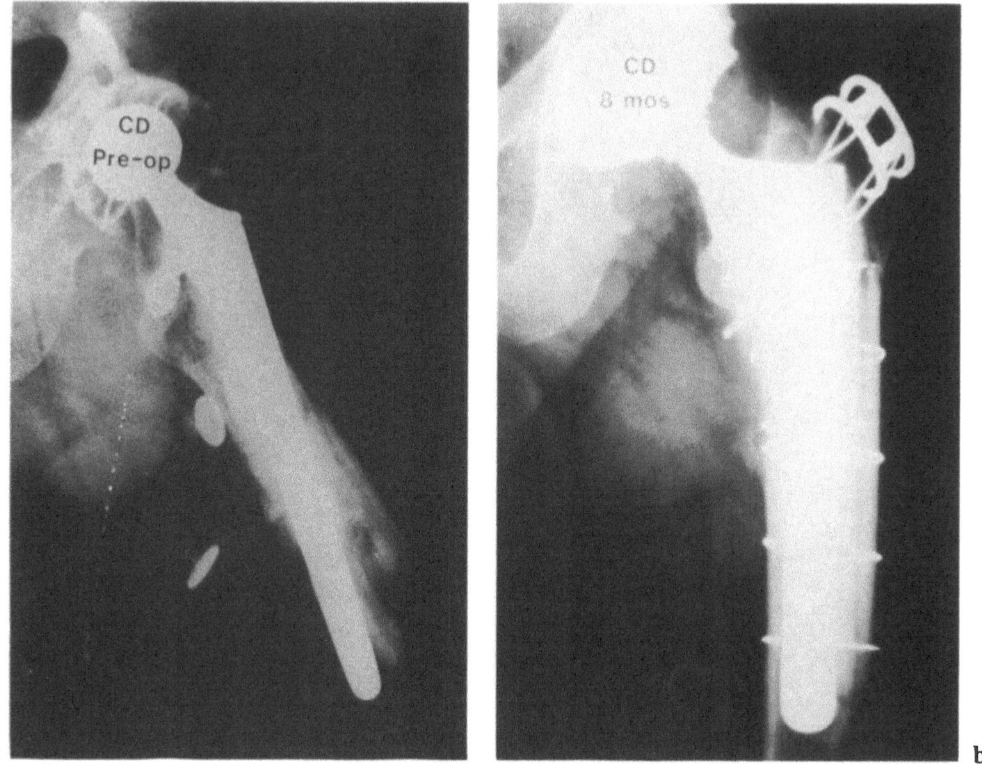

Fig. 14.10. **a** The patient after a primary total hip replacement. The femur is clearly fractured. **b** Reconstruction with a mid stem and allograft struts secured with Dall wires, cables and sleeves.

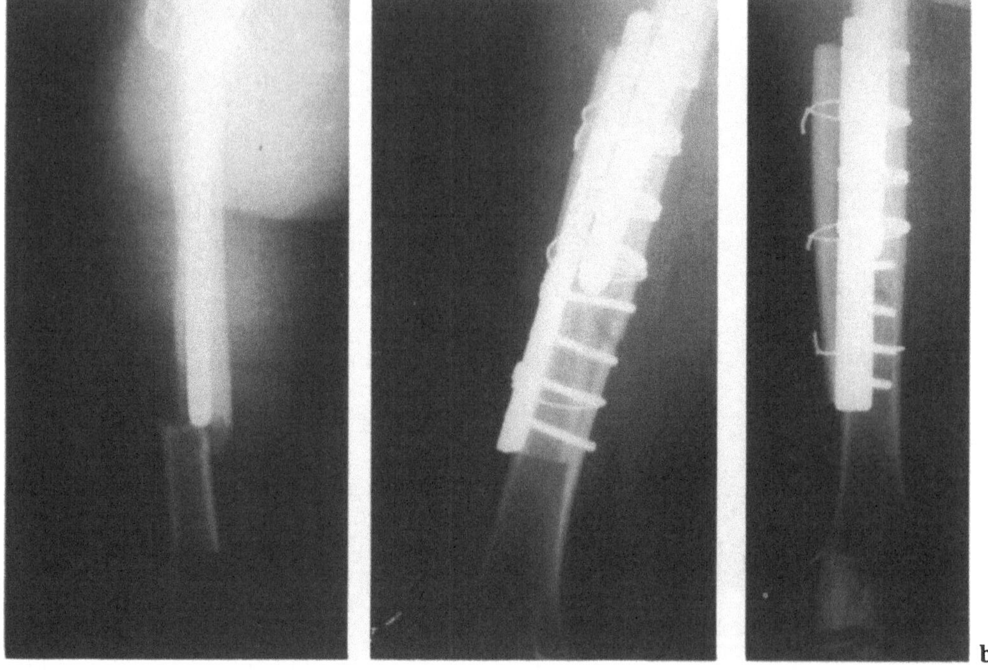

Fig. 14.11. **a** A fracture at the tip of a long stem. **b,c** Ogden plate used in conjunction with an allograft cortical strut.

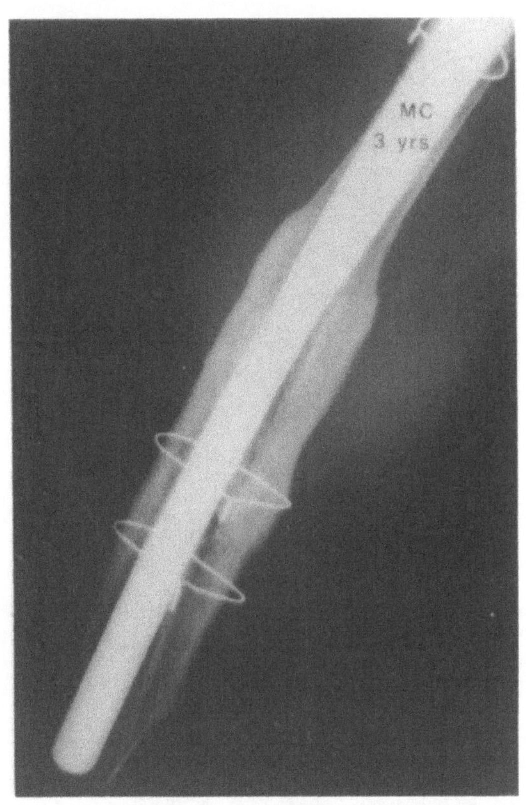

Fig. 14.12. The fracture has united with marked remodelling of the allograft struts.

healed completely and the patient is asymptomatic (Fig. 14.12).

Case 15

The patient's stem had started to migrate out of the side of her femur. She was revised with a longer curved stem but without effectively solving the problem. At three years nine months she had unabated thigh pain. We were concerned that she would fracture through the cortex so a lateral allograft strut was put on the femur which incorporated very well. Her thigh pain disappeared immediately post-operatively. We have used this technique to treat and cure thigh pain on several occasions since; the prerequisite being a proximally stable prosthesis (Fig. 14.13a,b).

Case 16

Fixation of the struts is important. We used Dall cables and sleeves. Using "plastic bands" that

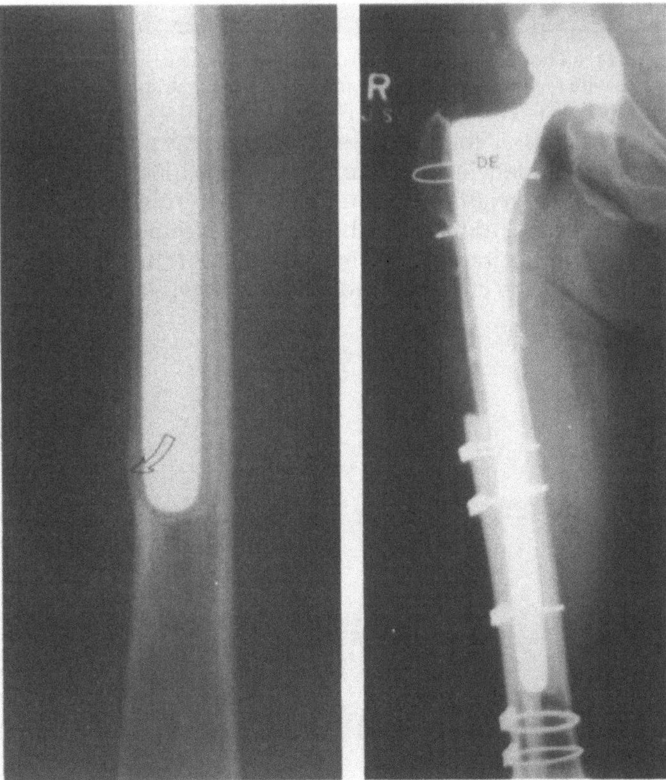

Fig. 14.13. a The stem tip migrating through the lateral cortex associated with thigh pain. **b** Post-application of cortical strut grafts. The thigh pain resolved.

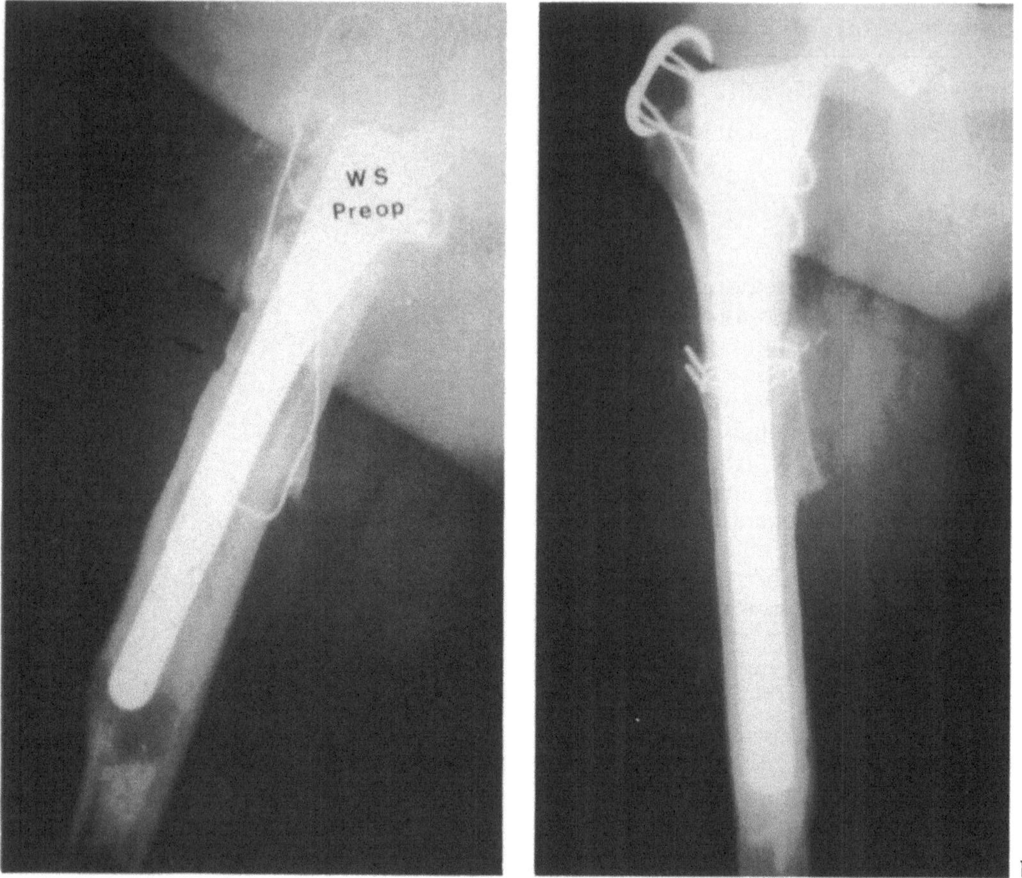

Fig. 14.14. a Pre-revision radiographs. Arrows show where plastic bands had segmented the femur. **b** Post bone grafting and cerclage wire to junction of allograft and host. Union was finally complete 18 months post-operation.

have a "memory" for cerclage is dangerous. Small notches were left in this femur by these bands. This patient, weighing about 135kg, has Fröhlich's syndrome. At revision, the femur was found to be in five separate pieces, because each of the plastic bands had cut right through bone down to prosthesis. We managed to obtain an excellent fit and apposition and thought union was certain. A delayed union resulted which healed after bone graft and cerclage cables. In retrospect this should probably have been done at the time of the initial surgery (Fig. 14.14a,b).

15 Bone Grafting in Revision Total Hip Surgery

V.M. Goldberg

Surgical technique is a critical determinant of the success of bone grafting in revision total hip replacement. The host bed must be appropriately prepared and the scar tissue excised. Soft tissue coverage is important in determining the outcome of the bone grafting procedure. Infection should be prevented by meticulous surgery and prophylactic antibiotics. The host bed provides circulation and nutrition to the soft tissue and bone. The selection of the bone graft is also important to provide the clinical function required.

Fixation and apposition of the graft to the host bone are important considerations in determining the outcome of bone grafting. Table 15.1 summarises the clinical factors of importance in healing of bone grafts (Goldberg 1991).

Table 15.1. Clinical factors in healing of bone grafts

Preparation of host bed
Eliminate scar
Good skin coverage
Avoid infection
Circulation and nutrition of surrounding tissues and host bone
Selection of graft
Apposition of graft to host bone
Fixation

Pre-operative Planning

The extent and location of the bone deficiency present in the failed total hip arthroplasty must be defined to adequately plan the surgical procedure. Radiographic assessment includes standardised pelvic radiographs with magnification markers, with the hips in neutral rotation and abduction, obliques of the pelvis (Judet views), and anterior-posterior and lateral views of the femur. CAT scans may be helpful and in some circumstances 3D reconstructions of the pelvis and/or femur may provide the best picture of the bony deficits. If the socket has migrated significantly, arteriography or an intravenous pyelogram may be helpful diagnostic studies. The location and extent of the bone loss will usually determine the surgical approach to the pelvis and proximal femur.

Planning for the socket usually requires bone stock assessment of the anterior and posterior columns and the medial wall and superior dome. The approximate socket size and type and amount of bone graft should be determined. A similar analysis should be done for the femur. Bone quality, size and shape of the canal, and the size and position of cortical defects should be determined. Leg length discrepancies are assessed and a plan provided to reproduce a centre of joint rotation as close as possible to the anatomical state. The surgeon must have available the appropriate bone graft to reconstruct the deficiency. This usually requires access to a bone bank with an adequate supply of cortical and cancellous bone. Both cemented and cementless components should be available. Oversized sockets and long stem femoral components may be useful.

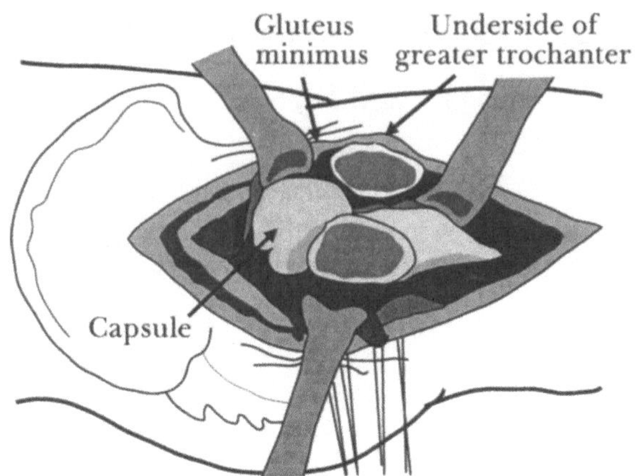

Fig. 15.1. Schematic view of the surgical approach providing extensive visualisation of the acetabulum and proximal femur.

General Techniques of Reconstruction

The restoration of bony integrity is critical in providing optimum prosthetic fixation, whether cemented or cementless. The fixation and containment of the graft is also important in providing the best environment for bone graft healing. The surgical approach that I have used when an extensive exposure is required is illustrated in Fig. 15.1. The greater trochanter is osteotomised, preserving the gluteus and vastus lateralis attachment in continuity. The entire construct with the bone is slid anteriorly, providing excellent exposure of the entire proximal femur and the acetabulum. The repair of the trochanteric bone is made easier by balanced soft tissues and a bone to bone interface.

Classification of Acetabular Defects

The types of acetabular bone deficiencies which aid both surgical planning and outcome assessment are shown in Table 15.2.

Table 15.2. Classification of acetabular component loosening

Type I	Intact acetabular wall Minimal bone thinning or loss
Type II	Marked thinning of acetabulum Massive acetabular enlargement
Type III	Bone loss: superior, anterior, posterior, central or combination leading to instability
Type IV	Massive acetabular collapse with fracture or extensive bone loss

Specific Acetabular Reconstruction Techniques

Cementless fixation of the socket should be considered if the metal shell is supported on 50% of its surface by bleeding, viable host bone. If this is not possible, consideration should be given to using cemented components. In either case the entire surface of the cup should be supported by bone if at all possible. It is important to have available sockets of many sizes. The use of extended rim liners and in some extreme problems constrained components may be necessary. When cementless fixation techniques are used, under-reaming the acetabulum improves the press fitting of the component.

Type I and II acetabular defects may be reconstructed using morsellised cancellous bone, either autogenous or frozen and/or freeze-dried allogeneic bone graft (Fig. 15.2). The graft is placed into the medial defect and shaped using an undersized hemispherical reamer in the reverse mode. Type III and IV defects are usually structural losses of the columns and require cortical structural bone grafts fixed firmly to the host bone. The posterior column and superior dome of the acetabulum are critical in obtaining a stable socket. If cementless liners are used for the reconstruction they also may act as the internal fixation system (Fig. 15.3). Type IV defects, where both columns and the medial wall are deficient, may require a total acetabular allograft in order to repair the socket (Fig. 15.4). Cement must be used in this circumstance. If necessary the socket may be translated up to 2cm proximally if this provides better bony

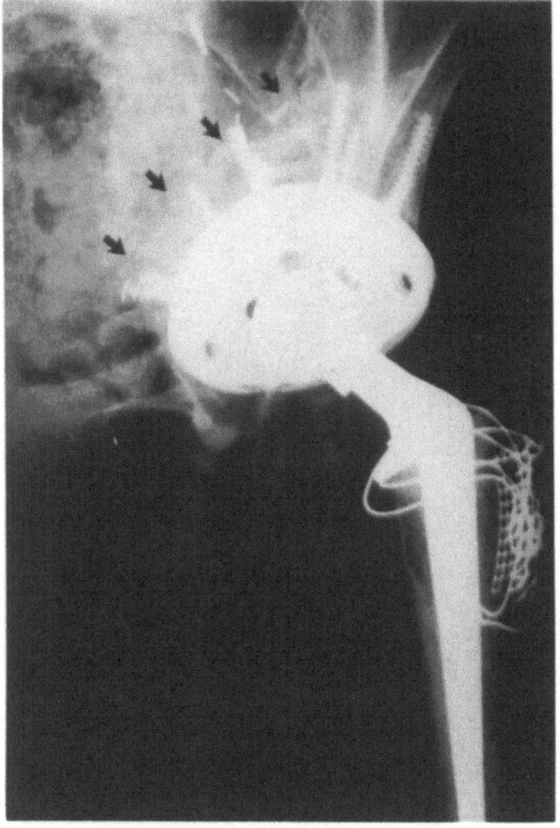

Fig. 15.2. Radiograph two years after reconstruction of a Type II acetabular defect with massive autograft. Arrows illustrate the extent of the incorporated bone graft.

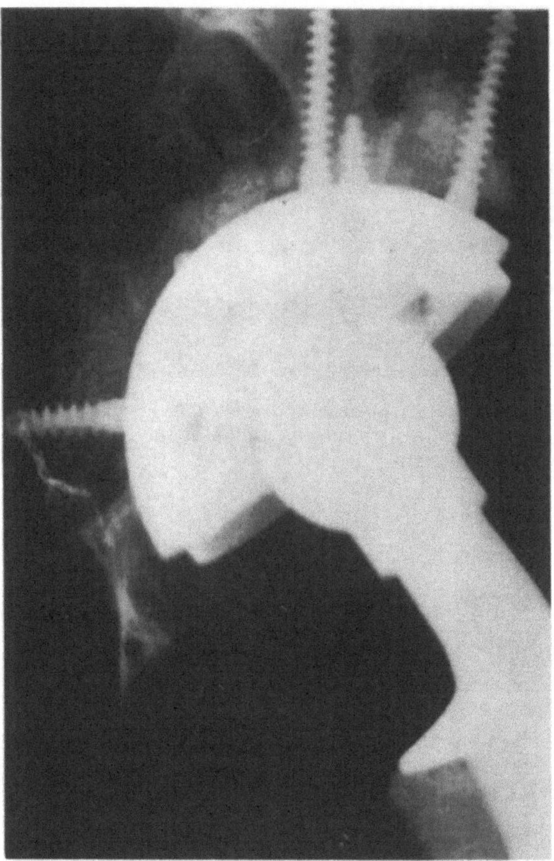

Fig. 15.3. Radiograph two years after reconstruction of Type III defect using the hemispherical metal liner as internal fixation device.

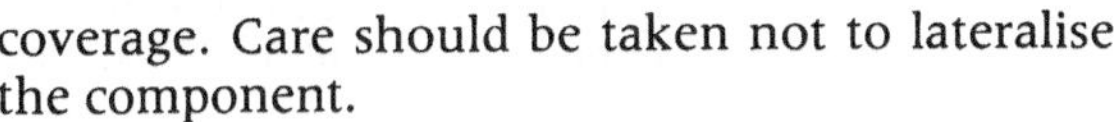

coverage. Care should be taken not to lateralise the component.

Classification of Femoral Defects

A qualitative classification of bone loss associated with failed femoral components is summarised in Table 15.3.

Specific Femoral Reconstruction Techniques

Cementless fixation of the femoral component is an appropriate technique if the clinical circum-

Table 15.3. Classification of femoral component loosening

Type I	Cement/prosthesis interface failure <50% thinning of proximal cortices Circumferential wall intact
Type II	Cement/bone interface failure >50% thinning of proximal cortices Canal enlargement Circumferential wall intact
Type III	Posterior-medial bone loss Unstable
Type IV	Proximal circumferential bone loss Loss of all columns

stances enable the surgeon to obtain an intrinsically stable implant interfaced with a surface of viable host bone. In elderly patients, or when intrinsic stability cannot be restored, cement fixation should be strongly considered. In either case, preservation of the soft tissues is critical. Wide exposure of the proximal femur is helpful in enabling the surgeon to remove all of the poly-

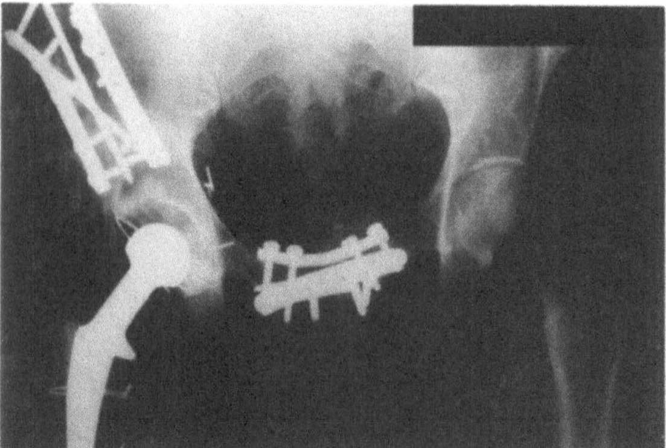

Fig. 15.4. Radiograph of pelvis demonstrating total acetabular graft two years after surgery. Union at the allograft–host junction is present.

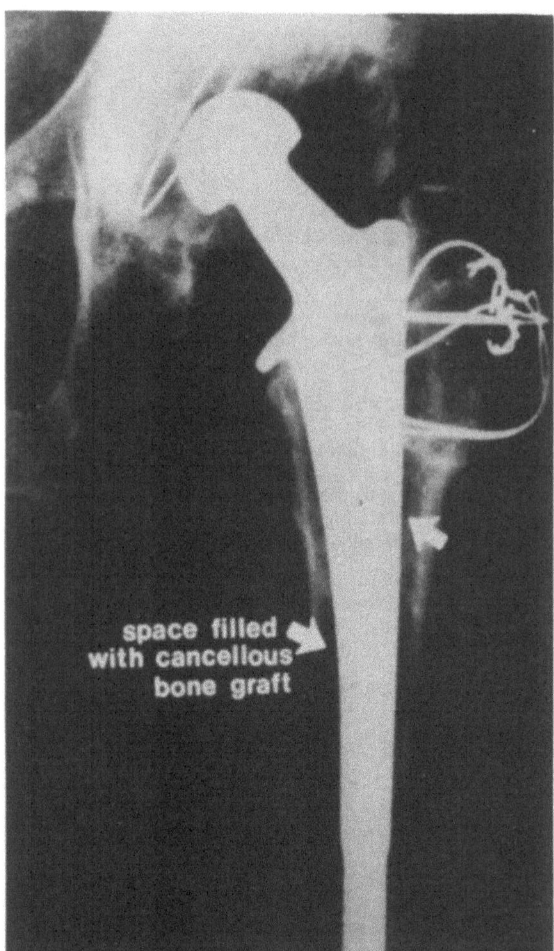

Fig. 15.5. Radiograph demonstrating femoral reconstruction of Type II bone deficiency using morsellised autograft and allograft cancellous bone graft.

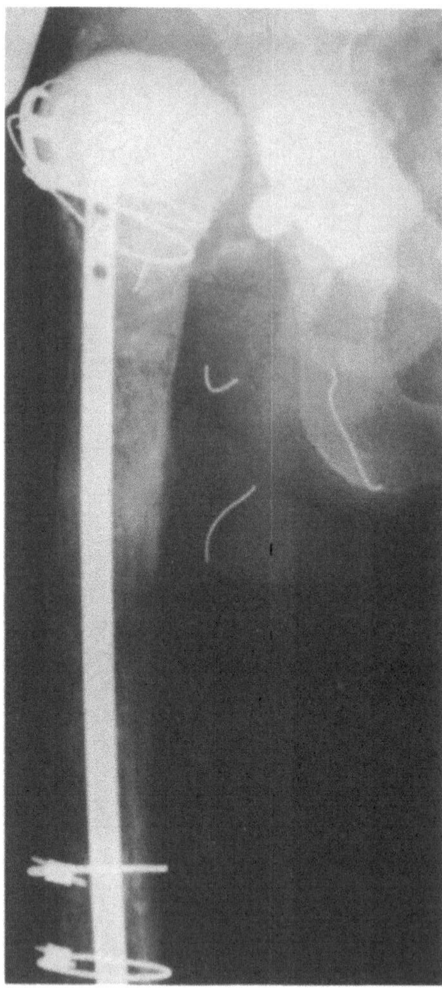

Fig. 15.6. Radiograph illustrating the intramedullary rod and spacer technique in reconstructing difficult bone loss problems.

methylmethacrylate. Autologous bone graft should be used as much as possible. When structural requirements are necessary, frozen allografts are useful. Freeze-dried bone does not have the structural integrity to function as a load bearing material. Type I and II bone deficiencies usually may be reconstructed with morsellised packed cancellous graft (Fig. 15.5). Any cortical defects should be repaired by cortical grafts well fixed to host bone. Type III defects require cortical bone to reconstruct the femur. Type IV proximal bone loss may require an entire proximal femoral allograft with a long stem component preferably fixed with cement only to the allograft. Recently when we have had extensive bone loss in a young patient, a two-stage procedure has been done. The first stage debrides all of the cement and non-viable soft tissue and then packs the proximal femoral canal completely with morsellised cancellous autograft and allograft. An intramedullary rod is used with a spacer to provide interim stability (Fig. 15.6). At the definitive procedure usually three months post-grafting, an uncemented component is used to reconstruct the femur. This type of approach may provide a viable alternative when difficult bone loss problems are seen in young active patients.

Results

The clinical scores of revision total hip replacements requiring bone grafting, a minimum 36 months after surgery, are summarised in Table 15.4. The functional results deteriorate the more significant the bone loss. Many of the patients have had multiple hip procedures so that deficient soft tissues are clearly additive to the bone loss problem.

Table 15.4. Femoral and acetabular typing vs. Harris hip score

	Number of hips	Pre-operative average	Post-operative average
Femoral component			
Type I	7	39.1	93.0
Type II	23	51.2	83.4
Type III	23	45.5	79.8
Type IV	3	35.8	78.2
Acetabular component			
Type I	14	44.0	83.7
Type III	18	49.4	83.3
Type III	7	39.3	76.4
Type IV	3	27.4	64.0

Conclusions

The extent and location of bone loss in revision total hip replacement is critical in determining the outcome of the procedure. CAT scans and 3D reconstructions are excellent techniques to define the problem and prepare for surgery. Complete and careful planning of the reconstruction is essential for a satisfactory result. Appropriate instrumentation, a wide choice of components and adequate bone graft must be available. If cementless fixation of the implant is selected, intrinsic stability of the components must be achieved and viable bleeding bone must be interfaced with the porous surface. Progressive conservative weight-bearing with prolonged crutch and/or cane use must be included in the postoperative rehabilitation. Finally, patient expectations must be reasonably assessed and cautious optimism communicated by the physician.

Reference and Further Reading

Goldberg VM (1991) Bone grafting in revision total hip arthroplasty. Instructional course lectures AAOS

16 The Use of Massive Proximal Femoral Allografts in Hip Surgery

C. Delloye and A. Vincent

Failures of hip arthroplasties with structural bone deficiency at the femur are becoming more frequent due to an ageing population. Femoral bone deficiency, whether in amount or quality, may result from an extensive osteolysis secondary to an aseptic loosening. Deficient bone may also result from a fracture of the bone around the prosthesis, or resection of a tumour. An allograft can be one of the surgical options to reconstruct the lost or weakened bone.

Material

During the past three years, 21 patients underwent a hip arthroplasty procedure which was associated with a massive proximal femoral allograft. The indication was revision arthroplasty for 17 patients, primary malignant bone tumour for two and pathologic or recurrent fractures for the remaining two patients.

The mean follow-up period is now 21 months. In the revision arthroplasty group, there were 15 women and two men. Their mean age at the time of allografting was 67.8 years, which is an older mean age than in other series. Most of them had undergone several revision procedures, with an average of 2.3 – the maximum being six surgical procedures.

Method

Reconstruction around the hip with a massive allograft is a major surgical procedure to reconstruct the upper end of the femur. We used a sterile-procured allograft that had been deep-frozen at −80 °C. The insertion of the gluteus tendons had been preserved on every allograft. The mean length of allograft was 15cm (± 6.5cm). This type of surgery took about five hours with an average blood loss of 3.5 litres. The mean stay in hospital was 34 days.

The reconstruction was carried out by an allograft alone (Fig. 16.1) or, more rarely, wrapped around by the residual host femur. A long-stem prosthesis was used to neutralise the bending forces. When a long allograft was used, a plate at the junction neutralised the rotatory forces. The prosthesis was secured in both bones by cement. In four patients, however, where the prosthesis was very well locked in the allograft, no cement was used in the allograft. Reinsertion of the gluteus tendons could be performed either by

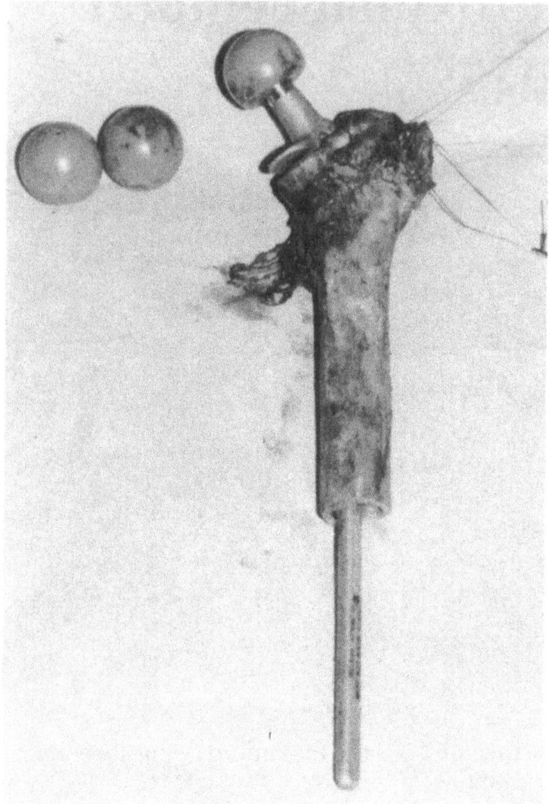

Fig. 16.1. Assembly of a prosthesis and an allograft on which the insertion of the gluteus tendon has been retained to facilitate the reconstruction. A modular prosthesis with large head size is preferable to the use of a conventional long-stemmed prosthesis.

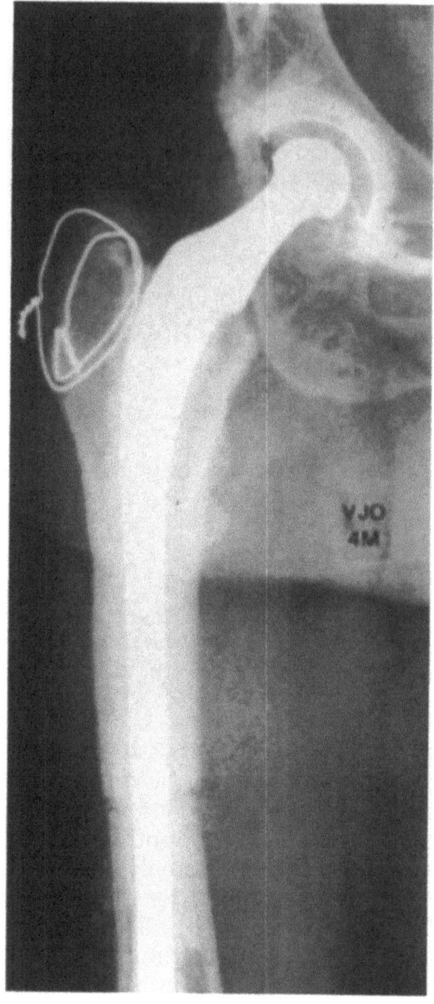

Fig. 16.2. Reinsertion of the host trochanter to a femoral allograft.

reinsertion of the trochanter, or by direct sutures on their counterpart which had been left attached on the allograft.

Patients began their post-operative rehabilitation around the fourth day, using two crutches or a walking frame for six weeks. Pendular walking was encouraged to avoid excessive demand on the gluteus. At six weeks, patients were allowed to walk with one crutch for one month. After this period most of the older patients used a cane outdoors.

Case Reports

Case 1

A 64-year-old patient had had two previous operations before being allografted. The acetabulum was re-shaped by a femoral head, while the femoral allograft was secured by a long-stem prosthesis cemented in both bones. The trochanter was reinserted by wires (Fig. 16.2).

Case 2

A 62-year-old male patient had had several hip arthroplasties. The primary pathology was avascular necrosis of the femoral head caused by severe alcohol consumption. His third prosthesis had rapidly loosened and become very painful (Fig. 16.3). A 29cm allograft, a long-stem prosthesis that was not cemented in the graft and a plate were used for reconstruction (Fig. 16.4).

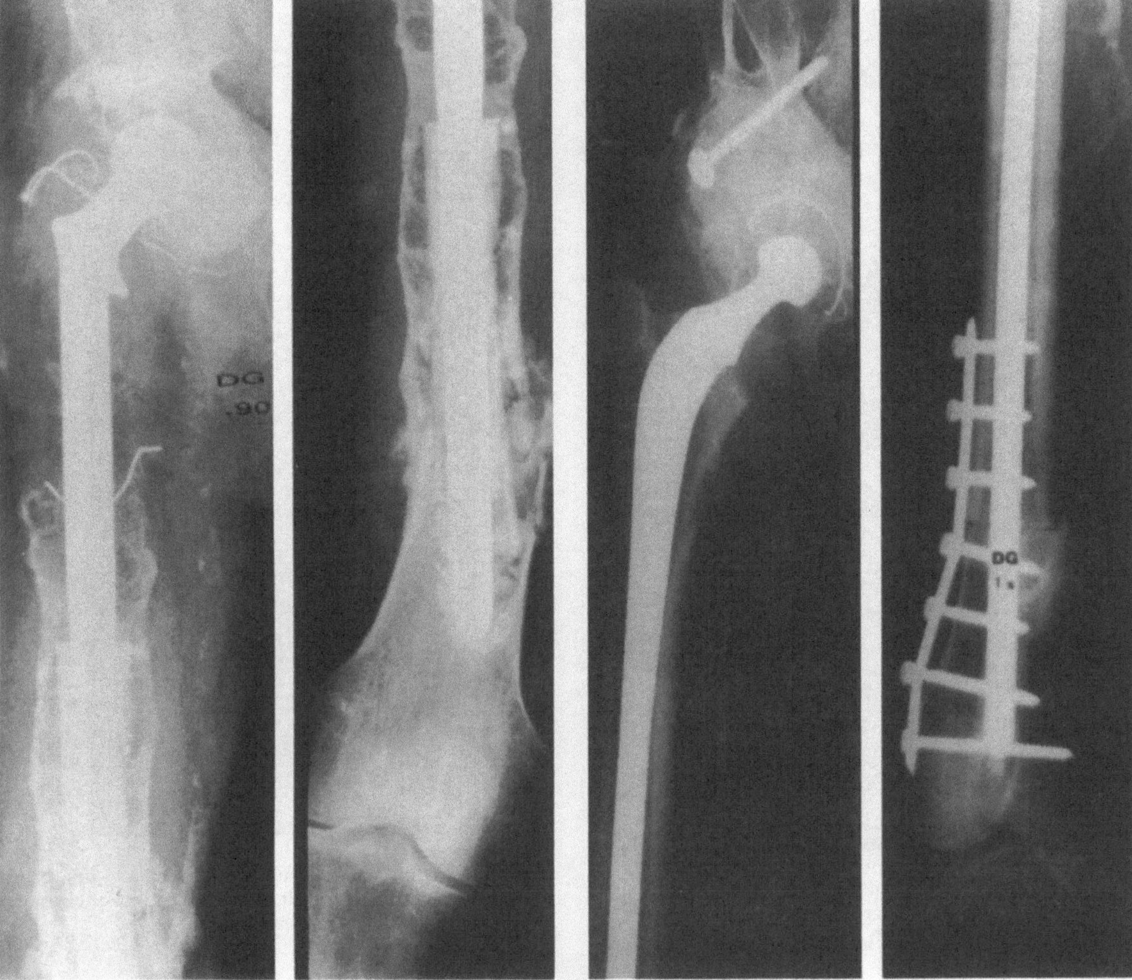

Fig. 16.3. A second revision hip arthroplasty that rapidly failed, requiring a third revision with allograft.

Fig. 16.4. Reconstruction of the hip shown in Fig. 16.3, with allograft and prosthesis.

Case 3

A 28-year-old male patient presented with an osteosarcoma. The anatomy and hip function were restored by a prosthesis and an allograft with a direct tendon suture (Fig. 16.5).

Case 4

A second type of reconstruction can be used in patients with aseptic loosening of a hip prosthesis. The allograft is adjusted to the size of the host bone and driven into the host. Provided that the allograft is long enough, a standard prosthesis can be implanted and cemented in the allograft, while cerclages fix the whole reconstruction. An example of such a reconstruction is seen in a 62-year-old patient (Fig. 16.6). There is apparently a bony union between the graft and the host bone.

Results

Functional results were graded according to the Merle d'Aubigne hip score system. Excluding those with less than six months of follow-up, 16 patients had been followed for an average of 21 months. Seventy-eight per cent had a good or excellent hip function, while 21% of patients had a poor or fair function. The patient with a poor function due to a gross failure of the reconstruction has had a further operation with a good result. Pain, motion and walking had respectively

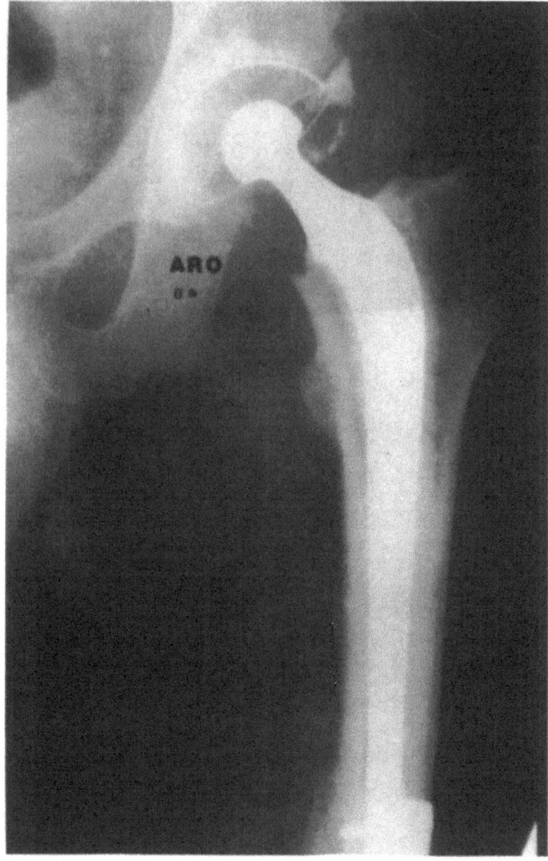

Fig. 16.5. Limb salvage surgery of a proximal femur by a combined prosthesis and allograft. Gluteus tendons were reinserted directly on the allograft.

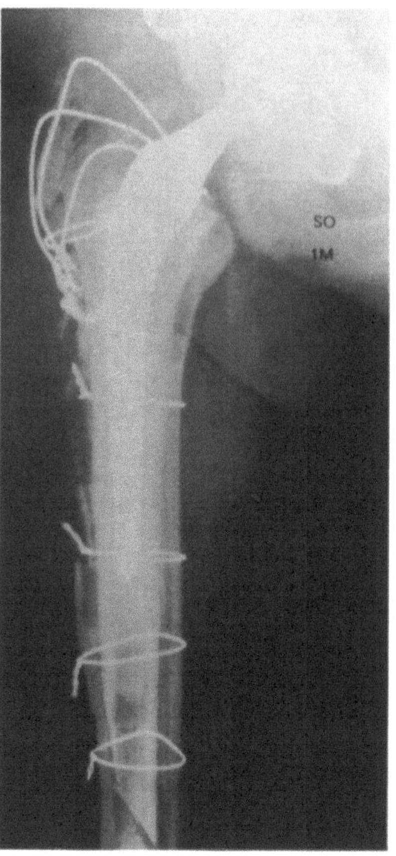

Fig. 16.6. In this patient with an aseptic loosening of the proximal femur, the allograft was inserted into the residual femur by-passing the weak area.

a score of 2.6, 2.5 and 2.4 points. Post operatively they scored 5.3, 5.5 and 4.6 points. Walking had the lowest post operative score (Fig. 16.7). Gluteus insufficiency in this old population was a common finding even after surgery. Most patients required the use of a cane for walking.

In evaluating the functional results, we also have to consider the general and local conditions in this rather old population. Most of them had a contralateral arthroplasty in good or fair condition which could impair the function.

The function obtained in young individuals who required an allograft after a tumour resection was excellent. In particular, the direct suture of the gluteus tendons left attached on the allograft was found very effective and apparently long-lasting. For example, two years after reconstruction following tumour resection in an 11-year-old girl. This young patient had good function standing on her operated lower limb (Fig. 16.8).

The radiographic fate of these allografts has been encouraging. No trochanteric non-union or gross resorption has been observed, except in some patients where a small resorption under the collar of the prosthesis was seen.

Union at the junction has been slow to develop, often with persistence of the osteotomy line. Callus has been favoured by the preservation of the remnants of the proximal femur in non-tumour conditions. Three non-unions were observed.

Some important complications were experienced in 37% patients. There were three deaths, two of which were more or less directly related to the operation. The first was a cirrhotic patient who died from a coagulating failure on the sixth day. The second patient, 82 years old, died at home, probably from an embolism. The third died from a neoplasm. There were also three dislocations, two fractures of the host femur, two infections, two non-unions and one gross failure.

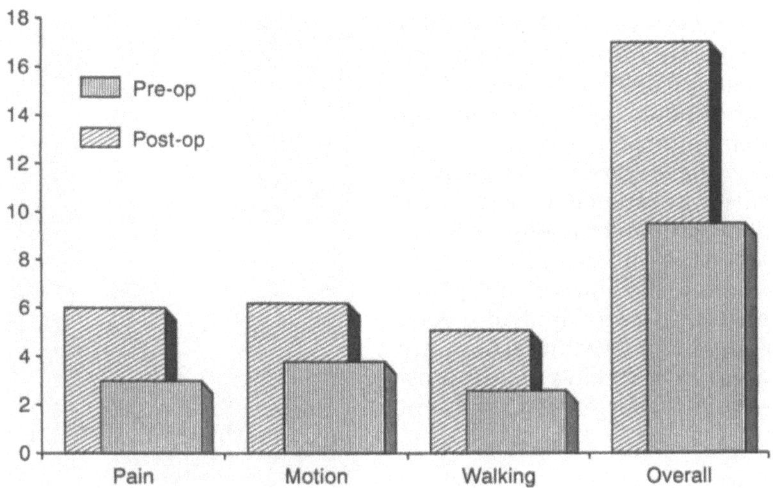

Fig. 16.7. Chart showing the pre- and post-operative (last follow-up) hip scores according to Merle d'Aubigne.

The first complication always occurred within the first four months. Once this critical period has been passed, no complications have occurred for the first time.

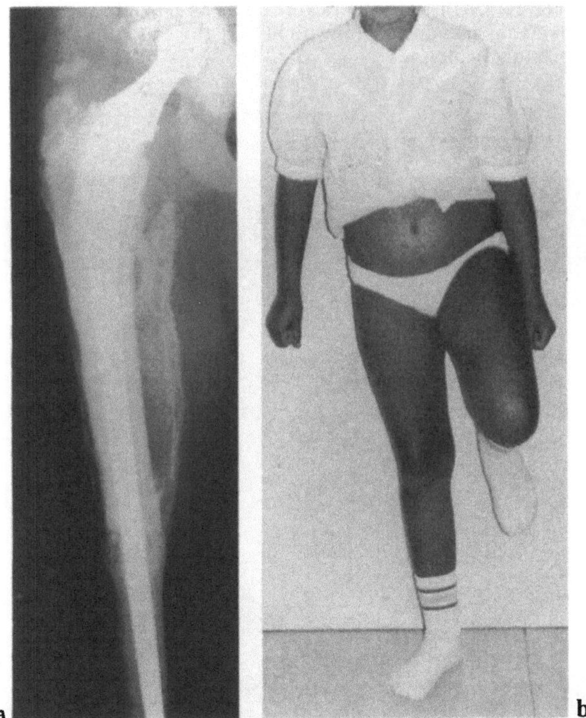

Fig. 16.8. Radiograph (**a**) and clinical function (**b**) of a young patient after a proximal femur reconstruction. Tendon reinsertion on its counterpart left on the great trochanter. There was normal hip function and normal walking performance without a limp.

Dislocation has been the most frequent local complication, occurring in three patients who had had at least three previous local operations. The number of previous operations was found to be the most critical factor in the dislocations. The hip always dislocated anteriorly. Two patients had multiple dislocations which required an augmentation procedure of the acetabulum to cover the head. To reduce this complication, the use of a thinner neck to avoid a cam effect or the use of a large head should be encouraged. The availability of different neck length or head thickness should also minimise this complication (Fig. 16.1).

Fracture of the host femur occurred twice after trauma in osteoporotic patients. Plating was performed without further complications (Fig. 16.9) Gross failure of the whole construct occurred in one patient.

Discussion

What is the value of the residual host femur that is to be placed around the allograft? We believe that the radiological appearance cannot distinguish union between allograft and host bone.

A 66-year-old patient died from a neoplasm 15 months after surgery. There was an apparent union between both bones (Fig. 16.10a). A transverse microradiograph, however, clearly demonstrated a large demarcation, fibrous interface between the viable bone and the allograft (Fig. 16.10b). We believe, therefore, that it is not always possible to demonstrate radiological union between both bones. The microradiograph also showed no remodelling activity in the allograft 15 months after surgery. Should we favour a press fit or a cement fixation of the prosthesis in the host femur? Up to now we have relied on cementation because our patients were rather old – an average of 68 years – usually with an osteoporotic bone, especially in the endosteal area. Is press fit insertion a long-lasting fixation in this population? In the patient in Case 2 who had several arthroplasties for necrosis of the femoral head, we experienced a progressive sinking of the prosthesis into the neck of the allograft, with shorten-

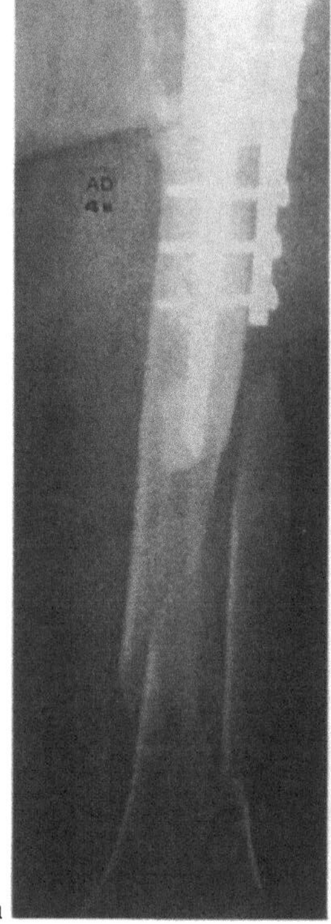

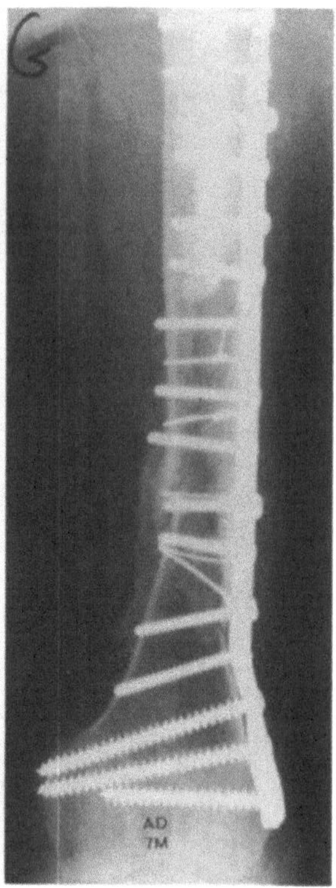

Fig. 16.9. a Distal femur fracture under an allograft, after a fall. **b** Plating was uneventful with healing three months after the fracture plating.

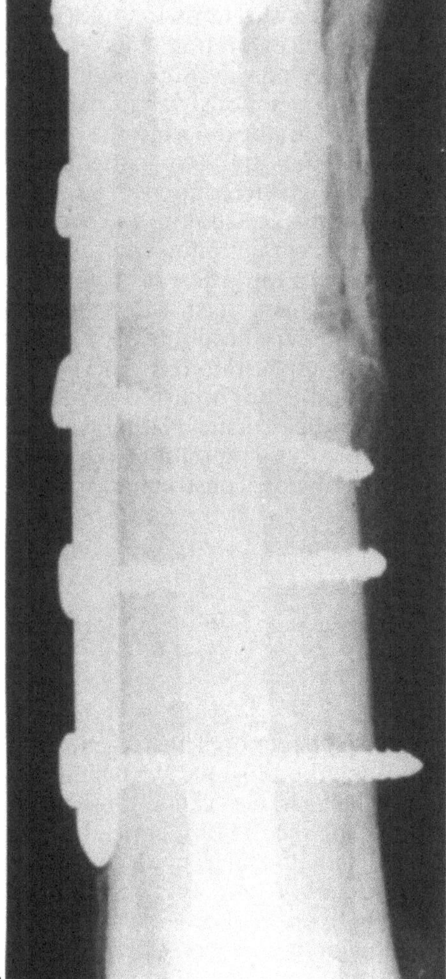

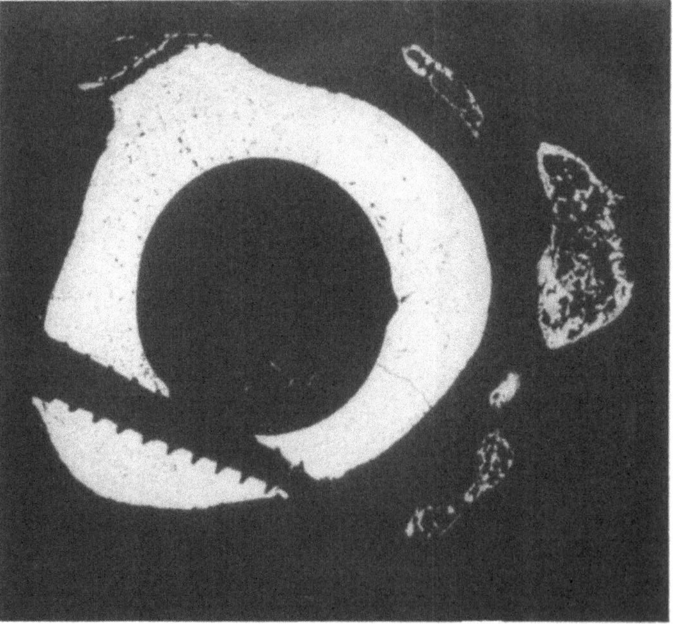

Fig. 16.10. Radiograph (**a**) and microradiograph (**b**) of a diaphyseal junction autografted 15 months previously. The cross-sectional microradiograph shows clearly the space between the allograft and the autograft.

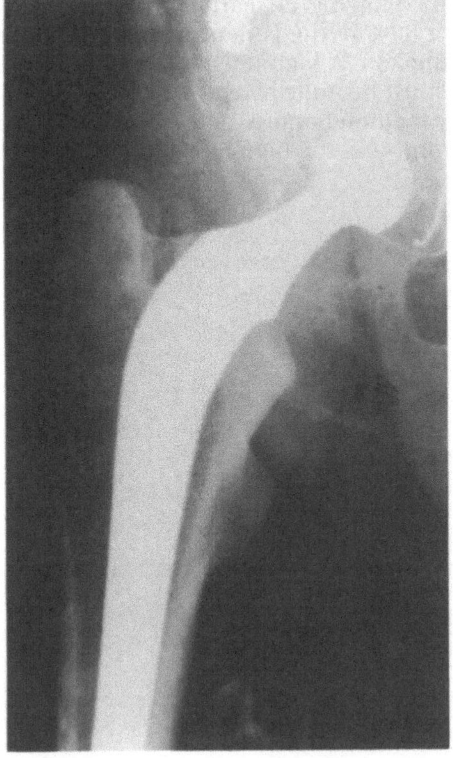

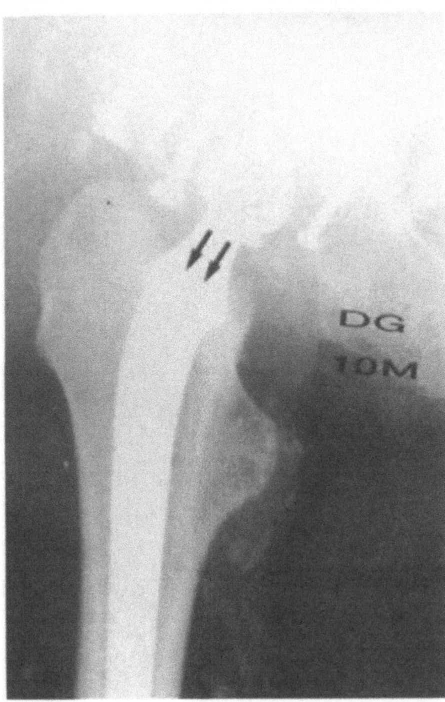

Fig. 16.11. Progressive sinking of a prosthesis that was not cemented in the allograft, with subsequent shortening and dislocation.

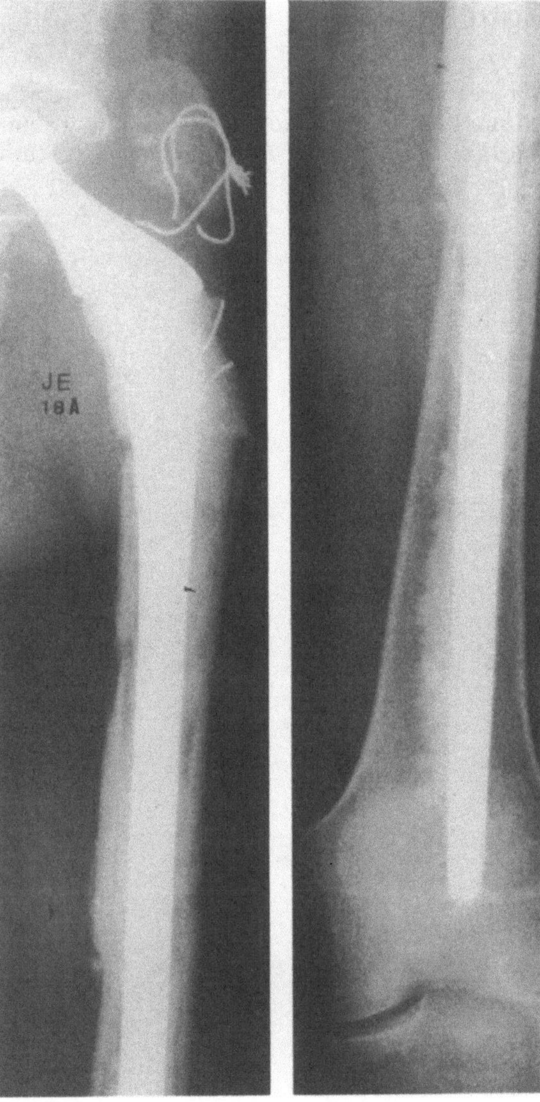

Fig. 16.12. Long-term aspect of a prosthesis and an allograft, 18 years after implantation. There is no loosening and no pain at present.

ing and several episodes of dislocation (Fig. 16.11). In order to avoid that, cementation in the allograft is recommended.

It is interesting to know that allograft and prosthesis can form a sound, long-lasting partnership as in a patient who had a prosthesis for 18 years (Fig. 16.12). Although there was minimal resorption along the medial cortex and a trochanter avulsion, there was no loosening and the patient is doing well.

Conclusion

Massive allografts are very valuable in treating substantial femoral bone loss or weakening about the hip. They provide an immediate structural support allowing an early functional recovery. Almost 80% of the patients had a good or excellent function. A high rate of complications, however, must be anticipated in very old patients in poor condition because of the long average operating time and the high blood loss.

17 Impaction Cancellous Grafting in Cemented Revision Hip Arthroplasty

G.A. Gie, J. Templey and R.S.M. Ling

Aseptic loosening of cemented femoral components with design shortcomings or where poor cementing technique has been used are significant problems in the long term. Radiological loosening of cemented femoral components in some series is reported to be as high as 30–50% at 10 to 12 years (Stauffer 1982, Wejkner and Stanport 1988, Maloney et al. 1990).

When loosening occurs the changes are progressive and often accompanied by significant loss of proximal bone stock and widening of the medullary canal. This osteolysis is not restricted to cemented components. Harris, Galante and others have discussed the frequency of localised osteolysis in uncemented femoral components which are radiographically stable and have reported 14 cases in a retrospective review of 474 primary total hip arthroplasties. Others have also recently described osteolysis in uncemented components (Santavirta et al. 1990, Tallroth et al. 1989).

Although it may be possible to cement in a new component without graft, the loss of bone stock makes this technique unsatisfactory with early appearance of radiolucent lines at the cement/bone interface. Callaghan, Salvati and others found that of 136 replaced femoral components followed for two to five years, only 32% had no radiolucent lines in the initial post-operative radiographs and 26.5% showed progressive radiolucencies. In an earlier study, only 8% of hips showed no initial post-operative lucencies (Callaghan et al. 1985, Pellicci et al. 1982).

When using cemented stems in revision surgery, Amstutz (1982) found that 37% of patients demonstrated immediate post-operative cement/bone radiolucencies that increased to 83.9% at an average follow-up of 2.1 years. Early experience of using cement in femoral revision arthroplasty, therefore, gave little grounds for optimism in the long term.

Hungerford and Jones (1982) advised against the use of cement in any revisions but reported a 10% failure rate with minimum follow-up of only two years in uncemented revision. It would appear that uncemented stems have their own problems in revision surgery.

Slooff (1984) reported improved clinical and radiological results in cemented revision using allograft bone chips to reconstitute bone stock loss in the acetabulum. In an effort to improve results on the femoral side, this technique was extended to the femur. It was felt that milled bone grafting might improve bonding between cement and bone and, at the same time, restore bone stock loss.

Methods and Materials

Technique

After removal of the stem of the femoral implant all the cement, fibrous membrane and granulomatous tissue, the cavity is thoroughly irrigated. An acrylic plug is inserted distally to occlude the cavity. It is placed 1cm beyond the anticipated stem tip or 2cm beyond the most distal area of bone lysis, whichever is the most distal. Segmental defects in the femoral cortex are then sealed by a thin wire mesh held in position with cerclage wires. Allograft bone chips are then packed firmly into the canal, initially with a distal impactor and then using an over-sized stem. This process is repeated until the grafts reach the cut surface of the femoral neck. The impacted stem should be so tight in the canal that it is impossible to remove by hand. Trial reduction is then carried out and if satisfactory, a smaller stem is cemented in position using cementing techniques as employed in primary surgery.

Initially, patients were kept in bed for three weeks post-operatively, and this is still the case when there is severe loss of bone stock. With increasing confidence in the procedure and reasonable bone stock however, mobilisation has been started progressively earlier and many patients are now mobilised at two to three days. Touch weight-bearing for the first three months is essential.

As impaction of the graft was sub-optimal in a number of cases and varus positioning of the stem not uncommon, cannulated instruments have been designed which have overcome these problems.

Clinical Material

At the time of this review, 72 patients have undergone cemented revision surgery with femoral grafting. The average age was 69 years (range 46-87 years). There were 42 males and 30 females. Five patients had died, one was lost to follow-up, two were on holiday abroad and of the two who were unfit, one had fallen and fractured below the femoral shaft and the other had suffered a pre-operative cerebral vascular accident. In 19 cases, the follow-up was less than one year and in five cases only partial canal grafting was performed for localised osteolysis. We report, therefore, on the first 38 cases in whom the whole canal was grafted and where follow-up exceeded one year. The maximum follow-up to date is 38 months and the average 20 months.

Results

Complications encountered in the first 72 cases are listed in Table 17.1.

Table 17.1. Complications in the first 72 cases

Complication	*n*
Intra-operative femoral fracture	2
Femoral component pierced femoral shaft	1
Dislocation	2
Post-operative cerebral vascular accident	1
Post-operative femoral fracture	1

Clinical Status

The patients were assessed clinically according to the Charnley modification of the D'Aubigne-Postel scoring system (Charnley 1979). There were 10 Category A, 14 Category B and 14 Category C patients. The pre-operative and follow-up clinical status of the patients is set out in Tables 17.2 and 17.3. Overall clinical assessment has been comparable with that achieved in primary interventions using cement, with no apparent deterioration over three years. The degree of bone stock loss has not influenced the clinical result.

Table 17.2. Pre-operative status

	Pain	Function	Movement
Category A	2.2	1.66	4.0
Category B	3.1	2.0	4.1
Category C	2.7	1.86	4.0

Table 17.3. Post-operative status

	Pain	Function	Movement
Category A	5.7	5.3	5.5
Category B	6.0	5.07	5.5
Category C	5.78	3.57	4.85

Radiological Appearances

The X-ray assessments were carried out on a consensus basis by three orthopaedic surgeons. Radiolucent lines were sought at graft/host and graft/cement interfaces and were expressed as a percentage of the relevant interface affected. In 77.8% of cases no lines were evident. Where lines

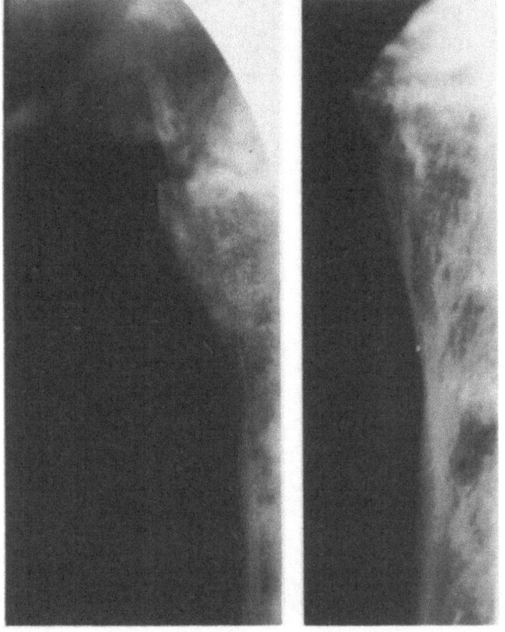

Fig. 17.1. a A post-operative radiograph showing a mass of milled autograft in the calcar region of the femur. **b** A radiograph three years later showing remodelling of the graft.

existed they averaged 20% at the graft/host interface and 17.5% at the graft/cement interface. No hip had lucent lines extending over more than 50% of these interfaces and none were progressive. The radiological appearances in fact continued to improve over the 38 month period. Stem subsidence within the cement mantel occurred in all but two cases. Most components subsided only 1-2cm but two subsided as much as 8mm. Subsidence could be correlated inversely with quality of graft impaction and cement insertion. Subsidence of the stem within the cement mantel was not associated with increasing pain or a lower clinical score. A similar phenomenon has been reported in primary total hip replacements (Fowler et al. 1988).

The late appearance of the graft was also assessed and is described in Table 17.4, but predominant features, trabecular and modelling alone (Fig. 17.1) and remodelling associated with cortical hypertrophy (Figs. 17.2, 17.3). Graft resorption occurred in four cases. The site of this resorption was invariably proximal to the lesser trochanter and only occurred if the graft was unsupported. This did not indicate impending failure. We have modified our technique and if the medial femoral neck is deficient, the graft is

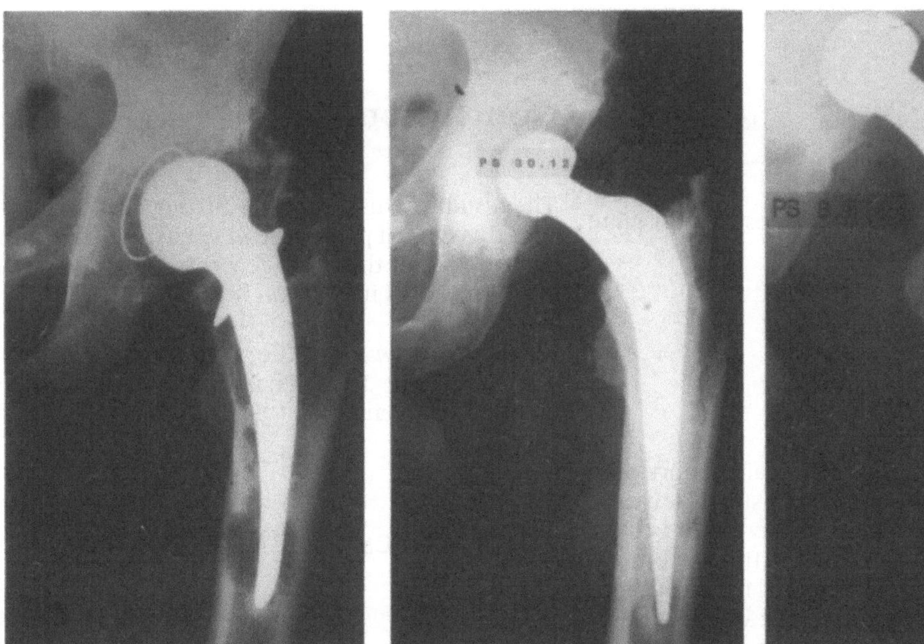

Fig. 17.2. a Radiograph of a McKee/Arden prosthesis with considerable endosteal lysis. **b** Radiograph 16 months post-operation. Despite varus positioning of stem, the implant is lying free and bone stock recovery is impressive. The medial cortex has healed completely and the only evidence of a cyst having been present distally is slight ectasia in the lateral cortex. **c** Three years post-operation, the radiograph shows normal bone stock and the interfaces remain perfect.

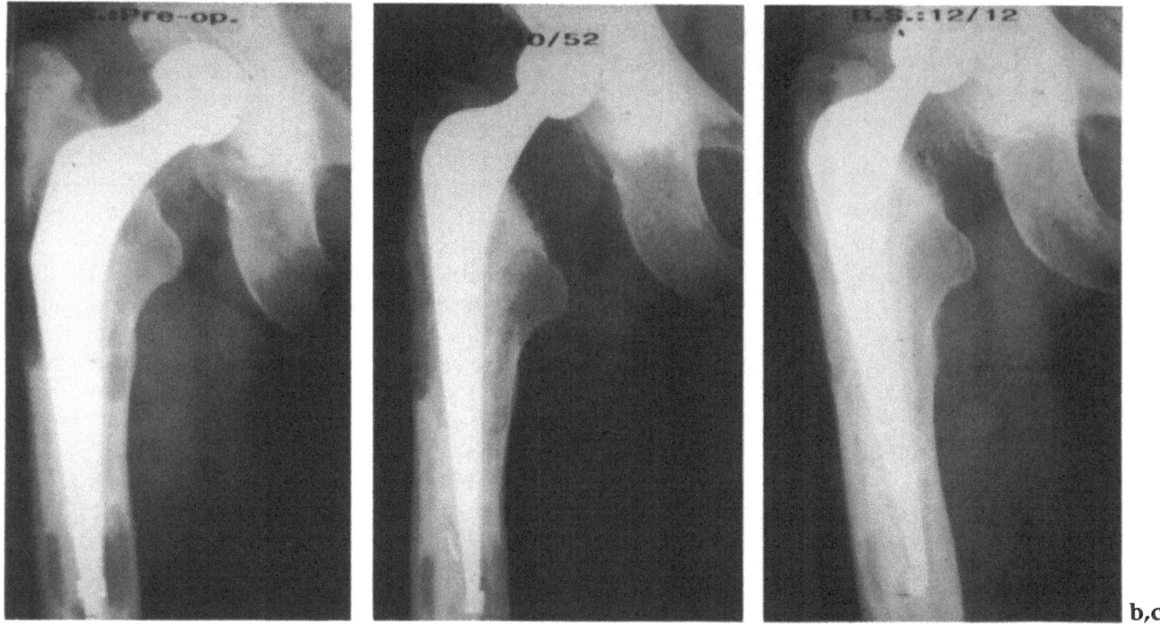

Fig. 17.3. **a** A pre-operative radiograph showing extensive lytic lesions and a cortical deficiency laterally. **b** A post-revision radiograph at ten weeks showing impaction grafting. **c** At one year, the radiograph shows the lateral cortex has reconstituted and the distal cyst healed.

Table 17.4. Later appearance of graft

	n
No change	1
Resorbed	4
Trabecular incorporation	5
Trabecular remodelling	7
Cortical hypertrophy	5
Cortical hypertrophy and trabecular remodelling	15
Not stated	2

supported by an acetabular reconstruction plate covered by a thin wire mesh.

Discussion

This chapter presents a preliminary report on a new technique in cemented stem revision. Despite the relatively short follow-up, it is clear that cases operated on using this technique behave in a totally different way from revisions where cement is used alone. The clinical results correlate favourably with primary cemented stems and the radiological appearances continued to improve over the three year follow-up period. There were no cases with progressive lucent lines or recurrence of bone lysis. Trabecular remodelling within the graft and reconstitution of deficient cortical areas were evidence of graft incorporation.

Conclusion

The results of this technique encouraged the hope that this type of procedure may have a significant part to play in the management of patients with bone stock loss in revision hip arthroplasty.

Acknowledgement. Mr Gie was generously supported by the John Charnley Trust to whom thanks are given.

References and Further Reading

Amstutz HC, Ma SM, Jinnah RH, Mai L (1982) Revision of aseptic loose total hip arthroplasties. Clin Orthop 170:21–33

Callaghan JJ, Salvati EA, Pellicci PM, Wilson PD, Ranawat CS (1985) Results of revision for mechanical failure after cemented total hip replacement, 1979 to 1982. J Bone Joint Surg (Am) 67:1074–1085

Charnley J (1979) Low Friction Arthroplasty of the Hip: Theory and Practice. Springer-Verlag, Berlin Heidelberg New York

Fowler JL, Gie GA, Lee AJ, Ling RSM (1988) Experience with the Exeter total hip Replacement since 1970. Orthop Clin North Am 19(3):477–489

Hungerford DS, Jones LC (1982) The rationale of cementless revision of cemented arthroplasty failures. Clin Orthop 170:12–24

Maloney WJ, Jasty M, Harris WH, Galante JO, Callaghan JJ (1990) Endosteal erosion in association with stable uncemented femoral components. J Bone Joint Surg (Am) 72:1025–1034

Pellicci PM, Wilson PD Jr, Sledge CB, Salvati EA, Ranawat CS, Poss R (1982) Revision total hip arthroplasty. Clin Orthop 170:34–41

Santavirta S, Hoikka V, Eskola A, Konttinen YT, Paavilainen T, Tallroth K (1990) Aggressive granulomatous lesions in cementless total hip arthroplasty. J Bone Joint Surg (Br) 72:980–984

Slooff TJJH, Huiskes R, Van Horn J, Lemmens AJ (1984) Bone grafting in total hip replacement for acetabular protrusion. Acta Orthop Scand 55:593–596

Stauffer RN (1982) Ten year follow-up study of total hip replacement. J Bone Joint Surg (Am) 64:983–990

Tallroth K, Eskola A, Santa Virta S, Konttinen YT, Lindholm TS (1989) Aggressive granulomatous lesions after hip arthroplasty. J Bone Joint Surg (Br) 71:571–575

Wejkner B, Stenport J (1988) Charnley total hip replacement. A 10–14 year follow-up study. Clin Orthop 231:113–119

18 A Method to Estimate the Initial Stability of Cemented and Non-cemented Hip Stems Using a Bone Grafting Technique

B.W. Schreurs, R. Huiskes, T.J.J.H. Slooff and P. Buma

The predominant cause of major bone stock loss in failed total hip arthroplasty is aseptic loosening of the prosthesis, both on the acetabular and the femoral sides. Morsellised trabecular bone grafts are used in our orthopaedic department to reconstruct the bony defect on the acetabular side (Slooff et al. 1984).

The aim of this study was to investigate the feasibility of using a similar technique on the femoral side, in combination with cemented or non-cemented prostheses. A method is described to estimate the initial stability, which seems of critical importance to the incorporation of the grafts, and hence the long-term success of the method.

Methods and Materials

In this in-vitro experiment, femora of the goat were used. These femora contain hardly any trabecular bone inside the intramedullary canal, and have a very hard and smooth endosteal cortex similar to the sclerotic bone often seen during revision surgery.

The femora and grafts used were freshly frozen. The chip-like trabecular grafts were of sternal origin. After thawing, the femur was prepared in the usual way including canal lavage. Then an appropriately sized bone cement plug was screwed on a metal rod and introduced into the canal. The space between this rod and the cortical bone (2–4mm) was filled with grafts in the retrograde fashion. By using a special set of instruments, consisting of metal tubes sliding over the central rod, the trabecular grafts were compressed. In this way an intramedullary wall of bone chips was created. Then the rod was removed leaving a central cavity surrounded by bone grafts. Impaction of the graft created a firm construction. A collapse during removal of the rod was never seen.

Two types of prostheses were inserted. In one group prostheses were inserted in combination with bone cement (Mathys type 2.30.702) (Fig. 18.1a). In the other group non-cemented

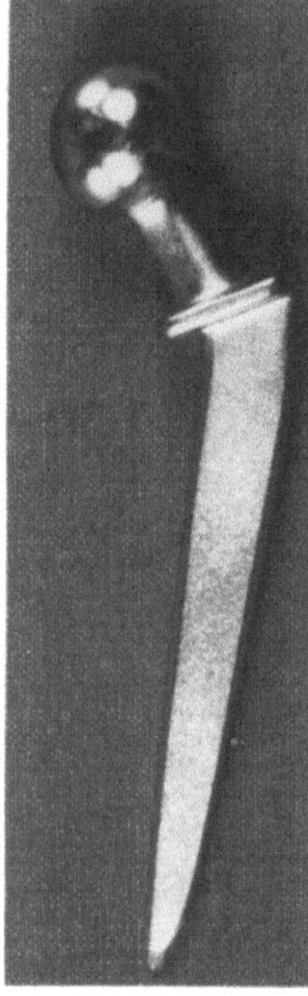

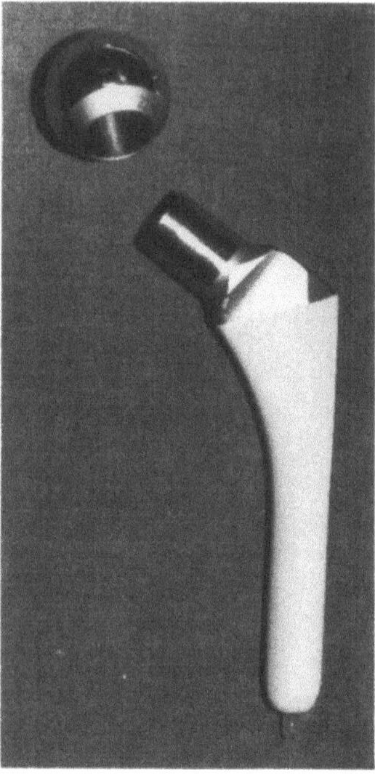

Fig. 18.1. a Femoral prosthesis used with cement. **b** Full hydroxyapatite-coated titanium prosthesis.

titanium prostheses, fully coated with hydroxyapatite, were inserted (Fig. 18.1b).

To estimate the initial stability of the stems immediately after implantation, we investigated the possibility of using roentgenstereophotogrammatic analysis (RSA) (Selvik 1974). RSA is a very sensitive and specific method to estimate (micro) movements (Karrholm 1989).

Because of this technique, both prostheses had a tantalum pellet attached to the tip prior to insertion contained in an acrylic strut glued to the metal. After insertion of the stems, bones were wrapped in physiological saline-drenched gauze bandages and kept for 24 hours at 4 °C. Then the femora were resected just above the condyles and embedded in polymethylmethacrylate (PMMA). Tantalum pellets were inserted proximal and distal on the medial and lateral side of the cortical bone, and two small PMMA rods containing tantalum pellets were glued to the proximal-medial and proximal-lateral parts of the prosthesis.

The prosthesis/bone structures were then loaded in an MTS loading device. A study comparing hip forces in sheep, dog and man concluded that the maximal hip forces in dogs and sheep were directed to the antero-medial region of the femoral head (Bergmann et al. 1984). When testing the stability of stems it is important that the applied loading results in both bending and rotational forces simultaneously (Schneider et al. 1989). To obtain such a physiological load on the femoral head, the femora were tilted 15° in the lateral direction and internally 45° relative to the vertical position.

The maximal load on the femoral head of sheep during gait was estimated at 110% of bodyweight (Bergmann et al. 1984). The load applied with the MTS machine was stepwise from zero to 200, 500 and 800N (Fig. 18.2). A group of goats operated on in a concurrent in vivo study had an

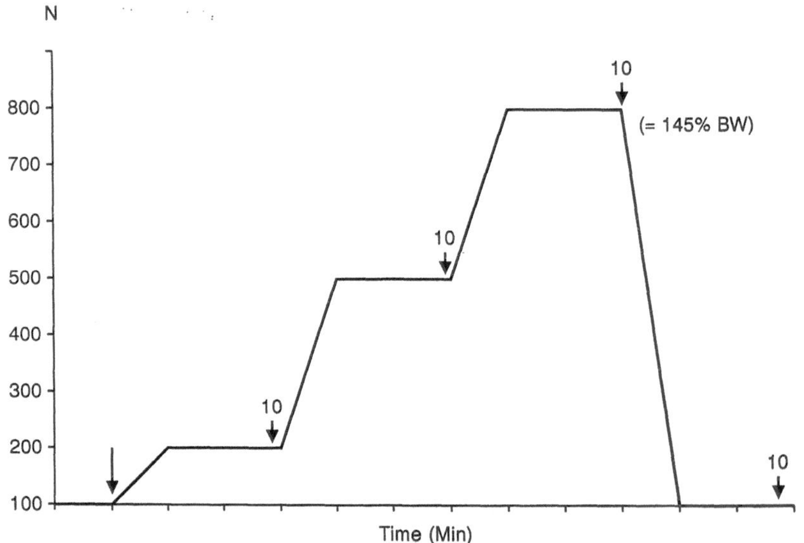

Fig. 18.2. Loading schedule for intramedullary grafts. Each arrow indicates a roentgenstereogram. Each loading time lasted ten minutes.

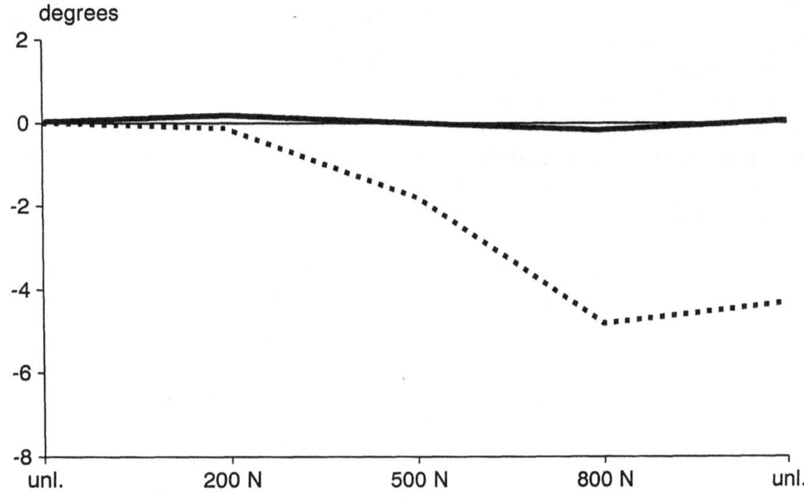

Fig. 18.3. Axial rotations of a cemented (continuous line) and a non-cemented (dotted line) prosthesis.

average weight of 55kg, so the maximal load used (800N) was 145% of body weight. Each loading period lasted ten minutes. As already mentioned, the 3D displacements of the prosthesis relative to bone (three rotations and three translations) were measured using RSA. Before loading, ten minutes after application of each load, and again ten minutes after final unloading, stereoroentgenographs were taken.

These were evaluated on a digitiser, and the 3D pellet positions were determined with the RSA computer system. Relative translations along and rotations around the coordinate axes were calculated. To increase the accuracy of the method, all roentgen films were measured five times, and averaged.

Results

In both stems most rotations occurred around the axial y-axis. Especially in the non-cemented stem, the rotation increased with load. After unloading, the major part of the rotation proved to be permanent. Rotation in the tested cemented specimen was small (Fig. 18.3).

In both stems, most translations occurred in the axial y-direction, resulting in subsidence of the prosthesis (Fig. 18.4). There is some elastic recovery after unloading.

The major goal of this part of the study was to determine if RSA could be applied successfully in

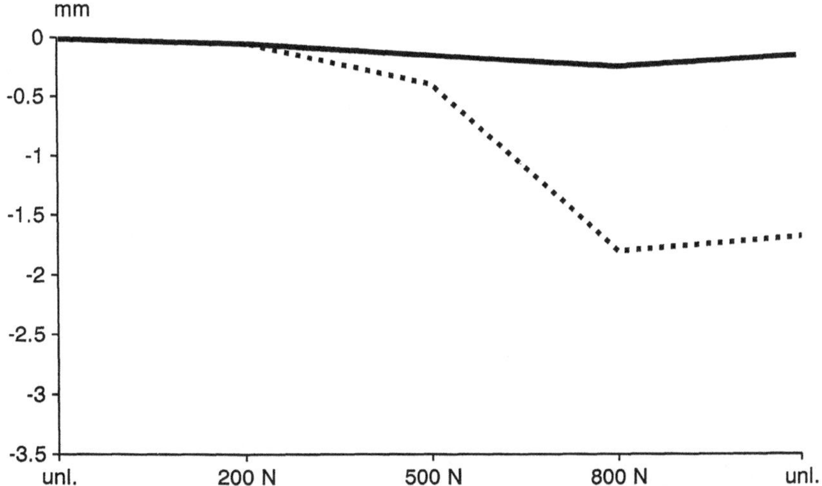

Fig. 18.4. Subsidence of a cemented (continuous line) and a non-cemented (dotted line) prosthesis.

testing the stability of the stems. It was of critical importance if the pellets, and especially the distal pellet fixed to the tip of the stem, would be fixed well enough to the prosthesis to perform RSA. The standard deviations for the displacements were estimated at 0.036mm for the translations and 0.07° for the rotations. It can be concluded that RSA is a suitable method to estimate the 3D movements of the stems relative to bone.

The preliminary results of an in vivo study in goats using this grafting technique are presented in Chapter 3.

Acknowledgement. The authors would like to thank Ton Bijlaart, Huub Peters and Willem van de Wijdeven for their assistance during the experiments.

References and Further Reading

Bergmann G, Siraky J, Rohlmann A (1984) A comparison of hip joint forces in sheep, dog and man. J Biomechanics 17:907–921

Karrholm J (1989) Roentgenstereophotogrammetry. Acta Orthop Scand 60:491–503

Schneider E, Eulenberger J, Steiner W, Wyder D, Friedman RJ, Perren SM (1989) Experimental method for the in vitro testing on the initial stability of cementless hip prosthesis. J Biomechanics 22:735–744

Selvik G (1974) A roentgenstereophotogrammetric method for study of the kinematics of the skeletal system. Thesis, University of Lund, Lund, Sweden

Slooff TJ, Huiskes R, van Horn J, Lemmens AJ (1984) Bone grafting in total hip replacements for acetabular protrusion. Acta Orthop Scand 55:593–596

19 Reconstruction of the Femur in Total Hip Replacement Using Multifilament Cerclage Cables Fastened with Crimp Sleeves

D.M. Dall

It was the challenge of fixation of the greater trochanter, which was shared with equal enthusiasm by Charnley, which prompted me to search for a more reliable technique than the standard monofilament wiring techniques. Working in Cape Town with Miles, a mechanical engineer, we realised the advantages of multifilament cable and by 1978 had developed a system for trochanteric fixation using this material (Dall and Miles 1983). As a result of the success of this fully developed system, it was a logical step to use multifilament cables for cerclage fixation.

Numerous cerclage materials are currently available. Monofilament wires of stainless steel and chrome cobalt alloy are commonly used. Multifilament cables are available in stainless steel, chrome cobalt and titanium alloy. Parham bands are made of stainless steel and Partridge straps of nylon.

Vascular Effects of Cerclage

Charnley (1957) condemned cerclage techniques because he said it "strangles the bone". Cerclage was felt to interfere with the periosteal blood supply and cause cortical ischaemia and necrosis which resulted in lysis, interfering with bridging callus of fractures and resulting in non-union and re-fracture (Jones 1986) (Fig. 19.1). Rhinelander (1968) was the first to study the vascular effects of cerclage. He found that the periosteal blood supply consisted of innumerable small vessels, none of which ran longitudinally. Furthermore, in fracture callus all vessels are directed perpendicular to the outer surface of the cortex. In his experiments, he found that wire loops, being round, had minimal contact with the bone and did not block the periosteal vessels (Fig. 19.2). He stated that Parham bands, being wider, acted differently.

Kirby and Wilson (1991) studied the effect of circumferential bands on cortical vascularity and viability. Cerclage 18-gauge wire, steel and nylon bands (varying in width from 2.5 to 10mm) were studied using micro-angiography and histology on dog femora sacrificed at 1, 4 and 15 weeks. They found no evidence of complete cortical devascularisation under any of the bands at any time interval. Furthermore, there was no difference if the cerclage was applied over or under the periosteum. In principle, I think that either the

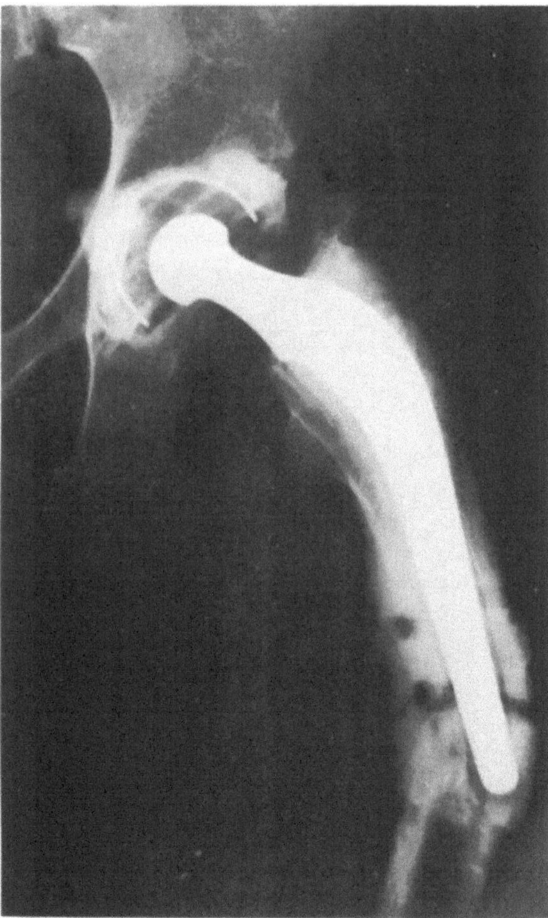

Fig. 19.1. Fracture through the lytic area deep to the Partridge bands two years after insertion. (Reprinted with permission from Jones (1986).)

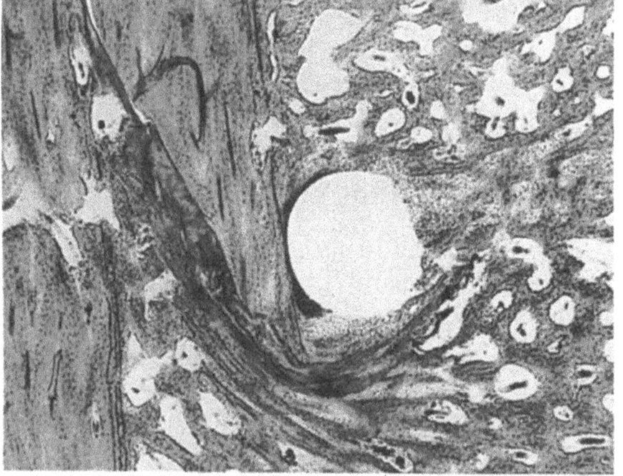

Fig. 19.2. Photomicrograph of periosteal callus at four weeks showing the result of tight fixation of the tip of radial fracture fragment by encircling wire loop (haematoxylin and eosin, ×45). (Reprinted with permission from Rhinelander (1968).)

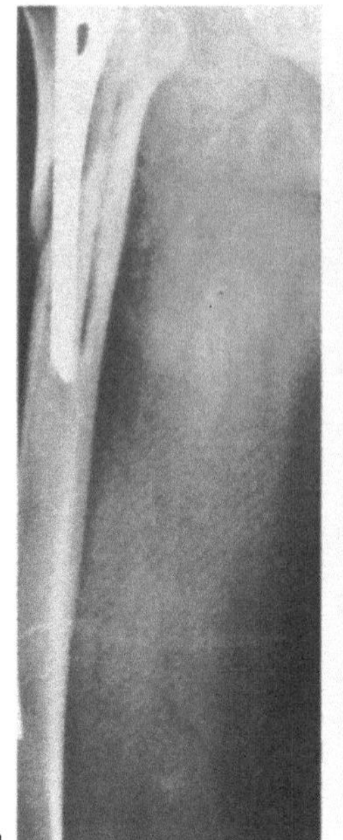

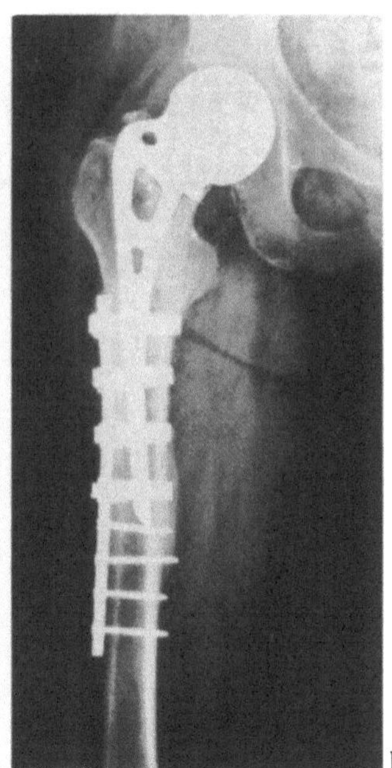

Fig. 19.3. **a** An 80 year old man with a displaced proximal femoral fracture three years after insertion of Austin Moore prosthesis. **b** Follow-up radiograph after treatment with an Ogden plate and Parham bands showing satisfactory alignment and healing. (Reprinted with permission from Wang et al. (1985).)

Parham bands or Partridge straps can be used effectively for the correct indications. Wang et al. (1985) reported three patients who were successfully treated for a fracture about the tip of a femoral stem using Parham bands proximally with a long plate and cortical screws distally (Fig. 19.3a,b).

Mechanical Properties of Cerclage

The mechanical strengths of the various multifilament cables currently available are compared with 18-gauge stainless steel and chrome cobalt monofilament wires in Figs. 19.4 and 19.5. It can be seen that the mechanical properties of both stainless steel and chrome cobalt multifilament cables are superior to monofilament wires of the same materials and significantly superior to titanium cable. Furthermore, the yield strength of crimped cables is significantly higher in chrome cobalt and stainless steel than in titanium (Fig. 19.6).

Wilson (1988) compared Parham and CPC (circumferential pneumatic compression) bands

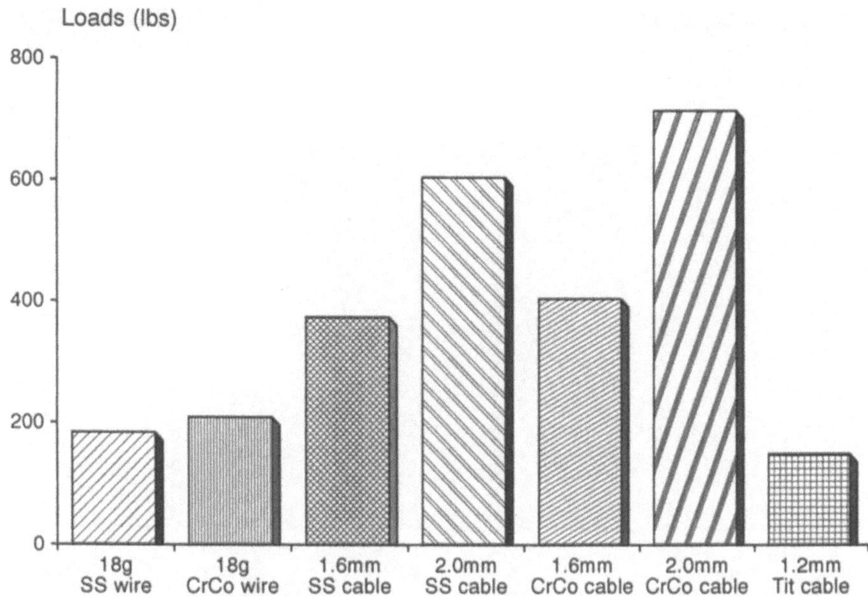

Fig. 19.4. Ultimate tensile strength of monofilament wires compared with multifilament cables. 18g, 18-gauge; SS, stainless steel; CrCo, chromium cobalt; Tit, titanium.

with wires and found that neither band produced knot-slip resistance as great as 1.2mm twist-knotted stainless steel wire. Kirby and Wilson (1989) compared nylon bands with wires and found that when soaked in saline for 24 hours, the knot strength of the two largest sized bands dropped to less than that of 1.2mm twist-knotted stainless steel wire. It appears, therefore, that monofilament wire is definitely stronger than steel or nylon bands in terms of cerclage fixation.

Invaco et al. (1990) showed that if the wire is double-wrapped, more than double the strength is achieved. This was confirmed by Taylor et al. (1991) in their excellent study on static tension

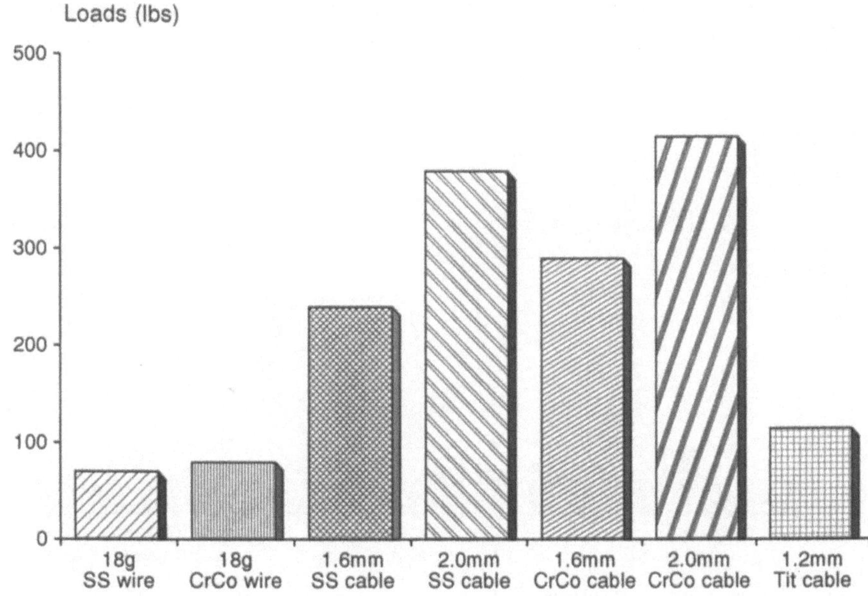

Fig. 19.5. Yield strength of monofilament wires compared with multifilament cables. 18g, 18-gauge; SS, stainless steel; CrCo, chromium cobalt; Tit, titanium.

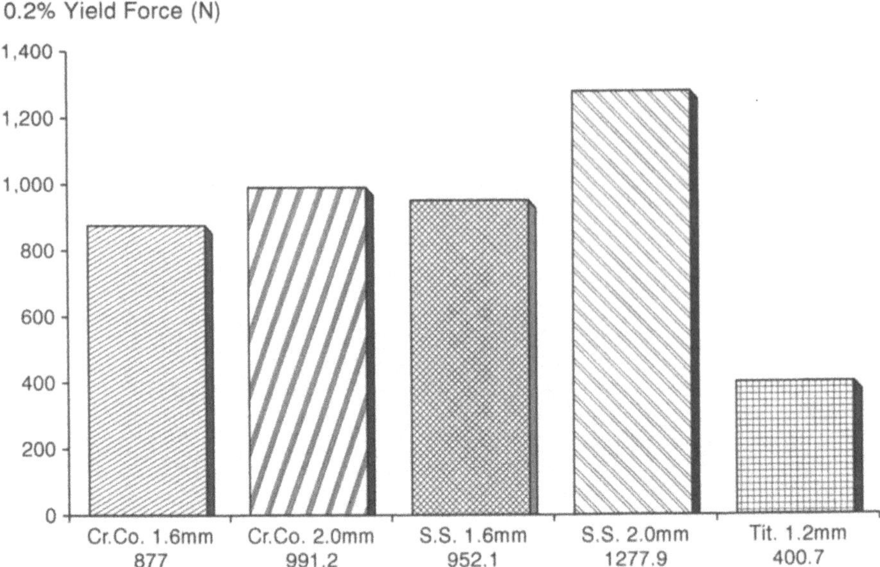

Fig. 19.6. Yield strength of crimped cables. 18g, 18-gauge; SS, stainless steel; CrCo, chromium cobalt; Tit, titanium.

and mechanics of failure of cerclage systems in total hip replacement. Furthermore, they showed that 1.6mm chrome cobalt cable fastened with a crimp sleeve was significantly better than 18-gauge monofilament stainless steel wire even if the latter was double wrapped. This superiority of the multifilament cable was demonstrated to occur not only in the peak compressive forces generated during tensioning but also in the final forces achieved (Fig. 19.7). Furthermore, the cable/crimp system required significantly greater force to produce 1mm displacement than all other systems (1.8–5.4 times greater, $P > 0.0001$). There is no doubt that multifilament cables fastened

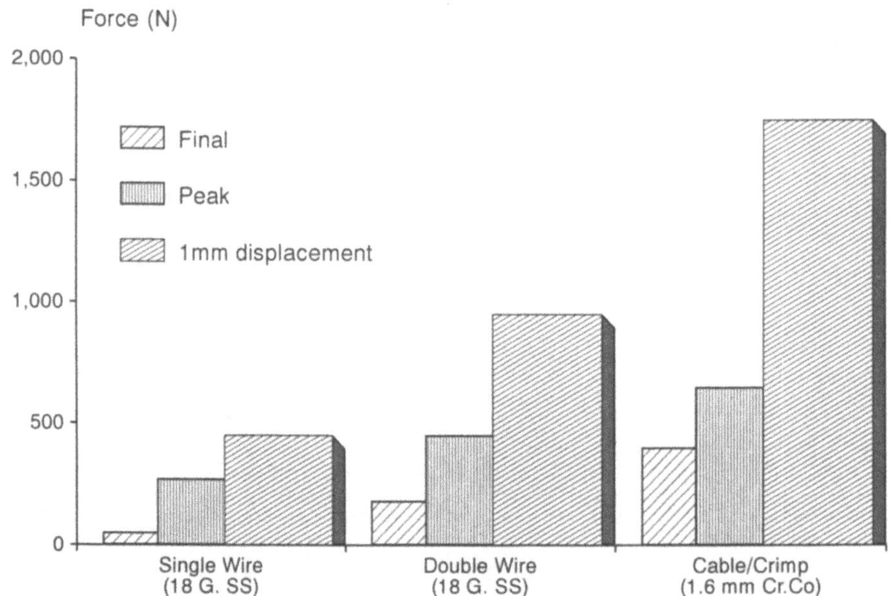

Fig. 19.7. Peak and final forces and 1mm displacement forces for cerclage systems of single and double 18 gauge stainless steel (18G.SS) wire compared with 1.6mm chrome cobalt (Cr.Co) multifilament cable fastened with crimp sleeve. (Data extracted from Taylor and Hayes (1991).)

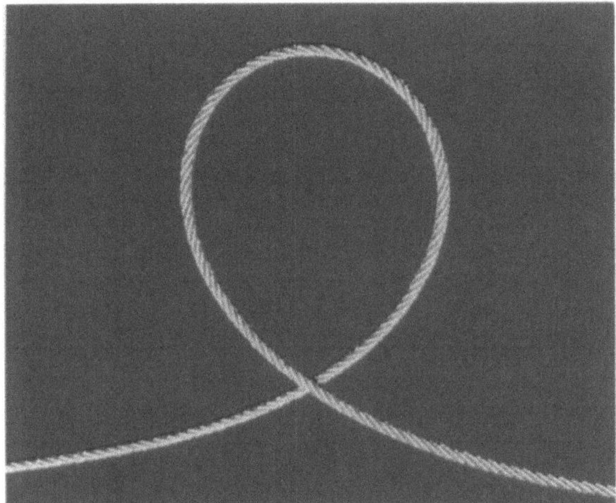

Fig. 19.8. Multifilament chrome cobalt cable.

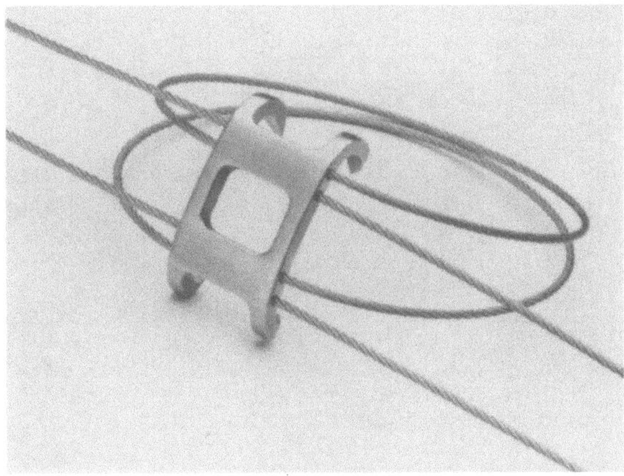

Fig. 19.10. Trochanter cable grip showing proximal and distal hooks with proximal and distal cables passed through holes in respective bridges. (Reprinted with permission from The Dall-Miles Trochanter Cable Grip System. Surgical Technique (brochure) 1989, HOWMEDICA, Rutherford, New Jersey.)

with crimp sleeves are significantly superior mechanically to other techniques used in cerclage.

Finally, multifilament cables are far easier to handle than monofilament wires (Fig. 19.8) The latter are very inclined to kink resulting in potentially weak areas (Fig. 19.9). Maintaining tension during fastening of the wires is difficult and twisting or knotting the wires is very technique-sensitive.

Multifilament Cables: Principles of Cerclage Technique

Multifilament cables cannot be fastened properly with a knot. It is essential to use a crimping technique for fastening. This is well illustrated in the application of the trochanter cable grip system where proximal and distal horizontal cables are passed through holes in the respective bridges of the grip (Fig. 19.10). After tensioning, the cables are fastened by crimping the bridges (Fig. 19.11). Cerclage techniques around the proximal femur, for example, employ the use of crimp sleeves (Fig. 19.12). Cerclage passers are used to place the cables around the bone, the ends are passed through the holes in the crimp sleeve and after suitable tensioning, fastened by crimping the sleeve. It is important to use a purpose designed tensioner to ensure an optimum plane of tension. Ordinary wire tighteners tend to be forced out of the optimum plane of tensioning when the arms are opened in a deep wound. A purpose designed crimping tool is also essential to match the size of

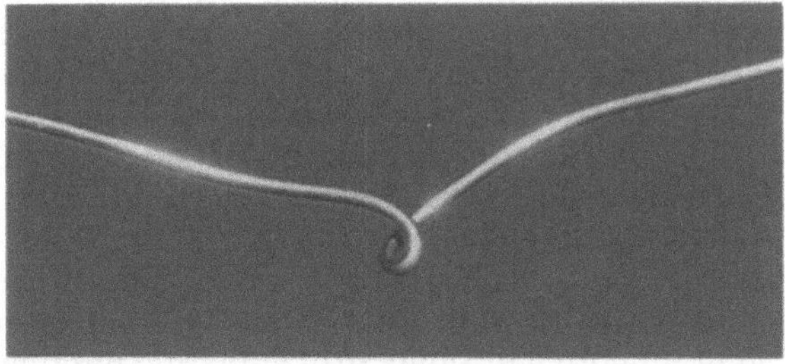

Fig. 19.9. Monofilament wire demonstrating kink.

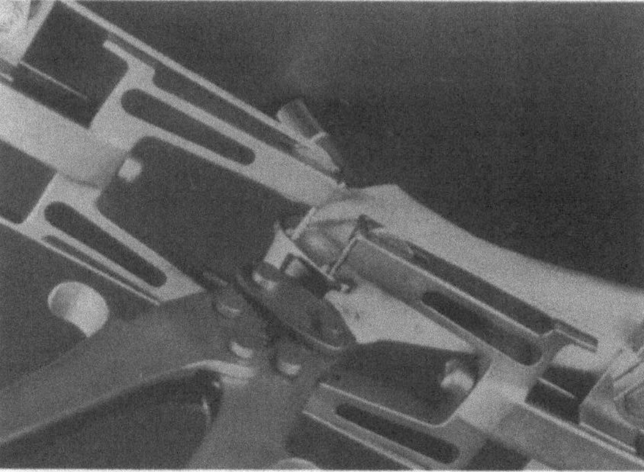

Fig. 19.11. A model showing fixation of the osteotomised trochanter with the trochanter cable grip system. Specially designed tensioners are used to tension the proximal and distal cables and the crimping tool is used to fasten the cables by crimping the bridges.

Fig. 19.12. Crimp sleeve illustrating how the cable is fastened by crimping the sleeve. (Reprinted with permission from The Dall-Miles Trochanter Cable Grip System. Surgical Technique (brochure) 1989, HOWMEDICA, Rutherford, New Jersey.)

the crimp sleeve exactly. Stainless steel or chrome cobalt cables of 1.6mm or 2mm diameter are recommended. Titanium cable is not recommended because of its inferior mechanical properties.

Clinical Application in Femoral Reconstruction

The main indications for this powerful cerclage technique are in revision total hip surgery associated with bone stock loss of the femur and/or fractures of the femur about the stem. It is an ideal way to fix allografts to the femur during femoral reconstruction. In addition, the technique is very useful in preventing or repairing longitudinal cracks which occur in the proximal femur during broaching or insertion of cementless stems.

Two cerclage cable passers have been designed to facilitate passing the cables – a large-diameter cerclage passer for the proximal region and the smaller-diameter passer for the femoral shaft below the lesser trochanter. If a trochanteric osteotomy is being used for exposure, the proximal and distal cables of the trochanter cable grip system can frequently be useful in fixing proximal allografts. If this approach is not being used, cerclage cables and crimp sleeves can also be used to secure the grafts proximally. The following examples illustrate the clinical applications.

Case 1 (Fig. 19.13a–c)

In this patient with a segmental defect at level I, fresh frozen allograft was fashioned to comprise the medial femoral neck, lesser trochanter and the medial cortex extending distally. The allograft was fixed proximally with the trochanter cable grip system and distally with cerclage cables and crimp sleeves.

Case 2 (Fig. 19.14a–c)

The patient had a level II segmental defect and a level III cavity defect resulting in a thin eggshell cortex after removal of the stem, cement and granulomatous material. Here a fresh frozen proximal femoral allograft fashioned like a "clothes peg" inserted into the remaining thin cortical tube of the femur was used. The allograft was then reinforced with cortical strut onlay grafts. Again, the trochanter cable grip system was used for proximal fixation. Cerclage cables and crimp sleeves were used to provide distal fixation.

Case 3 (Fig. 19.15a–c)

This patient illustrates the very useful application of this system in repairing femoral fractures about the stem. After removal of the existing loose femoral stem and cement, the fracture surfaces were exposed and carefully cleaned. The reduction was held temporarily with a single cerclage cable and the crimp sleeve applied at the mid point of the long oblique fracture. A long stem PCA without cement was used for the revision. Multiple cerclage cables and crimp sleeves were used to secure the onlay cortical struts thus providing internal fixation of the fracture. The X-ray shown at one year illustrates how well the onlay strut grafts can incorporate.

Case 4 (Fig. 19.16a–c)

This failed revision total hip arthroplasty with an uncemented stem and bipolar prosthesis presented significant problems. In addition to the gross lateral cortical defects of the femur, there was a transverse fracture below the tip of the stem. A fresh frozen allograft using the proximal two-thirds of a femur was fashioned distally to form a lateral tongue, which in turn was matched to a step cut made in a patient's femur distal to the fracture (to provide rotary stability). A longer stem supported by onlay strut grafts and multiple cerclage cables completed the reconstruction. In

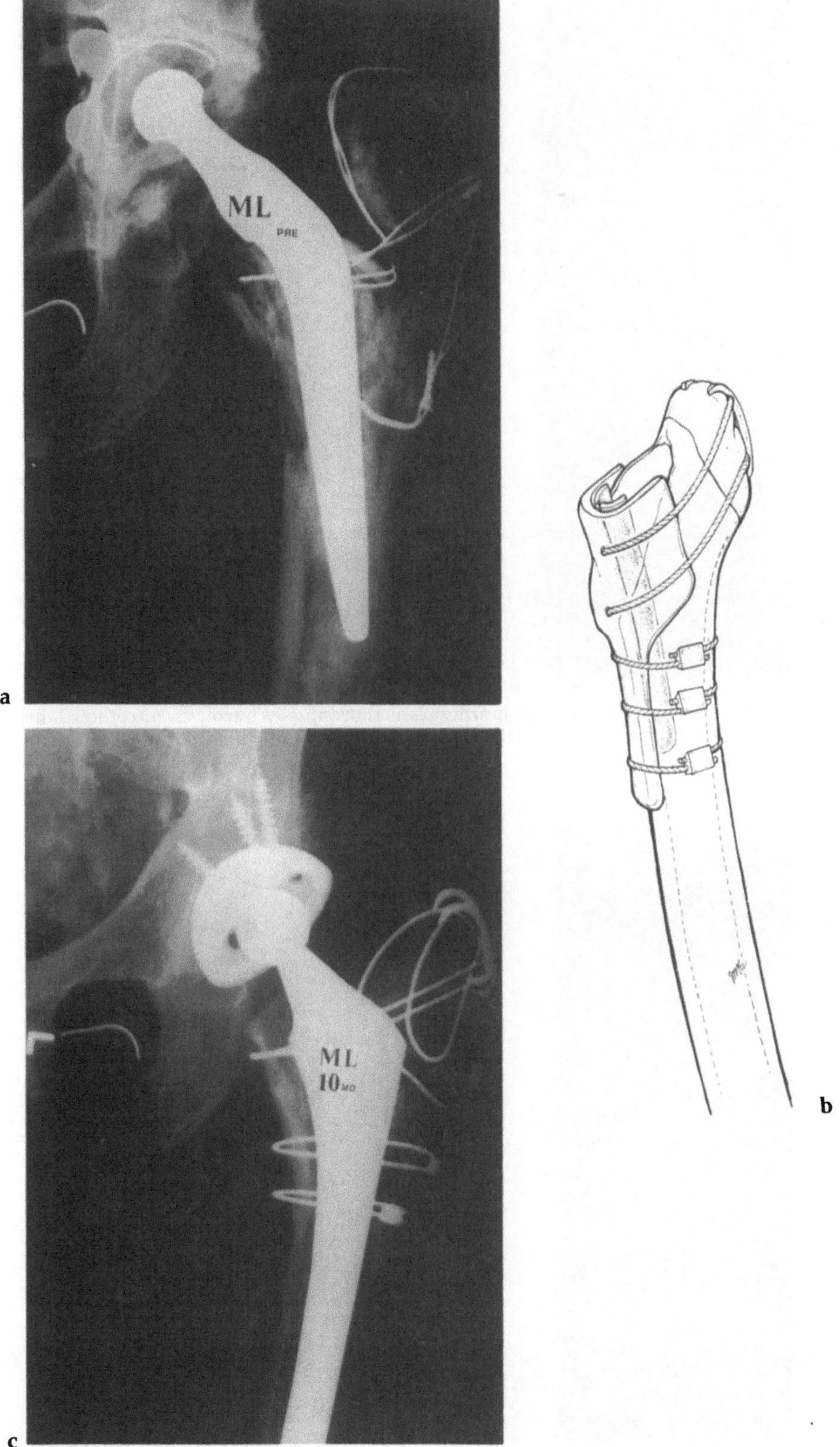

Fig. 19.13. **a** Pre-operative radiograph of failed cemented total hip arthroplasty showing deficiency of the calcar femoral and medial proximal femur. **b** Diagram of reconstruction. **c** Post-operative radiograph at ten months.

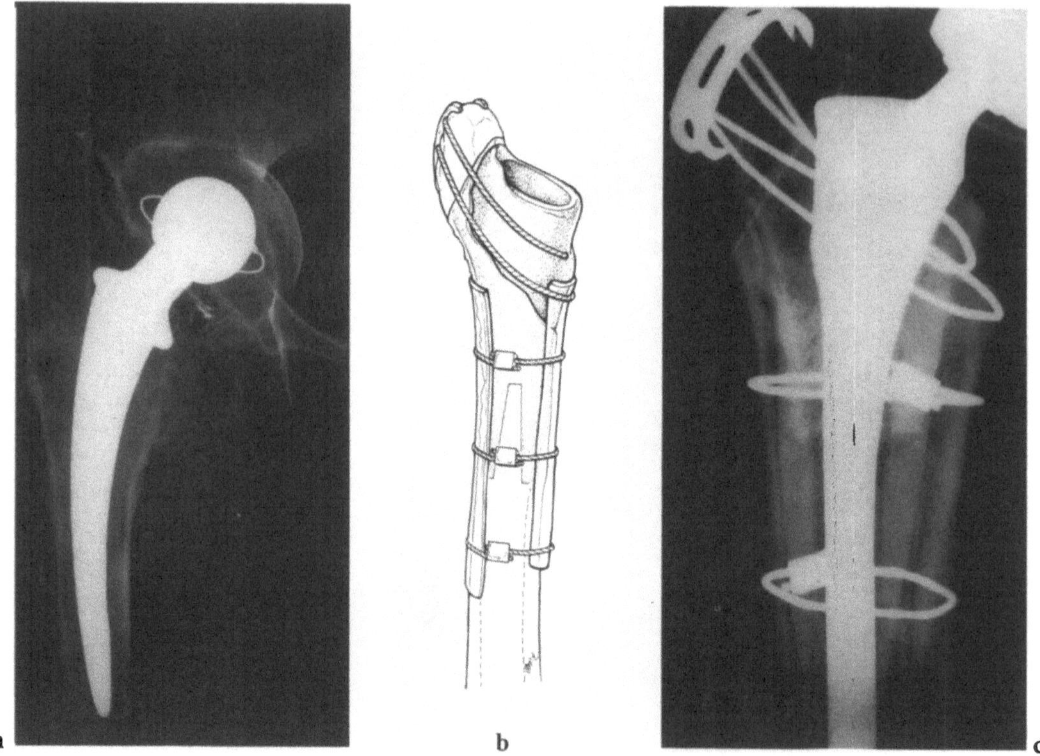

Fig. 19.14. a Radiograph of failed cemented total hip arthroplasty showing gross osteolysis and cortical thinning of the proximal femur and supero-medial migration of the socket. **b** Diagram of femoral reconstruction. **c** Post-operative radiograph. (Reprinted with permission from Dall (1991).)

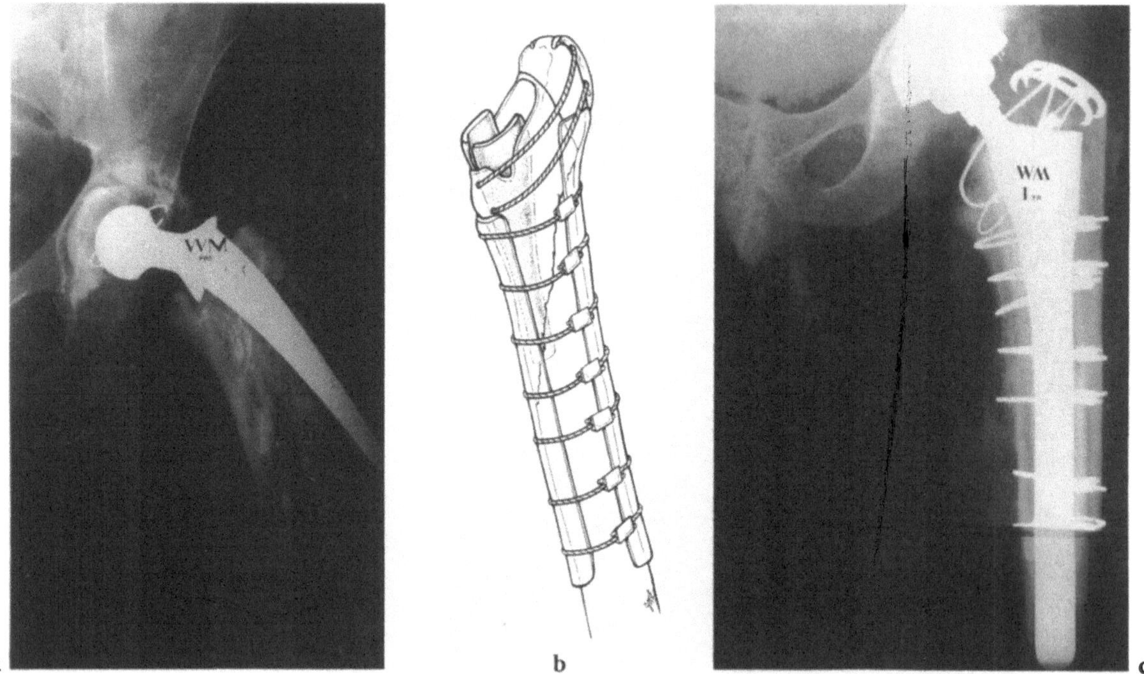

Fig. 19.15. a Failed cemented total hip arthroplasty with fracture of the femur about the femoral stem. **b** Diagram of reconstruction. **c** Post-operative radiograph at one year. (Reprinted with permission from Dall (1991).)

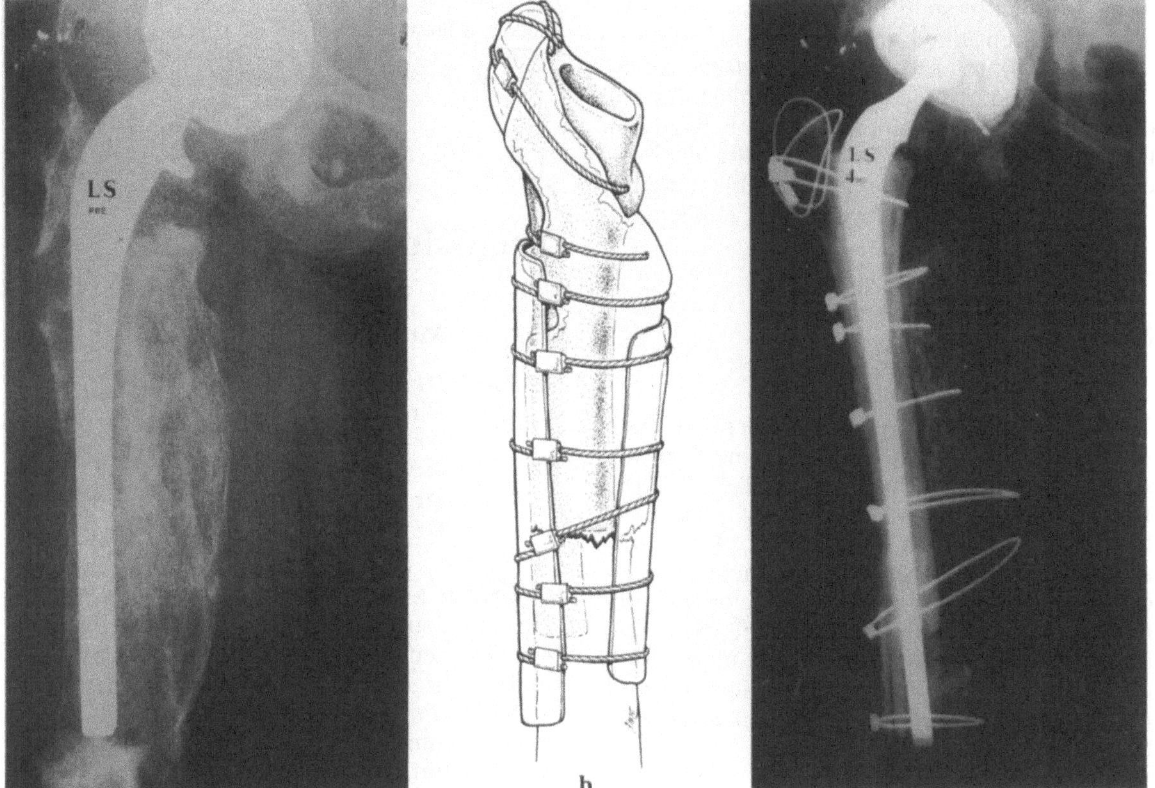

Fig. 19.16. a Failed revision total hip arthroplasty with uncemented stem and bipolar prosthesis showing gross lateral cortical defects in the femur and a fracture below the tip of the stem. **b** Diagram of reconstruction. **c** Post-operative radiograph. (Reprinted with permission from Dall (1991).)

this patient, an alternative method of trochanteric fixation is illustrated. Here a horizontal cable and two vertical cables were used.

Crack Fractures of the Proximal Femur

It is not uncommon to inadvertently produce a crack fracture of the proximal femur in attempting to obtain a tight proximal fit of cementless stems. Philip et al. (1990) have shown that this will reduce the rotary stability of the stem to 60% or less. It is therefore important that these fractures be supported with powerful cerclage using one or two mulifilament cables and crimp sleeves (Fig. 19.17). Prior to tensioning the cables, the stem should be extracted sufficiently to enable the cables to close the gap in the fracture on tensioning. While the tensioners remain applied to the cables, the stem is tapped down to the required

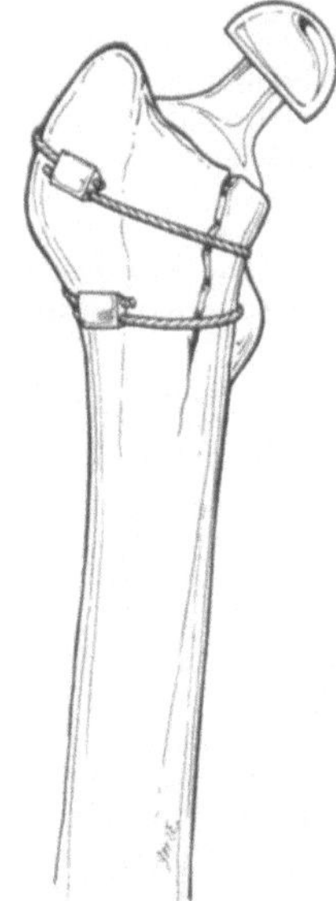

Fig. 19.17. Iatrogenic longitudinal crack fracture of the proximal femur fixed with two cerclage cables and crimp sleeves. (Reprinted with permission from Dall (1991).)

level. When the stem is seated, the cables are again fully tensioned and fastened by crimping the sleeves. I prefer to apply prophylactic cables, tensioned and crimped prior to broaching and insertion of the cementless stems. I believe that this technique significantly reduces the incidence of iatrogenic crack fractures and enables a tighter fit of the stem with improved rotary stabilty.

Summary

Cerclage fixation with multifilament stainless steel or cobalt chrome cables is mechanically superior to all other forms of cerclage fixation. The technique is very useful for fixation of femoral allografts and the repair of fractures about a femoral stem. Multifilament cerclage cables fastened with crimp sleeves provide the most powerful fixation of iatrogenic crack fractures of the proximal femur which might occur during broaching or insertion of the femoral component. They are also beneficial in preventing these fractures.

References and Further Reading

Charnley JS (1957) A closed treatment of common fractures. E and S Livingstone, Edinburgh, pp18-20

Dall DM (1991) Cable techniques for trochanteric and femoral allograft fixation. Techniques Orthop 6(3):7–16

Dall DM, Miles AW (1983) Re-attachment of the greater trochanter. J Bone Joint Surg (Br) 65:55–59

Incavo SJ, DiFazio F, Wilder D (1990) Strength of cerclage fixation systems: A biomechanical study. Clin Biomechanics 5:236–238

Jones DG (1986) Bone erosion beneath Partridge bands. J Bone Joint Surg (Br) 68:476–477

Kirby BM, Wilson JW (1989) Knot strength of nylon-band cerclage. Acta Orthop Scand 6:696–698

Kirby BM, Wilson JW (1991) Effect of circumferential bands on cortical vascularity and viability. J Orthop Res 9:174–179

Phillips TW, Messieh SS, McDonald PD (1990) Femoral stem fixation in hip replacement. J Bone Joint Surg (Br) 72:431–434

Rhinelander FW (1968) The normal microcirculation of diaphyseal cortex and its response to fracture: An instructional course lecture. The American Academy of Orthopaedic Surgeons. J Bone Joint Surg (Am) 50:784–800

Taylor JK, Hayes DEE (1991) Cerclage systems in total hip replacement. Static tension and mechanics of failure. Abstract, Orthopaedic Research Society 37th Annual Meeting, March 4–7

Wang G-J, Miller TO, Stamp WG (1985) Femoral fracture following hip arthroplasty. J Bone Joint Surg (Am) 67:956–957

Wilson JW (1988) Knot strength of cerclage bands and wires. Acta Orthop Scand 59:545–547

Discussion: Hip Problems

Chairman: **Mr Older**

The Panel: **Dr Chandler**
Mr Gie
Dr Goldberg
Dr Gross
Dr Loty
Dr Paprosky

Mr Older: As we have seen from our American colleagues, the operative procedures are technically very demanding and should not be undertaken by a surgeon who is unwilling to spend time to shape and secure the bone graft accurately. Speed is commendable as long as the result is excellent, but sacrificing quality for quickness and saving an hour in the operating room can cost weeks or months in the hospital stay or post-operative rehabilitation.

Reconstructive hip surgery, in the face of bonc deficiency, is more complex and of longer duration than a routine total hip replacement procedure so pre-operative assessment is paramount. The extent of the bone deficiency should not come as a surprise when we operate. Pre-operative detailed planning is essential – patient evaluation, radiographic analysis of the bone defects and the proposed reconstruction.

What of the acetabulum? Do we morsellise or do we put in a solid graft? Are the rim defects very important? How do we fix them? Do we use cement or not? What type of prosthesis do we use? Those are some of the questions that I would like to discuss.

Bone grafts should not be advocated if an equally effective solution is available. Is that true? Now that we have a larger component size of prosthesis for the acetabulum, maybe bone grafting in the future will not be required.

What of the femur? In the past, there is no doubt that the solution with femoral problems has been more cement and longer stems, but that meant that our cement was going beyond the isthmus of the femur; it would also produce severe osteopenia of the proximal femur.

Chandler has popularised a very good phrase – "the cemented long stem disease". How true, because the results will be earlier failure, second and third revisions more difficult than the first revision, and an increased need, therefore, for femoral grafting.

What of the grafts? Although autograft is preferable, rarely is there enough bone autograft available to provide that structural support for major deficiencies. Allograft is nearly always necessary. Should it be supplemented by autografts or not?

What of the prosthesis? Do we use a long one, a short one, a conventional one? What is the place of custom prostheses in revision surgery, particularly for the femur?

To cement or not to cement? What are Slooff and Ling doing? What are our colleagues doing in America? Can the two groups come together?

Then there is the whole problem of intra-operative and vascular complications and fractures below the stem of the vascularised femur, or loss of position at the junction. We have heard how important it is to have a step.

Then, of course, there are post-operative complications – high dislocation rates, infections. And how do we treat excessive ossification?

Next, there are the complications related to the graft: non-union, migration, resorption and fracture. What about the implant?

There is also the whole subject of rehabilitation. When do we start weight-bearing? What are the factors involved in that rehabilitation?

Finally, on the subject of human bone morphogenetic protein, does the work that Urist started years ago and that of Burwell and Mankin and others in the basic sciences, have any bearing on our future treatment of revision surgery?

The demands of the operative procedure are obvious to all of us, but we could start from the point of view of the pre-operative assessment. Perhaps Dr Chandler would comment on the importance of pre-operative assessment of the extent of the bone defect, and what particular methods of evaluation he has learnt over the years?

Dr Chandler: That is an important question: A person may have so many other disabilities that it may not be worth the price to have to go through the risk and dangers involved in such a major reconstruction.

I learn most from standard radiographs of both hips. If I am in any doubt, I will occasionally use CT scans and pure models.

I personally aspirate every hip. I may have an arthrogram. Vascular investigations are uncommon unless the acetabular component is within the pelvis. Gallium or bone scans rarely give useful information.

Mr Older: A good history, comprehensive radiographs and a series of ESRs are more informative than gallium scans.

Dr Goldberg: I agree with Dr Chandler that it takes time to do a proper investigation. It is very important to rule out infection. Wroblewski has talked about his experience of positive cultures as high as 40 or 50% in his revision work. I find that the scans are sensitive but not sufficiently specific.

My patients are always prepared for a two-stage procedure; to wake up with leg traction and the prospect of further surgery at a later date.

Mr Older: Do you mean by a two-stage procedure that you have taken everything out and you have given them a Girdlestone?

Dr Goldberg: Yes. A very good pathologist with whom I have worked for many years does frozen section biopsies of the membrane. I respect his level of suspicion concerning the number of polymorphonuclear leucocytes present. If the level is too high, I will back out at that point and wait for the permanent cultures. The best judgement with regard to infection is to always be suspicious, critically appraise all the investigations and combine this with clinical acumen at surgery. If there is a large bolus of cement or a protrusio, arteriography possibly with an intravenous pyelogram is indicated.

Mr Older: Dr Gross, may I ask you about your particular investigations to measure the size of a defect and the extent of grafting that will be required in the hip?

Dr Gross: We are very spoilt because we have the tissue bank right beside the operating room, so I tend to get lazy about doing exact measurements. Probably the best method of assessment is routine X-rays including Judet-views which will tell you whether it is an anterior or posterior column defect. If you are still confused about the defect, a CAT scan will reveal exactly what is missing. Then you have to make sure that you order the correct bone.

Mr Older: A metal implant can affect the CAT scan. How do you overcome this?

Dr Gross: You cannot. You may order a magnetic resonance scan and hope that there will be less scatter. When ordering the bone it is very important to make sure that you order enough! With an acetabular defect, you are probably better to order a whole hemi-pelvis and then use the iliac crest as morsellised bone to fill the cavities and whatever structural graft is required.

On the femoral side, when you are ordering a proximal femur you must ensure that you order a right hip for a right hip. You should also keep in mind whether the femoral prosthesis has a straight or curved stem. With a curved stem a proximal femoral allograft is better than a proximal tibial allograft. Bulk allograft must be done in two stages if you are concerned about infection. We do not do a primary Girdlestone. We fabricate a prosthesis that usually consists of two long Rush pins and a huge bolus of cement on top with a neck shaft angle incorporated into it. Then we put the Rush pins without cement down the femoral shaft, and the bowl of cement impregnated with antibiotic is placed in the socket of the acetabulum as a spacer; this is very stable. The patients do not need traction and can be out of bed within a day or two. It establishes a cavity so that subsequent operations are easier. It is more cost-effective than the post-operative care of a Girdlestone procedure in terms of days in hospital.

Dr Goldberg: We do basically the same thing, but what was your reason, Dr Gross, for selecting cefamandole? I avoid the penicillins, and we use tobramycin powder in our cement.

Dr Gross: We used cefamandole because for a long time it was impossible to get gentamycin in North America. The original work on cefamandole showed it was leaching out of the cement in adequate amounts and not weakening the cement. It seemed to work out quite well; gentamycin powder is now available.

Mr Older: Do you use antibiotic-loaded cement for all your revisions if you are using cement?

Dr Gross: Yes, for revisions but not in primary cases.

Mr Older: Dr Paprosky, do you use the same techniques that our colleagues have mentioned to make sure that you are dealing with aseptic loosening rather than infection?

Dr Paprosky: I agree with Dr Goldberg regarding history and suspicion. With the high degree of false negatives now being determined, I have stopped using scans as they are of little value. I only do pre-operative aspiration on 50% of hips. If there are two cultures or specimens intraoperatively that suggest infection, you stop and come back later.

The current work from Fitzgerald and others about antibiotics in the cement is that leaching probably only occurs within the first hour. I used it until about a year ago but we have stopped altogether.

I agree with using spacers; they maintain the cavity, and make it easier if further surgery is needed because all the scar tissue does not have to be dissected.

When using bone graft, I agree that it is important to have the right supply available. CT scans are going to help you for they differentiate between cortical bone, cement or cancellous bone; you have to know how to do these procedures and be prepared to do all these big complicated grafts.

Mr Older: Mr Elson, with all your experience of infected hips and revisions, please will you comment on deciding whether things are effective or not?

Mr Elson: I was interested in what was said about the spacer concept. We have been doing this for a long time. Occasionally, there is quite a violent reaction with very large cement plombes. I have had three or four patients where there has been an enormous collection of curious fluid looking like port wine which has developed around cement. There were tiny fragments of acrylic cement in the wall. It is quite clear that the friction of the soft tissues against the plombe of cement rubs off microscopic fragments of acrylic and produces this reaction. I am surprised that it has not happened more often if you are actually using it as an articulating head inside the bone.

On the subject of infection diagnosis, we have never done any grafting, autograft or allograft, until we were as sure as possible there was no infection. Suspicion of infection is one of the strongest indications for a two-stage procedure. In expert hands the single stage is reasonably safe, not as safe as a two-stage procedure, but it spares the patient from having two very major operations. It is wonderful to achieve this.

Our experience with staphylococcus is now very extensive. It is very treacherous, as is haemolytic streptococcus, which we originally thought easy.

I am very suspicious about the concept that you require more than one specimen to diagnose infection. In the operative field, when you do biopsies and take specimens you must first distinguish the difference between infection and contamination. Contamination is very high. There will be contaminated tissue which is not necessarily an infection in most revision procedures. I do not believe there is any necessity to have three out of five specimens. As far as the two-stage operation is concerned, do you do biopsies before the second stage, especially if you contemplate massive grafting?

On principle, I am very much against putting antiseptics into the human body in this kind of environment because I think this kills cells. It may kill bacteria but it must also kill cells, and the most important thing in these revision operations is to be left with healthy bleeding tissue. It is irrational to put in antiseptics and grafts soaked in povidone-iodine.

Cefamandole is not particularly good at eluting from cement. The highest are the aminoglycosides, it does not matter which one especially if you use Palacos. Streptomycin is a very valuable agent in some of these infections and streptomycin powder is available. Lincomycin is especially valuable for Gram-positive cocci.

I do not think we need worry too much about allergy. I would be interested to hear of a patient having an allergic response to antibiotic coming out of his cement.

Dr Goldberg: I never use aminoglycoside. With regard to delayed reimplantation, we tried one-stage exchanges but had a 40% failure rate without any organisms, so a two-stage procedure is now used routinely.

The one piece of advice that I have is that you need to involve the microbiologists. It is very hard to deal with these problems on your own. There are very complex and strange organisms, that require microbiological input in the specific culturing and use of antibiotics.

We use high dose intravenous antibiotics with peaks and troughs, usually for six weeks, then change over to oral antibiotics. I do not re-explore before three months. Looking at Fitzgerald's data, there is a clear point around three months in those patients where we have done re-exploration. Failure is more likely before three months.

I usually do a limited open biopsy, although I re-aspirate the patient after he has been off antibiotics for a minimum of ten days. I explore and investigate. At that point, if the cultures that I aspirated were negative, the exploration essentially leaves me with non-specific changes and no problems. Clinically you have a clean situation and I reimplant. Depending on bone stock circumstances, I use either a cemented or uncemented technique.

There have been a number of occasions where it has been right to back out and come back again after another course of antibiotics.

With regard to the issue of soft tissue coverage, we are a referral centre and therefore see two- and three-time losers as far as infection is concerned. The soft tissue coverage and particularly the blood supply to the area is deficient. We have worked with our plastic surgeon, and done a free tissue flap or turned a flap from the muscle on a vascular pedicle down into the area of the hip. We have done that six times and have been able to cure the infection, and go back and reimplant.

Dr Gross: We have the allograft brought into the room but do not take it out of its container until the joint has been exposed and we are absolutely sure that we are going ahead, otherwise a valuable specimen will be wasted. Once it has been opened, most tissue banks will not accept it back. It is best to leave it in its box surrounded by dry ice outside the operating room until you are absolutely sure you are going ahead. When we unwrap it, we thaw it out in warm povidone-iodine and put it into hydrogen peroxide as the final disinfectant. It is one of the well-documented agents for killing the AIDS virus. Then we wash everything off the graft with bacitracin and saline. We irrigate the wound with 50% povidone-iodine and hydrogen peroxide, and then wash out very carefully with bacitracin and saline.

Dr Elson: You have done the damage then, have you not? You have killed the cells. You produce the irritation and it is no good saying that you wash it out; the damage is done. Have you done cases without the povidone-iodine and without this washing? You talk of using bacitracin and saline. You should not really use saline. Saline itself, the tissue culture merchants will tell you, kills cells. If you get saline into a cut on your hand, it stings like anything.

Dr Gross: It is good when it stings!

Dr Elson: That is what they used to say about iodine. Tincture of iodine was said to be very good. I have seen Charnley irrigate a wound with tincture of iodine. I thought this was unwise. In any irrigation of a wound you should not use saline. The tissue people know this. If you must use bacitracin, use it in physiological saline. When I do these massive grafts they will be pure, and I do not think I will have any more sepsis than you.

As for hydrogen peroxide, there have been reports of cardiac arrest. We are now being dissuaded from using hydrogen peroxide in large wounds.

Dr Goldberg: At the last Academy meeting in our institution there were reports on 10,000 hips. We have had a number of cardiac arrests and problems with acute death in using large grafts with cement in elderly people. This occurred where a large bolus of cement had been used in elderly women with some cardiac history. We had three deaths within 18 months. The Mayo Clinic had an incidence of 0.7%. In revision with large grafts it was in the region of 3%.

Dr Gross: We should probably do a double blind study, regarding the question of cement in revisions. We have had one intra-operative death and similar cases have been reported recently. It is probably fat embolism from using cement.

If you are doing a revision in an elderly Caucasian female with great big, wide femoral canals that are full of fat – a common situation with such people – I do not think you should be quite so keen about pressuring the cement. With three bags of cement put into huge guns and slammed home, the patients may not be dead immediately but they cannot get them off the anaesthetic machines. That is a well recognised complication, and also one of the good arguments against cementing in a long stem.

Mr Older: It is not only the cement. The more major the revision, the greater the increase in the blood loss. I am sure we all have very good anaesthetist colleagues, but maintaining the circulation with the enormous blood loss during revision surgery has a great bearing on the successful outcome.

Dr Loty: We never use large amounts of cement, certainly not three bags because the cement would be loose in a few years.

I agree about the importance of X-ray, and like to have X-rays with measurement. It is very important to know what type of reconstruction we want to do. If we want to have a bed of allograft, then we must choose a small femur, whereas an onlay allograft needs a big femur. X-ray with measurement is very important in these cases. We usually perform a one-stage procedure in the infected cases. We only operate in two stages where the bacteria are resistant to all the antibiotics we normally use. In recent years we have had staphylococci for which only vancomycin is efficient and use the two-stage procedure in these patients. In all the other cases we perform in one stage with cement containing gentamycin and systemic antibiotics for six months. In infected cases, arteriograms are very important for they show where infection can be expected.

Mr Gie: In Exeter, we pressurise cement in every single case to an extent that those who have not seen the technique would hardly believe. My anaesthetist believes very strongly that cement is not a problem as long as the canal is thoroughly cleaned. He is happy for us to pressurise as long as we get rid of all the fat from the canal.

We frequently use three mixes of cement. I do not know whether the amount of cement being used is connected with the loosening rate. We believe that the reason problems are seen, particularly with osteolysis, is that the cement mantle is incomplete. The implants are too big and there is not enough cement. In the laboratory, if you tap thin cement with a hammer it will crack. The place where you will find cement fractures is where the cement gets extremely thin in the proximal end. We believe in lots of cement and small implants.

We do not cement long stem implants because you cannot put in the plug and do a decent cementing technique.

Dr Czitrom: I wonder whether the panel, who are all revision surgeons, would consider taking the experience of sepsis surgery as a model for their work? In hundreds of patients it has been shown that the radical debridement in osteomyelitis is the most important part of this procedure. Done properly you see the so-called "paprika" sign – bleeding bone everywhere and healthy soft tissue.

The purpose of staging is to get back the results of all those cultures and tissue samples taken at the time of surgery so staging is done on a short-term basis; the final procedure being done ten to 14 days after all the cultures are back. There is no point in delaying the second stage.

The incidence of positive cultures after the first debridement compared with cultures two weeks later at the start of the second stage is the same, so you do not get less positive cultures by leaving the wound full of gentamycin. The gentamycin does not protect from infection; it merely tides over the period while you are waiting for the final stage.

In cases of infection, magnetic resonance scans really do indicate the intra-marrow involvement if you use them to show you the extent of infection inside the canal.

Dr Chandler: MR scans have not helped me when doing a big implant where there is a great deal of marrow. In a definitely infected situation, I would always stage it.

We studied alignment rods in total knee replacements and found that the resting pressure of the marrow cavity is about 20 to 40mmttg. The pressures will rise to 1200mmttg. If you can really clean out the canal so that there is no marrow, you do not get marrow emboli. In the cementing of long stems you cannot clean the canal well enough, and that is very dangerous.

Mr Older: Do you believe in morsellised, solid, or both forms of graft together?

Mr Gie: I use morsellised graft only where possible, but I do find that the solid graft is sometimes required.

Dr Chandler: I think that the rim has to be reconstructed by a solid graft. I use morsellised bone only in circumstances that do not bear weight.

Dr Loty: We almost always use solid graft.

Dr Gross: I would re-emphasise that the first goal is no graft, the second goal is morsellised, then solid bone, and finally you decide whether you are going to make the bulk graft.

Dr Paprosky: You can use morsellised graft when you have a stable rim. If no rim, you can get that only with bulk graft. That has been proved over and over again.

Mr Older: Because it is an important issue, can we deal with the fact that in the cementless sockets you will not see a radiolucent line. This is quite an important point when we are comparing cemented with cementless cups. Could we have some discussion on that aspect?

Dr Paprosky: Initially we thought that these sockets did not migrate and that they were all stable because we did not see the line which we were used to seeing with cement when we measured them. You must measure every year with a standing AP pelvis radiograph. All superior and medial migration is measured. Six of our 35 structural allografts showed evidence of migration greater than 3mm to 4mm. These things looked perfect until we actually measured them. Unless you do that on annual follow-ups, you will not assess accurately.

Dr Slooff: When we started a new technique in revisions in the late 1970s it was very modern to use a metal-backed cup in primary arthroplasty. Because at that time it was impossible to obtain metal-backed cups in Europe, the only thing available was a very non-rigid wire mesh. Since we never like to change a good technique, we are still using it and our results with this mesh are acceptable.

When the grafts fail, there is a host bone/graft interface and there is never a problem. We have shown that the problem is at the cement/graft interface.

Mr Older: It behaves similarly to a primary failure?

Dr Slooff: Yes, we believe so.

Dr Gross: When we looked at the morsellised bone grafts, the results with the cemented cups in the roof reinforcement rings were exactly the same clinically as with the uncemented cup. There was no difference between the resorption of the graft, the loss of bone stock and migration. The only difference was in the incidence of lucent lines which was higher around the cemented cup. If you had a very young and active patient, you would be more hesitant to cement in a cup if an uncemented cup was available. We still use a great deal of cement, but we tend not to use it in the higher demand patient.

I would advise Dr Slooff not to use a metal-backed cup for reconstruction. Experience with a cemented metal-backed cup in the United States is very bad. I believe that you should stay with exactly what you have.

Dr Goldberg: If you look at several series of primary uncemented cups, retrieved for reasons other than loosening, at autopsy or in any other situation, the variation of bone ingrowth is anywhere from 17% to as high as 84%; a fair number of them are very low. If you take the low average, because you are dealing with bone that is relatively less vascular with more problems, you are as low as the 17%–20% level. What is the true level of bone ingrowth needed for a long-term stable implant, which will at least not migrate in a cementless situation?

Dr Gross: You are absolutely right. If we look at our porous-coated cups in revisions, where we have at least 50% contact, the amount of bony ingrowth will not be at all significant. I believe that we are looking at an asymptomatic fibrous type of encapsulation, a fibrous ingrowth situation.

Dr Goldberg: Do you not think that that will fail ultimately?

Dr Gross: I do not know. I do not think so. Some people might think that is an ideal situation in terms of stress shielding. I do not think they will fail as long as they do not migrate. We have not seen them migrating so far.

I hope I was not suggesting that, because we have contact at 50% host bone, all of that is ingrown with bone. I do not mean that. I mean that it is a stable interface.

Dr Goldberg: That is an important point. You may not necessarily have bone ingrowth but you have a stable interface. Whether the so-called "moving" interface is viable is still a moot point requiring a good deal of discussion.

I agree with Dr Gross. I think we are achieving stable fibrous interfaces, which seem to last eight or nine years. That principle has been extrapolated into the revision situation.

Mr Older: Can we talk about prostheses? Some of you have emphasised that as we are seeing larger implants put into the acetabulum, perhaps that is taking us away from grafting. May I have your comments, Dr Goldberg?

Dr Goldberg: I would like to bring the centre down. The larger diameter sockets are a help. I try not to cement the sockets.

Dr Paprosky: A large diameter socket can be fixed on the rim and you do not have to hold it with pit screws. It is basically supported by itself. That is the way to go.

Dr Gross: We would go to a large socket only to get peripheral support, but we have to remember that larger is not better. We have already gone through that cycle on the femoral side and we know that it does not work.

Dr Loty: I think the great use of the allograft is that it allows you to use in revision the same implant that you use in primary arthroplasty.

Dr Chandler: I would much rather use a larger component without a graft if I have stability. I do not feel as strongly about the centre of rotation as some of the other panelists. You can move upwards but I do not think you can move laterally.

Mr Older: Offset is terribly important in terms of the biomechanical engineering of the hip that is being inserted. We should not place the cup too medially as this reduces the offset, particularly in big men.

Dr Chandler: I do not think it is bad if the centre of gravity to the axis of rotation is a little short. The offset gives you a trochanteric lever arm. If you lengthen your abductor levers, there is less force to the hips. The two are compatible.

Dr Kocialkowski: Could you explain why the threaded cups perform less well than the PCA cups regarding allografts?

Dr Paprosky: Maybe we were lousy surgeons, but the American experience with threaded cups has been a disaster! Maybe it was design-related. When we put in a threaded cup, it looked and felt tight. We would "grunt" and the whole patient would move, and everybody would say "This looks great". But when we later analysed some of these in the laboratory, we found that all the threaded cups did was to expand the acetabulum. In the revision situations where there was poor quality bone, it was easy to expand the acetabulum with the threaded cup in the columns that existed. You did not thread or cut into the area. Universally, we found a large amount of migration. A lot just fell out.

The results with the porous cups have been so much better with grafts that the threaded cups have been abandoned.

Dr Gross: I believe one of the reasons is because the less you do to the allograft the better. If you just put in a porous ingrowth cup, you do not interfere with cortical integrity. With the threaded cup you have to cut into the cortical integrity. The less you do to a bulk graft, the better.

Mr Older: It has also been shown in the primary situation that one of the main reasons for their failure is the stress shielding you see with the threads.

I would like to re-emphasise your good advice to Dr Slooff about not going for something more rigid. Mesh is very much more flexible and much better. I felt that the bad news with cemented metal-backed sockets which is appearing in the American literature, was very predictable when they first came out. That is one of the things that concerns me about cementless sockets. A great many sockets with very thick metal backing are being used.

Mr Gie, please comment on the subject of femoral prosthetic size. There was a time when we put in the long stem and more cement. We are moving away from that now and try to put in a standard size prosthesis.

Mr Gie: I have never used a custom-made prosthesis or anything other than the standard prosthesis that we have on the shelf.

Mr Older: Could you comment, Dr Chandler, on the place of custom prostheses?

Dr Chandler: I had only one made for a patient and it was so absurd I could not use it. In the past three years I have been interested in using the joint medical stem which allows you to customise for the patient with several metaphyseal units for each stem. The metaphyseal units can be put in any rotation. It is an uncemented component.

Regarding long stems, I prefer the standard stem if I have stability and the femur is strong enough to take it. The massive chrome–cobalt stems are tremendously stiff and do tremendous damage.

Dr Loty: If the patient is young, we try to use an embedded allograft and standard prosthesis. We prefer not to take the risk in very old patients and we use a long-stemmed prosthesis.

Dr Gross: I do not worry about using a long stem. I have always felt that the stem should go beyond the weakest point. I am a firm believer in intramedullary fixation. The advantages of a straight stem are that you get three-point fixation, which is outstanding. The disadvantage is that you often penetrate the cortex at the distal end. The advantage of the curved stem is that you will not penetrate the cortex distally.

Mr Elson: Dr Gross, three-point fixation may operate in bending but be very poor in rotation. That is the advantage of the long curved stem.

On the matter of thigh pain, a prosthesis which is cylindrical, with or without flutes, will have a point at the tip where bone will be bending and pressing against it. Those of you who are divers will know that pressure on the sinus produces intense pain, due to bone deformation. The simple expedient of having a taper, a run-off, at the end is brilliant. I had never thought of that before. Has it actually been tried yet?

Dr Goldberg: Yes.

Mr Elson: We are talking about an uncemented prosthesis. It does not end on a corner but there is just a run-off.

Dr Paprosky: We have a long curved stem with a 1 inch taper on the end of it.

Mr Elson: Does it make any difference?

Dr Paprosky: We have noticed a significant difference.

Dr Gross: I know that it has been shown in a biomechanics laboratory that the curved stem, even in a short stem, is as good as a straight stem cemented. The problem in the operating room is that you have to start with a lot of anteversion in order to worm that curved stem down. You cannot have an absolutely perfect tight fit and get it down the correct way. What often happens is that it is not quite as tight as three-point fixation. I am not saying that I favour a straight stem but I know that when you put down a straight stem you are going to end up with three-point fixation. When you put down a curved stem you will probably be 1mm short of having an absolutely perfect fit. In order to get it down, it has to be that 1mm smaller. That is the problem. We would probably be using 50% curved and straight stems. We have had some problems with straight stems which have been revealed when we have taken X-rays at the end of the operation. The tip of the stem may have caused thigh pain; it may even have penetrated through. We have not seen that with the curved stem. It is my impression that if you have a good three-point fixation with the straight stem you have less thigh pain.

Dr Paprosky: I have never used the custom prosthesis in revision surgery. It is expensive. The reason that you have trouble inserting a curved stem is that you are putting it through a whole allograft and you cannot turn it.

Dr Gross: We do not use a curved stem with an allograft. We use a narrow straight stem prosthesis so that we do not have to ream the strength out of the allograft. In a revision hip without an allograft, we use a long curved stem.

Dr Paprosky: Whether you use a long stem or not, depends on where the bone is. If you do not have any bone above, you have to go further distally. If you have bone up above, then you can choose the short stem.

Dr Goldberg: I agree that the medical stem has a great deal of versatility. We have been using that in increasing numbers, although I am concerned about the fretting. I have been impressed with the lack of thigh pain in these patients. Metaphyseal fit is incredible and works extremely well. There is a close fit distally. The titanium modulus is low and it tapers off to some extent. You actually feel that you can squeeze the closeness in your hand. It is very effective in a difficult situation. It is available both straight and curved and we have used it once in an allograft.

Mr Older: Now to the subject of thigh pain.

Dr Hedley: There are various causes of thigh pain. The most obvious is a loose prosthesis. If the component is loose proximally, there are radiolucencies around the porous coating and there is an end-bearing pedestal and you have lost. There is no point in doing anything other than revise the stem. But if there is no end-bearing pedestal and there is radiolucence between the pedestal and the tip of the stem, then there is proximal stability. A number of those patients will have thigh pain. The thigh pain is due to micro-motion at the tip of the stem. If you strut the femur, you stiffen it and remove the micro-motion, but it has to be stable proximally.

Mr Older: You are saying that there is an elastic modulus mismatch?

Dr Hedley: Yes.

Mr Older: In that respect, you must have intrinsic stability of the implant as a prerequisite. How do you know when it is stable? Are there any hints that anyone can give on how you can tell that the implant is intrinsically stable?

Dr Hedley: I had an opportunity to compare long stems versus mid-stems, versus short stems – primary stems – all in essentially the same type of prosthesis which was collarless and proximally coated. The incidence of thigh pain is more pronounced in the mid-stem than in the primary stem. The length of the stem definitely plays a role; it is not purely cortical with a modulus mismatch. In the equation somewhere is obviously the stiffness of the stem, but there is also the length.

Mr Older: From what I remember, Hungerford said initially that he had less thigh pain with the mid than with the shorter stem.

Dr Hedley: I think I was the one responsible for that. I used 50 mid-stems for primary arthroplasty in the belief that because the mid-stem was bigger at the top end and I was filling the metaphysis then I would do better. I regret the day I did that. Some of those patients are four years from operation and they still have thigh pain. It was not a good idea.

Dr Paprosky: One of the things we found in our studies was the degree of canal fill. Using a mid-stem the greater the degree of canal fill, the less relative motion around the stem and the less tendency towards thigh pain.

In our experience, the better the canal filling in the stem, thigh pain was markedly reduced – adding a tapered tip so that you do not have the femur bending on the end. When there is a modulus mismatch and not a loose stem, the cortical stress also improved the modulus.

Mr Older: How do you know when you have an intrinsically stable stem?

Dr Paprosky: Whatever kind of reamer you are using, once that engages the cortices and it is basically the proper size, then you can jam that stem in. On intra-operative X-ray and revision, if the reamer fills the canal, you drive a stem in and it is tight as it is goes in and does not impinge against the front of the femur at the cortex - those are going to be tight. In the past a lot of them were undersized.

Dr Loty: The advantage of the cortical bone allograft is not in the choice between a medium or a long stem but getting back to the standard prosthesis which is usually used. If one puts a strut graft longer than the end of the prosthesis, then we should try to get back to the standard length and avoid the long-stem prosthesis.

Mr Older: Dr Loty, please comment about the level of irradiation and weakening the bone. There seemed to be a definite decrease in the strength of the bone above 350–370 Gy. What was the testing technique? Was it cortical bone? Did you do it in torsion? How did you test the bone?

Dr Loty: We had small samples of cortical bone from the human femur, sufficiently small to be easily tested on a bending machine with the same load that one would have in a mechanical study in industry. We first verified that the mechanical properties of the bone from where we took the samples from the femur did not change. Then we tested with two amounts of irradiation. After 270Gy we had 80% of control values, but after 370Gy of irradiation we had only 65% from control. We have very accurate measurements for each bone. We request a certificate from the company which carries out the irradiation, giving the minimal and maximal dose designation. If bones are irradiated, it is important to have a very accurate designation.

Dr Kocialkowski: Could I ask the panel if the host allograft junction should always be autografted?

Dr Gross: You have to pay very close attention to that junction because that is the key. We have had this discussion about long-stem femoral components. Long-stem femoral components by themselves are not dangerous unless you are cementing them into the host, or unless you are trying to get a press-fit distally. Dr Paprosky and I probably disagree here because he is of the Engh school that talks about press-fit distally. I believe that that de-stresses proximal bone and causes stress shielding, even of the allograft.

I believe that if you pay very close attention to the step cut and cement the allograft with autograft there is stability as soon as there is union. Then the long stem does not mean very much distally because there can be a fairly narrow long stem; a press fit is unnecessary and there are none of the disadvantages of a long-stem prosthesis.

We spend much of our time at the junction. Before we finish it has to be absolutely rigid, with the cerclage wires, the vascularised bone graft and whatever autograft is available. If the step cut is allowing 10° of rotation, then we put in little shim struts to control it. Before we started doing all that, we had five or six non-unions where we had to plate and bone graft to get them to unite; then we had a plate on the allograft. The best allograft is a non-violated cortical tube. Your goal should be to end up with something that at least comes close to that.

Mr Older: Have you had to go back in and use a vascularised graft at this step cut area?

Dr Gross: No. We use our plate and autograft.

Dr Slooff: What should be the fate of the graft? At the moment you are operating and using a massive allograft. First of all, this graft will be united against the host. After some years, however, when the graft has remodelled, I believe that the graft will also be weakened. What does the panel think?

Dr Gross: We take the patient's residual femur with the soft tissue attached to it and wrap it around the graft. Ideally we would like to achieve union at the host/graft junction, and creeping substitution of a millimetre or two. In other words, we would like little or no creeping substitution. Hopefully the host bone will reinforce the allograft. We do not want it to replace the allograft because it could not do so in the patient's lifetime; it will only weaken it and cause it to fracture.

What you have said is absolutely true, Dr Slooff. Eventually the allograft has to fail. Eventually a primary total hip has to fail. Everything is working against us. The problem is how long we can make them last.

I would like to think that we could get ten years, plus being able to do another revision, the additional revision being made easier by the allograft.

Dr Slooff: In that case, would it be better not to speak of "incorporation" of the graft?

Dr Gross: The only grafts where I refer to incorporation are the small grafts that we use around the knee for post-traumatic defects. It cannot be avoided there because there is cancellous bone against cancellous bone.

Dr Goldberg: Mr Gie, how do you pressurise? Do you use the gun and the standard technique?

Mr Gie: We now have a spout that is cone shaped which allows the tip of the gun to get to the end of the canal without damaging the mantle. We have a hemispherical seal which fits into the proximal part of the femur. We are modifying that to put on some flanges so that it can sit on the graft at the top and stop any leakage of cement.

Dr Dall: Putting in low viscosity cement, do you not find that it seeps into all the chips and that you end up by sequestrating them?

Mr Gie: If you saw the operations, you would be amazed how tightly the chips are packed and how stable the situation is. The only thing is that as you are increasing the pressure in the cement you are also increasing the pressure in the graft. The cement will go a little bit into the graft, but you never push it all the way through.

Dr Dall: Does it work well in practice?

Mr Gie: Yes, and the X-rays confirm that. When you finally impact it, you cannot pull the stem out. Despite it being highly polished you have to tap it out of the graft because it is so tight.

Dr Delloye: Can you use the same system, using ceramics?

Mr Gie: We did a few uncemented cases initially. The intention was to go back and cement them later once the graft had incorporated. Two or three cases were done in 1984 and they all subsided markedly. I feel that you should not have a collar, because then you will not load the graft in the manner in which we would like it loaded. If you do not have a collar, you are likely to have some instability, which is borne out by the experiments carried out in Nijmegen. I suppose you could use ceramics, if you wish.

Dr Paprosky: If the bone graft is packed in so tightly that it does not separate or leak you will not have any real penetration of the cement. My understanding of the cemented technique is that there is lattice work and bonding into the cancellous or interstices as in the use of primary bone. Are you not cementing into the same kind of smooth, basically sclerotic type of cylinder there?

Mr Gie: No, not at all. In one of the patients where I caused the femoral shaft to fracture during the procedure I had to take the whole prosthesis out. You could see that the cement had gone into the graft despite it being heavily packed. If you pressurise well enough, it does go in. There is no question about that.

Mr Older: Do you feel that the stem can subside without fracturing a mantle of cement? It seems unusual to me that that can occur without breaking the cement.

Mr Gie: When this was observed originally in the early 1970s, Mr Ling felt that the cement was probably fracturing. He believed that because of the tapered shape of the prosthesis, a secondary stability was occurring by a kind of wedge technique. Later on in the 1970s and early 1980s an opportunity arose to examine the cement mantles in a number of patients where the cup had become loose and the stem had been tapped out. The stem had subsided and the cement mantle was examined. In all the cases the cement was intact. We have not been able to identify any fractures.

Certainly experiments in the laboratory would show that you can subside a prosthesis within the cement. During my time in Nijmegen I managed with six or seven specimens to subside a 6 ° taper within an unconstrained cement mantle of up to 7mm before the cement mantle fractured.

Mr Older: Given the excellent papers from Exeter and Nijmegen on the obvious benefits of cemented fixation, I find it difficult to believe, given the conservative approach that cement obviously offers in revision arthroplasty, that filling up someone's femur with metal rather than cement, which seems to be what happens in cementless implants, can be of any long-term benefit to the patient.

Dr Dall: I am in basic agreement with the philosophy that if you are looking at a situation where there is significant bone stock loss your primary objective should be to try to regain bone stock. If you think that this can be done best the cemented way or the cementless way, it is your choice. I do not believe that anyone has the answer but Dr Gross's attitude, where in order to gain stability with a bulk allograft you cement into it, makes perfect sense because it is dead anyway. At the same time my feeling would be, if you are trying to gain some stability in the viable bone and can avoid cement, then avoid it. My philosophy is to avoid anything that will create further bone loss.

Dr Goldberg: I would agree on the issue of preserving bone and trying to reconstitute bone. The issue of "cement versus cementless" is still a raging argument. As far as primary hips are concerned, on the femoral side I probably cement in 50% to 60% of the stems. Even those who use cementless techniques are not totally convinced in that direction.

Mr Older: Mr Gie, how do you manage your patients post-operatively?

Mr Gie: We keep the patients in bed for three weeks and then we get them up using crutches, touch weight-bearing for three months. However, some of the more severe cases do stay in bed for six weeks. The standard practice is three weeks in bed, three months on crutches. Unfortunately, patients do not stick to that regime and we have seen them using only one stick.

Dr Goldberg: How can you afford to keep them in bed for three weeks? In the United States they have to be out in five days!

Mr Gie: We feel better keeping it to three weeks, but it does not achieve anything, except that there is a fairly high dislocation rate in my revisions. In about 80 revisions, of which I have kept 40 patients in bed for three weeks and another 40 up straight away, I have had no dislocations in those who have spent three weeks in bed, but a 15% or higher dislocation rate in those who got out of bed straight away.

Dr Goldberg: Why, with bone grafting, are you using functional ischial weight-bearing braces?

Mr Older: I do not think any of us yet have the final solution on how to manage our patients. I cannot keep my revision patients in bed for six weeks, for various reasons, some of which are financial. They are out of bed and non-weight bearing. I have a lot of experience with functional braces. I have felt, and it is perhaps empirical prejudice, that with a functional brace the prosthesis and graft is protected. It seems to work.

Dr Goldberg: It is very difficult to obtain load data from in vivo situations. You are probably not unloading the hip, which is good. To some extent, you are getting the patient's co-operation to go partially weight-bearing, to use crutches and maybe protecting the soft tissue.

Dr Learmouth: Obviously the rate of incorporation and consolidation of the graft will differ, depending on whether you use morsellised chips, solid block cancellous grafts or cortical grafts. How does that influence the rehabilitation?

Dr Gross: I start most of the patients on partial weight-bearing by three months and full weight-bearing by six months for bulk grafts. If it is a morsellised graft, I would start them on partial weight-bearing at six weeks and full weight-bearing at three months.

You do not want the bulk grafts to remodel; they could not remodel in a patient's lifetime. Ideally, you would like to see them lie there like inert lumps with union at both ends and maybe a little bone laid down on top to reinforce the graft.

If you are hoping for remodelling, the graft will fail long before it remodels. All you are waiting for is union with the bulk grafts. As soon as you have union, weight-bearing can start. It is guesswork with the morsellised grafts but I would say six weeks' partial and three months' full weight-bearing.

Dr Paprosky: We concentrate on primary fixation with these rock-hard, stable, distal femurs when we want to use a structural graft. We probably go a little faster than Dr Gross. They are on partial weight-bearing after the first month and use crutches for up to four months. I have to assume that this patient will walk. Patients do not listen. If you look out of the window, they are carrying their crutches over their shoulder!

I just assume that they will walk on it right away, so I let them partially weight-bear initially. I am a little more aggressive in my post-operative approach.

Dr Slooff: In the cases of cancellous grafts, our patients start weight bearing after six weeks. When it is a graft in cases of protrusio and when the graft is very nicely contained, then it will be earlier – ten to twelve days.

Mr Older: Patients with morsellised autograft of the acetabulum are non-weight-bearing for six weeks and then partial weight-bearing for six weeks; at three months they are allowed to fully weight-bear, but with crutches for a further three months.

Part III

The Knee

20 Allografts in Total Knee Arthroplasty

A.E. Gross

Our experience with revision arthroplasty of the knee is more limited than with the hip because there is a 4:1 ratio of total hip to knee total joint replacements in North America. Loss of bone stock is the main problem with knee revision surgery although less than that associated with the hip. The options for salvage are to use more cement, a custom prosthesis or autograft. An arthrodesis is difficult at the best of times and generally unacceptable in North America today. Excision arthroplasty is not as good in the knee as the hip although, like an amputation, it is an option. Reconstruction with cement is weak and may lead to more rapid bone loss. Metal wedges are probably too small for most revisions so we tend to use bone grafts which give better load transfer.

Ideally, we want the host bone to lay down new bone on top of the allograft but not actually replace it, which is why we use a great deal of cortical bone to ensure that will not happen. The surgical principle is to correct the biomechanical environment by restoring bone stock.

Principles of Reconstruction

We try to preserve host bone, restore the anatomy, always autograft at the junctions and avoid injudicious placement of screws. Bulk, cortical strut and morsellised allografts are used in knee revision. For bulk allografts either distal femurs or proximal tibias are used. Morsellised bone is packed into cavities but not used for support. Bulk grafts have to be fixed absolutely rigidly in contact with the host bone. Bone defects are either contained or uncontained. A contained defect means that there is an intact rim of cortical bone, whereas an uncontained defect extends right to the periphery. Contained defects can be corrected with morsellised bone if they are small enough. Unlike the hip, a large contained defect cannot be corrected only with morsellised bone; a metaphyseal/diaphyseal graft fixed by a stem or screws is needed to support the prosthesis.

Tibial Reconstructions

For a small contained defect on the tibial side, we use morsellised bone and then a conventional prosthesis. If morsellised bone is used in a large contained defect it will not support the prosthesis and then there will be subsidence. We take a proximal tibia and ensure that some of the cortex is left intact together with all of the cancellous bone, and fix it either with the stem alone or supported with screws and then insert the

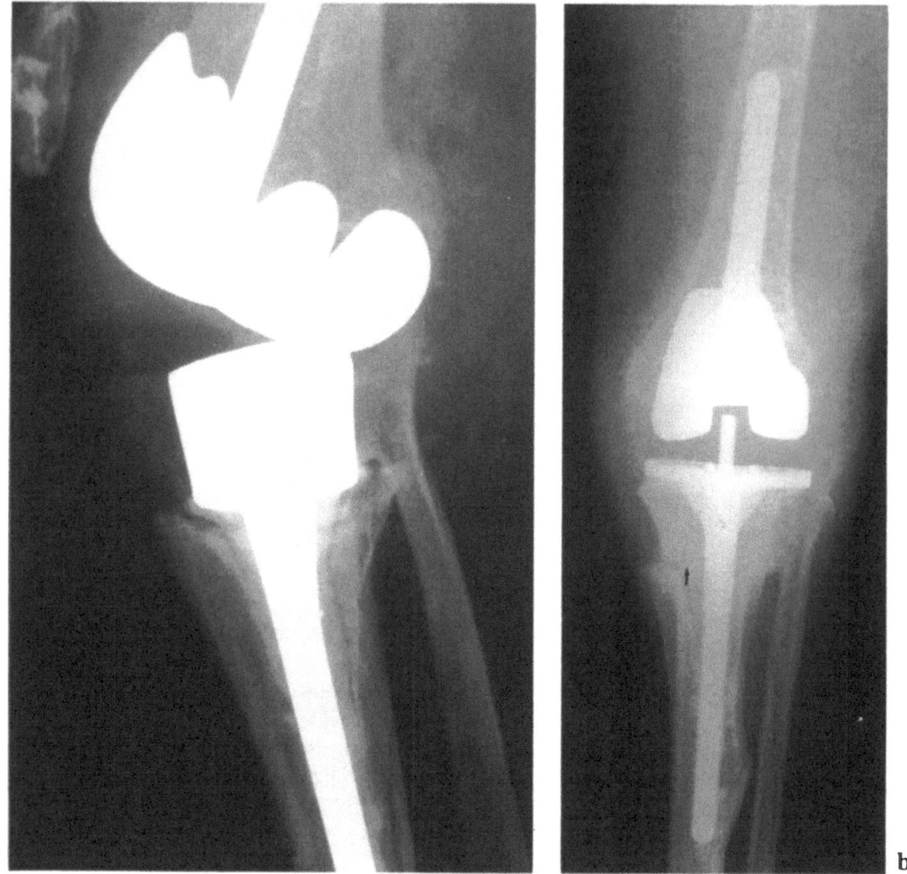

Fig. 20.1. a Failed total knee prosthesis with loss of bone stock on tibial side. **b** Modular Insall Burstein knee prosthesis has been inserted on a large tibial allograft. Radiograph is at four years' follow-up.

prosthesis. We only use cement on the surface so that it is against the allograft. In order to do this type of graft a modular system is needed, either the Insall modular system or the Johnson & Johnson, so that various length stems and diameters are available. Uncontained grafts have to be fixed back in position. The anatomy must be reproduced as though it were a virgin joint. The graft must be fixed before you start your instrumentation so that you can see the void and then put in the graft. We use a combination of screw fixation with a modular stem for a large uncontained defect on the tibia.

Femoral Reconstructions

Small contained defects on the femur are filled before doing the instrumentation in the usual way as with the tibia. A metaphyseal/diaphyseal graft is used for a large contained defect which is telescoped in before the prosthesis is implanted.

We restore the anatomy of the uncontained defects on the femoral side with a graft secured by screws, do our instrumentation and then put in the implant. The use of modular implants has simplified and rationalised knee revision surgery (Fig. 20.1a,b). Prophylactic anticoagulants are given to most of our patients having revision surgery.

Results

Thirty-two deep-frozen irradiated allografts have been used for revision reconstructions in 20 knees. Sixteen knees required femoral grafts and 16 required, tibial grafts. Grafts were done on both sides in 12 knees.

Initially, only long stem hinged prostheses were available, but since then we have used the condylar III, PCA revision and the PFC modular prostheses. We have had three clinical failures using our scoring protocol for a success rate of 85% at an average follow-up of 4.2 years (range 2–7 years).

Radiographic Results

We have had only one patient with a morsellised bone graft that failed. The morsellised bone was well incorporated but the prosthesis subsided at six years. The patient had a good clinical result but definite subsidence was present. Rim contact is essential. Without it, a bulk allograft becomes necessary.

There have been a few radiographic failures. In the worst instance the tibial bulk allograft fractured completely, fragmented and subsided. Lucencies have occurred but have not been a problem, primarily because we use surface cement for the allograft only.

Conclusions

Reconstruction to maintain structural integrity, stability and correct alignment has resulted in a clinical success rate of 85% in difficult cases. The correct biomechanical environment is needed for allografts to succeed.

21 Bone Grafting in Knee Revision Surgery

V.M. Goldberg

The technical aspects of revision total knee (TKR) surgery are similar to the primary arthroplasty. Mechanical alignments of the knee and component position are critical considerations. Component fit and bone implant contact are also important. In order to accomplish these goals, modular knee systems are used as well as custom components when indicated. Osseous deficiencies which require bone grafts for reconstruction must be adequately defined during the pre-operative evaluation. In addition to standard radiographs, CAT scans and 3-dimensional model reconstructions may help to delineate exactly the bony deficits.

An appropriate classification of the bony defects in both the primary and revision TKR is important for outcome studies. Bone loss in the primary TKR may be classified by the extent of bone loss on either the tibial plateau or femoral condyle. Type I bone loss is less than 5mm in depth and 50% of the surface. The Type II loss is usually greater than 5mm of depth and 50% of the surface. It may be treated by changing the bone resection line to accommodate the loss or by adding bone cement and/or bone graft. Type II usually requires augmentation with either bone cement, grafts or modular components. Customised components may also be useful. Modular wedges or bone graft should be used when there is a 10mm defect or more. The joint line should not be shifted more than 8mm as this has been shown to result in a poorer functional outcome (Goldberg et al. 1988).

The bone loss associated with revision TKR must be quantified in order to help predict the outcome of the surgical procedure. Type I bone loss is focal loss of metaphyseal bone but with the cortical rim still intact. This is usually seen after failure of a unicompartmental arthroplasty. Type II defects are extensive metaphyseal defects of more than 50% of the bone surface but with the cortical rim still intact. Cortical preservation is important because it serves as a base of support for the component. Type IIIA bone loss is when both the metaphyseal and cortical bone is deficient, but usually confined to a single compartment. Type IIIB bone loss involves significant metaphyseal and cortical bone deficiency, usually bicompartmental. This latter deficiency usually requires bulk allograft or customised components to reconstruct the structural loss.

Treatment Options

Options for treatment of the bone loss depend upon the extent of the deficiency as well as considerations of the age of the patient, functional status and state of the soft tissue. Bypassing the defect if small enough, with deeper resection

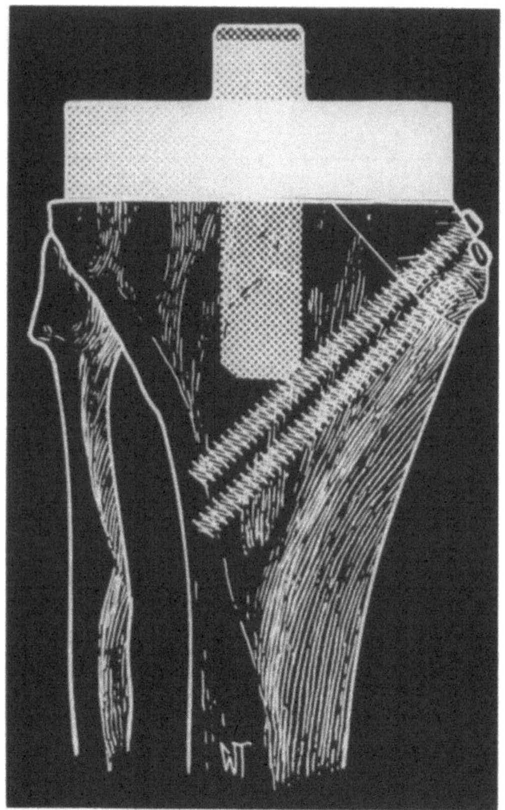

Fig. 21.1. Schematic illustration of autograft technique with the distal femoral resected bone to reconstruct Type I and Type II bone defect in primary total knee arthroplasty.

lines and thicker components with or without a thicker cement mantle is an effective technique. However one must remember to maintain the joint line to within 8mm of the pre-operative state. Reconstruction of bone loss seen with the primary TKR should use local autogenous bone graft whenever possible. Ample material is usually readily available.

Type I angular deformities present in primary TKR can be treated by bypassing the defect and using modular wedges. Type II deformities with a 5–15mm defect and with 50% or more of the plateau involved usually require modular wedges. Reinforced polymethylmethacrylate has been reported as a viable alternative to reconstruct this defect, although autogenous bone grafts may be a better biological alternative (Ritter 1986). Bone loss greater than 15mm in depth requires bone graft or customised components which use modular wedges. The technique of using bone grafts for this latter problem includes minimal bone resection in order to maintain the joint line and firm fixation of the autograft to its host bed. The distal femur is an excellent source of bone to reconstruct tibial metaphyseal and cortical bone deficiencies. The graft is fixed initially with K-wires while the standard tibial bone resection is performed. Cancellous screws are used for final fixation whenever possible. The implant may be fixed using either cemented or cementless systems depending upon the age of the patient and the extent of the bone loss. Stemmed components are useful to maintain the stability of the component (Fig. 21.1).

Revision TKR is more complex. Sufficient autogenous bone graft is often unavailable so that allograft must be used. The fixation of the allograft to the host bone is paramount to ensure the incorporation of the graft. Type I and II defects may be reconstructed using morsellised cancellous allograft, either frozen or freeze-dried. The bone is densely packed to provide structure and the cortical rim is preserved. These contained defects usually heal satisfactorily with bone allografts. The components may be fixed with or without cement depending upon the extent of bone loss and the clinical circumstances. A radiograph seven years after reconstructing a contained tibial Type I defect with frozen cancellous allograft shows excellent healing of the graft (Fig. 21.2). A significant Type II defect reconstructed with a bulk frozen allograft fixed within the boundaries of the cortical rim is seen in Fig. 21.3. The components used were semi-constrained and have functioned satisfactorily for three years.

With massive allografts, stems should be used to bypass the surface and provide load sharing. Supplementary autograft should be used whenever possible at the allograft-host interface and/or at the bone implant interface. Type IIIA bone loss is much more difficult when reconstructing the failed knee arthroplasty. The technique for problems with Type IIIA retains the soft tissue sleeve and uses long-stem components. Bulk allograft and autograft are required. The fixation options depend upon the clinical circumstances with either cement or cementless systems being used. The bone graft appears successfully incorporated in a radiograph of a knee two years after frozen allograft reconstruction of a Type IIIA defect (Fig. 21.4).

Type IIIB bone loss usually requires the use of bulk allograft interfaced with stemmed, partially constrained components (Kraay et al. 1992). Preservation of as much soft tissue as possible is important in providing the best environment for a successful clinical outcome. A radiograph taken five years after reconstructing a Type IIIB defect of the femur is shown in Fig. 21.5. The patient remains ambulatory but uses a cane full time.

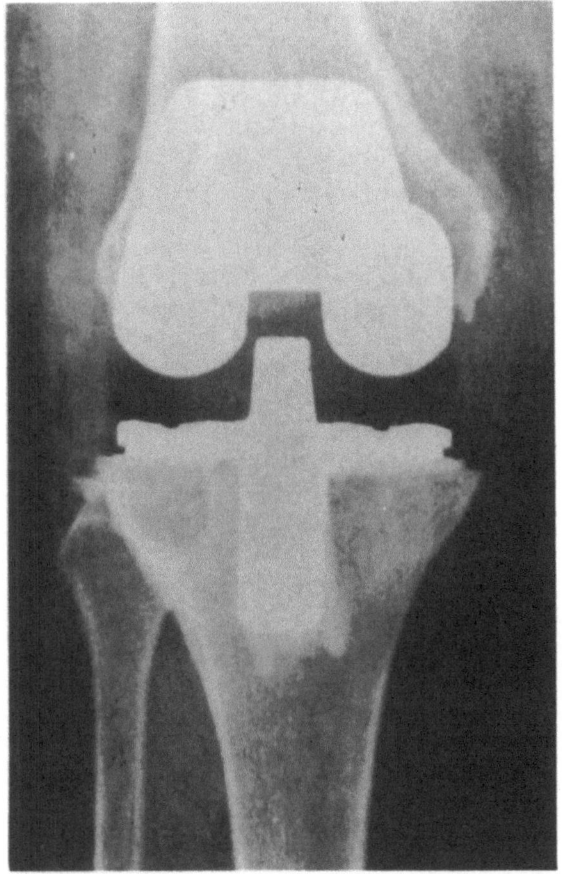

Fig. 21.2. Radiograph after reconstructing a contained tibial Type I defect with frozen cancellous allograft.

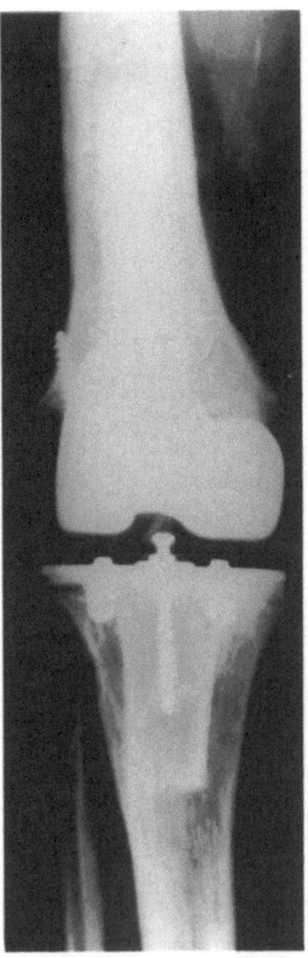

Fig. 21.3. Radiograph demonstrating a significant Type II defect reconstructed with a bulk frozen allograft fixed within the boundaries of the cortical rim.

Results

Where autografts were used, all primary Type I and II defects healed and the clinical results have not differed from non-grafted primary TKR. A summary of the clinical results of reconstruction for revision TKR using bone grafts is seen in Table 21.1. All of the allografts used to repair Type I and II contained defects in the revision TKR healed radiographically and functioned satisfactorily with a minimum follow-up of 40 months. The average knee score 40 months after surgery was 84. Of the twelve bulk frozen allografts used to repair Type IIIA defects, one partially resorbed 36 months after surgery. The average functional knee score was 79, and all patients are ambulatory. One of the seven bulk frozen allografts used for a Type IIIB problem partially resorbed. All of these latter patients use canes or crutches but remain ambulatory. Their functional score averaged 71, but they are relatively pain free.

Table 21.1. Clinical results in bone grafts for revision total knee arthroplasty (minimum follow-up 40 months)

Defect	*n*	Average knee score	
		Pre-operative	Post-operative
Type I	24	32	86
Type II	10	25	78
Type IIIA	12	33	79
Type IIIB	7	22	71

Conclusions

The long-term outcome of bulk allograft remains in doubt. If we compare the knee allografts to the

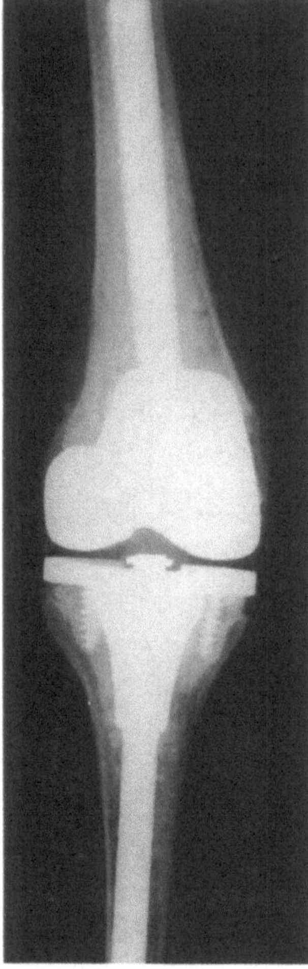

Fig. 21.4. Radiograph of a knee two years after frozen allograft reconstruction of a Type IIIA defect.

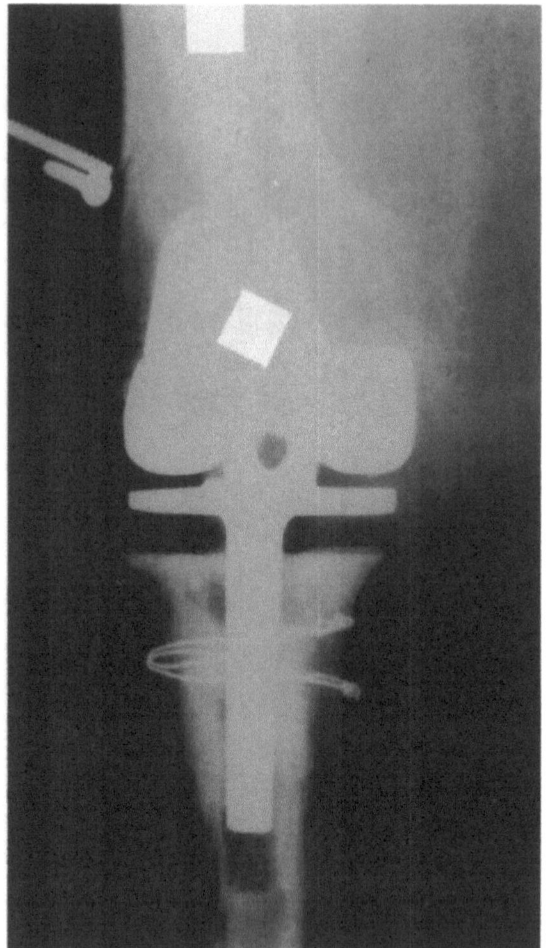

Fig. 21.5. Radiograph five years after reconstructing a Type IIIB defect of the femur.

allografts used for reconstruction of failed total hip replacements, the early enthusiasm must be tempered. However, because the bone grafts used to reconstruct failed TKRs are usually under some load and are well fixed to the host bone, fewer failures may be seen in the long-term follow-up.

The surgical principles of total knee replacement are similar whether or not bone grafts are used. Mechanical alignment of the extremity, size, position of the components, maintenance of the joint line and secure fixation of the components are mandatory for a successful reconstruction. The selection and fixation of the graft are important technical considerations which may determine the ultimate outcome of the bone graft and the knee arthroplasty. The components should be either semi-constrained or unconstrained. Linked hinge-like components are rarely used. Finally, careful pre-operative assessment and planning are critical to prepare adequately for the difficult surgical problems associated with bone loss in primary and revision total knee replacement.

References and Further Reading

Goldberg VM, Figgie MP, Figgie HE III, Sobel M (1988) The results of revision total knee arthroplasty. Clin Orthop Rel Res 226:86–92

Kraay MK, Goldberg VM, Figgie MP, Figgie HE (1992) Distal femoral replacement with allograft/prosthetic reconstruction for treatment of supracondylar fractures in patients with total knee arthroplasties. J Arthroplasty (in press)

Ritter MA (1986) Screw and cement fixation of large defects in total knee arthroplasty. J Arthroplasty 1:125–129

22 Bone Allografts Around the Knee Joint

A.K. Hedley

Minor Bone Loss

Filling in defects where there is an intact cortical wall is defined as a minor bone loss and called filler graft. We have learned from our experience of the hip that if there is a cavity defect with an intact wall, revision is relatively simple.

In reconstructing the knee, it is important to maintain the kinematics. An unconstrained prosthesis can generally be used because the ligaments are present, as they usually are with minor bone loss. In particular, many older patients' defects can be filled with cement.

The options are solid bone, particulate bone or a combination. Cement can be used in the elderly, with or without wedges attached to the prosthesis. Personally I prefer bone.

Case Reports

Case 1

The patient had an allograft femoral head to fill the defect. The graft healed completely and the patient was functioning well at six years. When first introduced, this prosthesis was hailed as the solution to "the problem" and made the cover of *Scientific American*. In my opinion, it is the cause of the problem. The defects remaining after removal of spherocentric prostheses are enormous.

The defects can be filled with a mixture of cancellous bone and solid pieces. These will often heal very well if the knee is stable and the ligaments are present; it is reconstructible despite the size of the defect.

It is important that minor grafts are protected with a metal base plate, particularly on the tibia. This should probably be augmented by a stem, but if the rim is intact some minor grafts do not need a stem for support. The key is the metal base plate which distributes the load and protects the graft from focal overload.

Case 2

The patient had rheumatoid disease with considerable bone loss and infection in the tibia, necessitating excision and packing with antibiotic beads. This particular defect was filled with two femoral heads held by screws. A femoral head was also placed in the femur securing the epicondyle. These grafts were still doing well five years later (Fig. 22.1a,b).

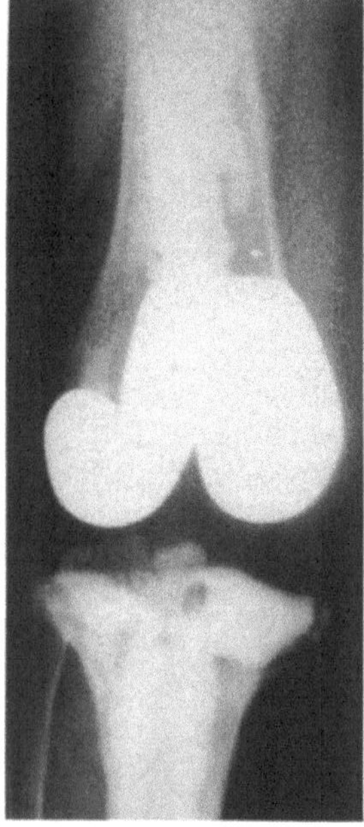
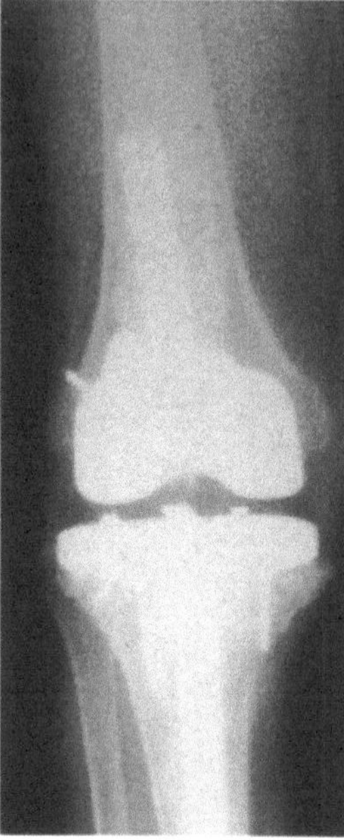

Fig. 22.1. a Radiograph showing infection around tibial component. **b** Five years post-operation.

Case 3

The patient had a fractured femur with considerable bone loss on both the tibia and femur but, nevertheless, intact collateral ligaments (Fig. 22.2a). Even with this much bone loss we regard it as a "minor defect" because it does not call for constraint from the prosthesis. All cement was removed leaving an "ice cream cone" of tibia.

This patient had substantial grafting of the femur using femoral heads. An allograft tibia was inserted into the tibia to correct the bone loss (Fig. 22.2b). The allograft was carefully shaped to fit into the defect and splinted with a stemmed metal base plate to protect the graft. Ideally, it should have been a little wider and the stem perhaps shorter.

At a year the patient was doing well (Fig. 22.2c). Radiographs showed the graft was a little smaller. At three years the patient presented with functional instability due to some subsidence of the graft, and the prosthesis had subsided a little due to resorption (Fig. 22.2d). Note the metal base plate close to the screws.

We believe the issue was twofold; firstly subsidence caused functional instability. Because the prosthesis was unconstrained, it needed revision. As the subsidence was on the tibial side it could be corrected by adding a thicker plastic.

The second issue concerns the fate of the screws through the metal base plate versus screws placed remote from the base plate. If the screws go through the base plate, because of the subsidence there would be gross movement between the screw and the metal base plate creating metal debris. The track record of metal to metal contact has never been good and should be borne in mind when multiple screws are put through base plates.

Case 4

In this patient there was a measurable distance between the screw head and the base plate at six weeks. Later, the graft had not disappeared but seemed to have shrunk. At three years, the patient had a re-operation and the tibial base plate was removed. Because there was only fibrous tissue and no bone ingrowth, the base plate was removed easily. An area of the graft was relatively vascular.

Microscopy showed a cellular reaction with fibrous tissue inside the allograft. A wave of new bone was being laid down on the dead trabeculae. Viable new bone could be seen with normal marrow spaces. At a higher magnification, there was cellular reaction with osteoblast and remodelling – typical creeping substitution along the osteoid seams.

Through the years, the top of the large piece of allograft seemed to be replaced with new bone moving from the edge towards the centre in a "centripetal" fashion, all the while providing mechanical support. I believe that there are many lessons to be learned from this. The question of constrained prostheses on big grafts will recur and we believe the basic principle is to use the least constraint possible.

Major Bone Loss

Major bone loss is a little different. Ligaments are absent with many major bone grafts; either the whole condyle or the entire metaphysis having gone with no ligament attachments. If there is to be long-term function of the knee when ligaments are absent, a so-called soft tissue "sleeve" is preserved. The knee will not function like a normal knee and may need to be braced. In an elderly patient, you may wish to consider slightly more constrained prostheses which will accom-

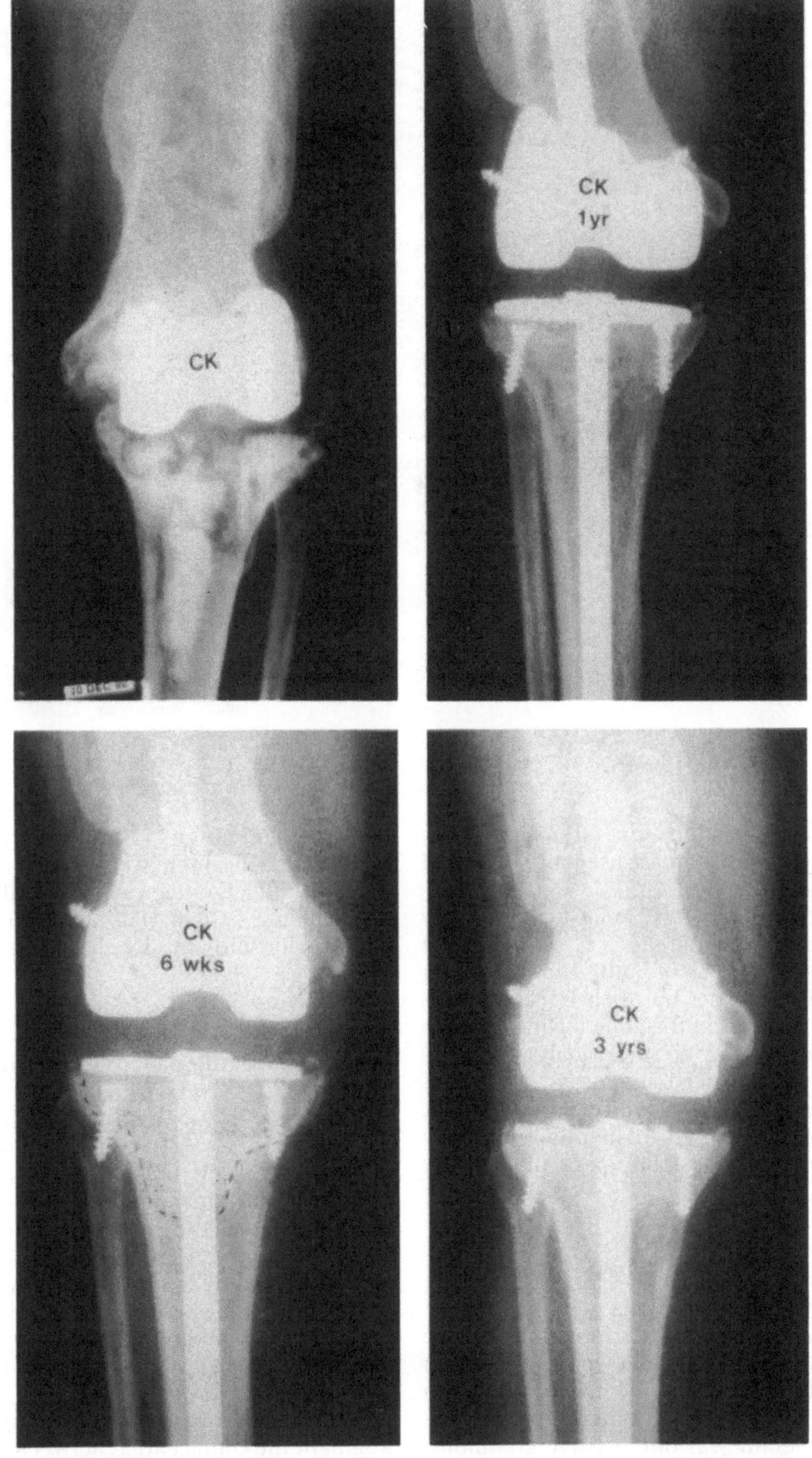
CK
CK
1yr
CK
6 wks
CK
3 yrs

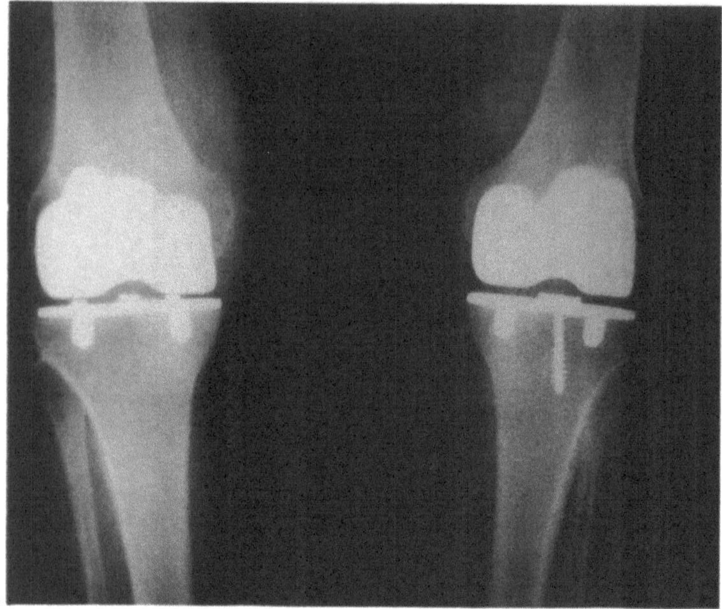

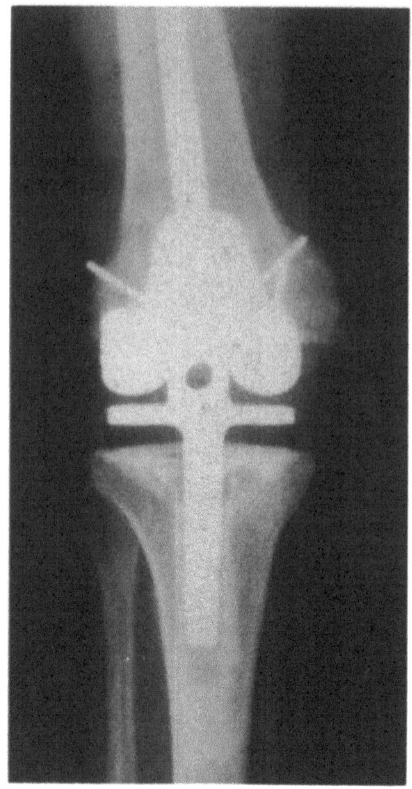

Fig. 22.3. a Radiograph of bilateral total knee prosthesis. **b** Post-operative radiograph.

modate absent ligaments. In all major losses where large pieces of graft are used the support of stems is indicated.

Many of these patients are indeed salvage cases requiring massive allografts – a whole distal femur or proximal tibia, or both. Another option is segmental prostheses but their track record is not good. Most of us at this time incline towards allografts because they do surprisingly well.

The major grafts need supportive prostheses with some constraint such as a Total Condylar III. Rotating hinges are used frequently and satisfy some of the criteria for stability, particularly in the elderly.

Case 5

This patient had bilateral total knees (Fig. 22.3a). One side was fine, the other not. One condyle gradually "washed-out", followed by the other side. A cavity appeared at the end of the femur which looked tumorous. When opened, the entire distal femur was one massive granuloma caused by plastic debris. This was treated using a "femur inside her femur". The medial side of the knee was so tenuous in terms of attachment of her collateral ligament that we resorted to a rotating hinge. She is doing very well with this but is a low demand patient (Fig. 22.3b). Failed hinges also provide fairly massive bone loss.

Case 6

When opened, this patient's knee was filled with "muddy sludge" (Fig. 22.4a). By the time we had curetted out the cement and all the sludge we were left with an "ice cream cone" in the femur and another in the tibia. There was considerable bone loss with some doubt about the histology. I therefore packed the cavities with antibiotic beads while we prepared an allograft. I then operated again and he now has a "femur inside his femur" and a "tibia inside his tibia". The grafts fit well and the patient is progressing extremely well (Fig. 22.4b). No change was seen on the last X-rays at two years (Fig. 22.4c).

The principles are cement into the allograft but augment the contact areas between the allografts and the host with bone paste, preferably autograft.

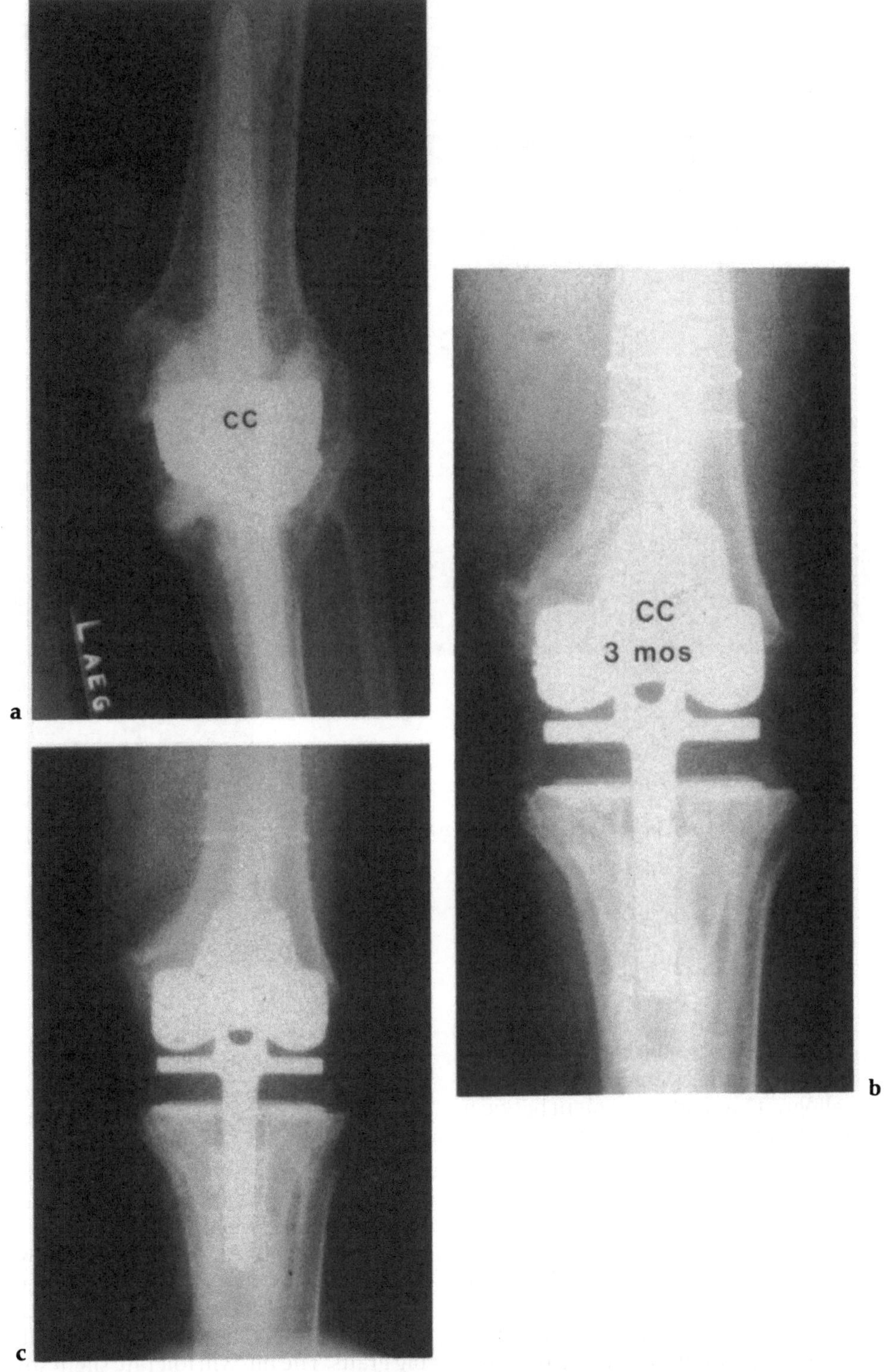

Fig. 22.4. a Pre-operative radiograph. **b** Post-operative radiograph at three months. **c** Post-operative radiograph at two years.

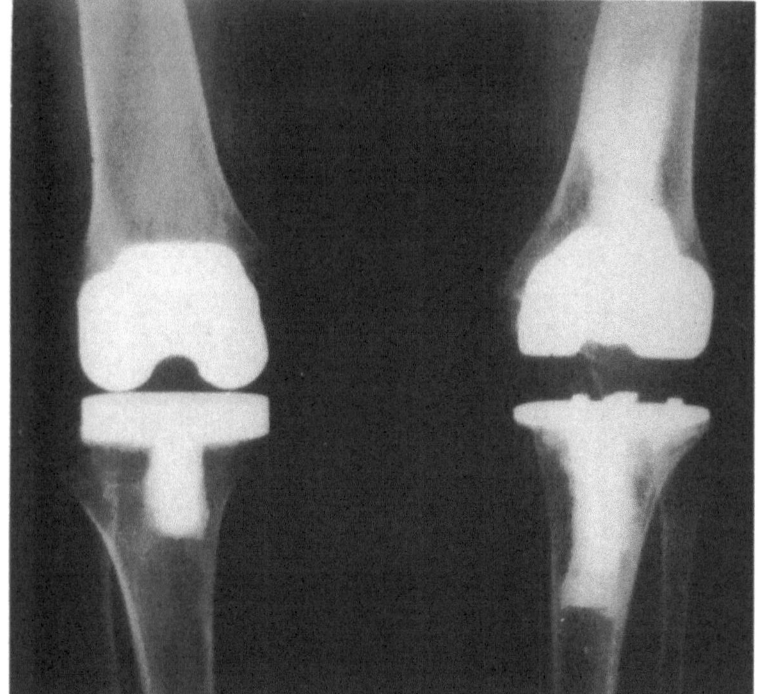

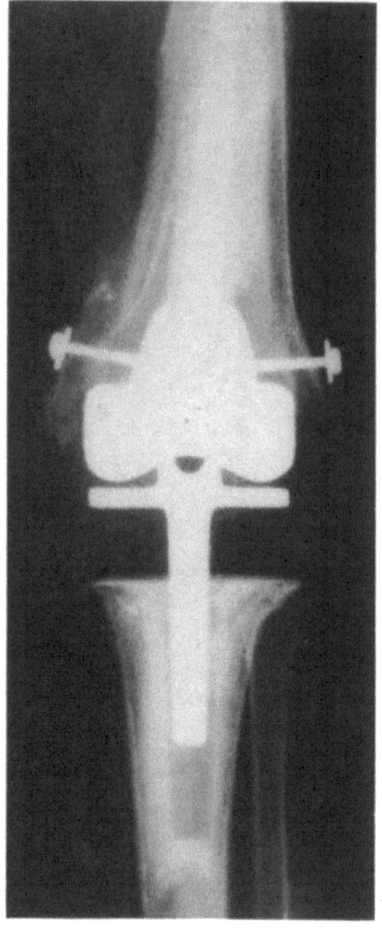

Fig. 22.5. a Radiograph showing left knee prosthesis infected. **b** Post-operative radiograph.

Case 7

The patient had had two revisions and became infected (Fig. 22.5a). This was revised after an interval with implanted antibiotic beads with a "femur inside the femur" cementing the prosthesis to the allograft but with no cement between the allograft and the host (Fig. 22.5b).

Case 8

This patient had a spherocentric prosthesis and was left with no cortical rim or even an "ice cream cone" but with a tremendously devastated distal femur (Fig. 22.6a). A "femur inside a femur" was used but with one exception. The patient had a femoral stem coming down from above so cortical struts were placed on the inside of the femur, bridging either side of the ends of both stems (Fig. 22.6b).

Conclusions

Techniques for these major or massive allografts are very demanding. If one does not use a constrained device such as a Total Condylar III or rotating hinge, a second operation may be necessary. In many revision circumstances with massive bone graft, and where the device is not constrained, any subsidence will require re-operation to achieve functional stability.

We are exploring a relatively new area and believe that the demands for "getting the mechanics right" are far greater in the knee than the hip because the hip can tolerate some subsidence of big grafts. The hip will tolerate slight subsidence of the prosthesis, but this is not so in the knee. If there is subsidence of one or other of the components with a non-constrained device, millimetres can render that knee functionally unstable. Patients have difficulty getting out of

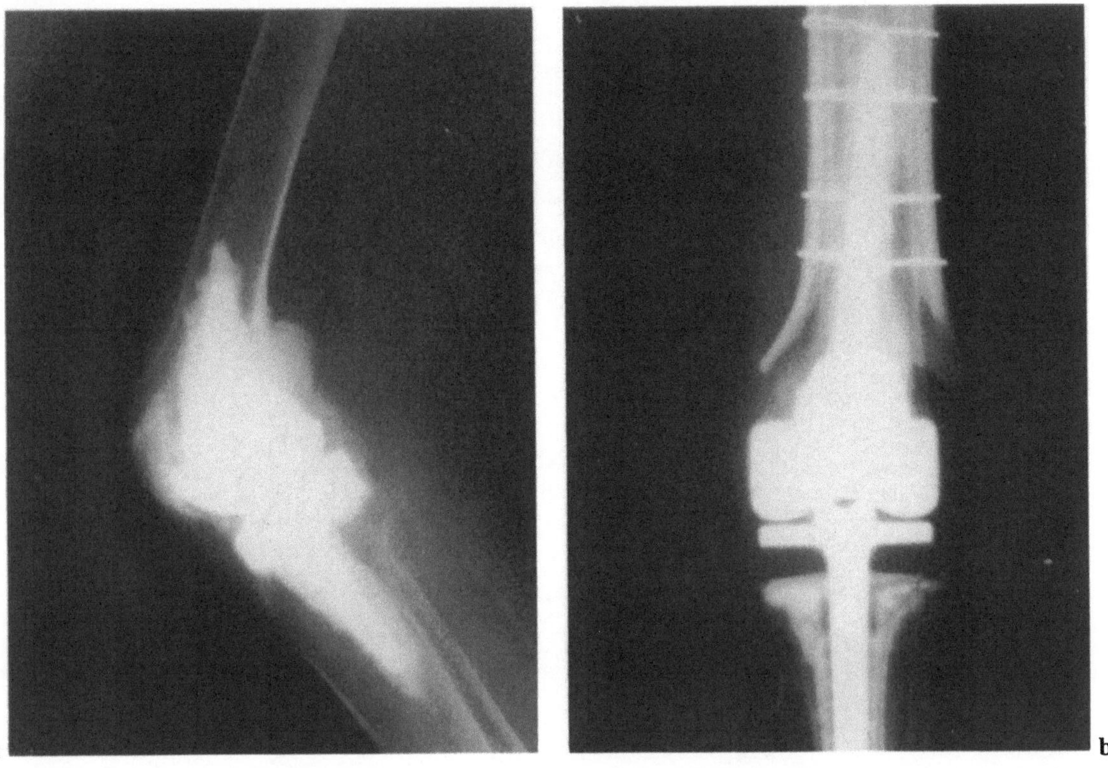

Fig. 22.6. a Pre-operative radiograph. **b** Post-operative radiograph.

chairs and going up and down stairs, and may need to be braced and use crutches or canes.

This is a new but very exciting field. So far, some of these reconstructions have been very gratifying for the patient. They are generally preferable to amputation, bearing in mind that for some of these patients arthrodesis is not even a consideration.

23 Structural Bone Grafting Technique in Total Knee Replacement Surgery

H.P. Chandler

There are several options when dealing with bone deficiency in total knee reconstructions. Cement used by itself is inadequate for structural weight-bearing defects that are more than 1 cm in depth. Cement, reinforced by screws, can be used to fill small defects in elderly patients. There is a definite place for wedges and revision components. Small contained defects can be filled with morsellised autograft or allograft bone, but for major bone loss I prefer structural weight-bearing bone grafts. The advantages of such grafts are that the bone architecture can be re-established and provides more bone stock for subsequent revisions, if these are necessary. If cement is used, the interface between the components and the graft is inert, at least initially.

Factors Affecting the Outcome of Structural Weight-Bearing Grafts

Sources of Graft Material

Autograft from the iliac crest or from local bone resection from the femur and tibia is preferred, but not enough for structural support. However, when morsellised, autografts can be used to fill contained defects. I have used femoral heads in the past for structural knee grafts, but the quality of such grafts may be poor. In recent years, I have more frequently used fresh-frozen tibial or femoral allografts harvested from healthy donors. Not only is the bone of better quality, but segments of the allograft bone corresponding to the defects in the host can be used and the anatomy of the host and the donor bone is similar.

Surgical Exposure

Wide exposure is necessary in these massive defects if major reconstruction is to be considered. I do not hesitate to perform osteotomy of the tibial tubercle to obtain this exposure if it is necessary.

Trabecular Orientation

In life, the trabeculae of the donor grafts were aligned to the forces to which the bone was subjected. If autografts or femoral heads are used as structural grafts, it is important that the trabeculae are realigned in the axis of weight-bearing forces. Allograft segments taken from ipsilateral knees are used to fill corresponding defects in the host bone. If the trabeculae are malaligned they will fracture and the graft will routinely fail.

Fixation of the Graft

It is important that the graft is completely supported by host bone. Care must be taken to have an accurate graft-host fit. With complex geometries, a methylmethacrylate mould of the defect is sometimes helpful in shaping the graft. With such complex grafts, a thin layer of finely morsellised autograft or allograft bone can be used at the graft-host interface to fill any small incongruities. With major segmental deficiencies, it is easier to change complex geometry to flat surfaces using an oscillating saw. In these circumstances, the distal graft-host junction should be at right angles to the weight-bearing axis to ensure that this junction sees only pure compression forces and not shear forces. With contained defects, the graft can occasionally be stabilised to the host bone by press fit alone, but in most circumstances additional fixation is necessary.

In the tibia, the graft should temporarily be fixed to the host bone by two or three 3/32 wires that are used in a retrograde fashion through the graft proximally and through the cortex of the host tibia distally. These wires are withdrawn distally enough so that they will not interfere with conventional proximal articular cuts using standard tibial jigs. One or two cancellous screws in each plateau are then countersunk through the graft and should engage the cortex of the more distal tibia. These lag screws should be roughly parallel to weight-bearing forces, and no threads should be in the graft so that it can impact against the host. It is important to orientate these screws in the graft in areas that will not compromise later fixation of the tibial component. It should be emphasised that screws bear no weight and are used only to hold the graft to the transverse buttress of the host bone.

Although I have successfully used PCA resurfacing-type tibial trays (Howmedica, Rutherford, NJ) in conjunction with large structural grafts, I now prefer a stemmed component with a cemented base plate but without cement used for the stem. The uncemented stem allows the graft to impact against the host bone but still provides additional resistance to bending or torsional stresses.

In the femur, grafts are also temporarily fixed to the host bone by retrograde wires until the articular surfaces can be cut by conventional jigs. Often the geometry of the femoral component contains the graft without further fixation, but if there is any question, cancellous countersunk screws can be used on the femur as well to secure the graft. As with the tibia, care should be taken to avoid placing screws in areas that might interfere with femoral pegs or stems. Because the geometry of the femoral component makes it more stable than the tibial tray, conventional resurfacing femoral components can more frequently be used, but with major grafts, a stem may be necessary. Cement is required less frequently on the femoral side than on the tibia if there is major contact with living bone. However, if cement is used, it is preferable to cement the articular surfaces but not the stem, so that the graft can be impacted against the host bone by weight-bearing forces.

Incorporation of the Graft

Cancellous bone (used exclusively at the distal femur and proximal tibia) incorporates by apposition of new bone on old trabeculae and therefore gets stronger if the graft trabeculae are orientated in line with weight-bearing forces (Springfield 1987). As with fracture healing, piezo-electric forces are probably important and therefore so is early weight-bearing (Bassett and Becker 1962). Cortical bone gets markedly weaker as it incorporates and, if the graft is large enough, the cortical portion should be protected with an intramedullary stem.

Case Reports

Case 1

This 75 year old man had a medial unicompartment replacement because of osteoarthritis. Three years later, he presented with a painful knee with recurrent varus deformity. X-rays showed a major tibial defect resulting from a loose tibial component (Fig. 23.1a,b).

Because of limited motion and a varus alignment, a tibial tubercle osteotomy was used for exposure. Although the defect was contained, the remaining medial rim was not strong enough to support the tibial component if morsellised bone had been used alone. The irregularities of the defect were smoothed with a high speed burr and then a methylmethacrylate mould of the defect was made. Sterile aluminium foil was first placed in the defect to prevent adhesions of the cement to the host bone (Fig. 23.1c). The methylmethacrylate mould was then used to shape the graft to the complex geometry of the defect (Fig. 23.1d). A very thin layer of finely morsellised autograft bone was placed in the defect to fill any small incongruities and then the graft was firmly impacted into the defect. A conventional tibial cutting jig was used to cut the graft and the opposite compartment, and then two countersunk screws were used to lag the graft to the host. A cemented PCA resurfacing component (Howmedica, Rutherford, NJ) was used. The femoral component was uncemented.

The patient used crutches for six weeks and then a cane for a month. At two years, nine months, the graft seemed to be doing well (Fig. 23.1e). At that follow-up visit, he had flexion to 120° no pain, and walked without a limp. It is now six years since reconstruction and he reports by telephone that he continues to function at the same level.

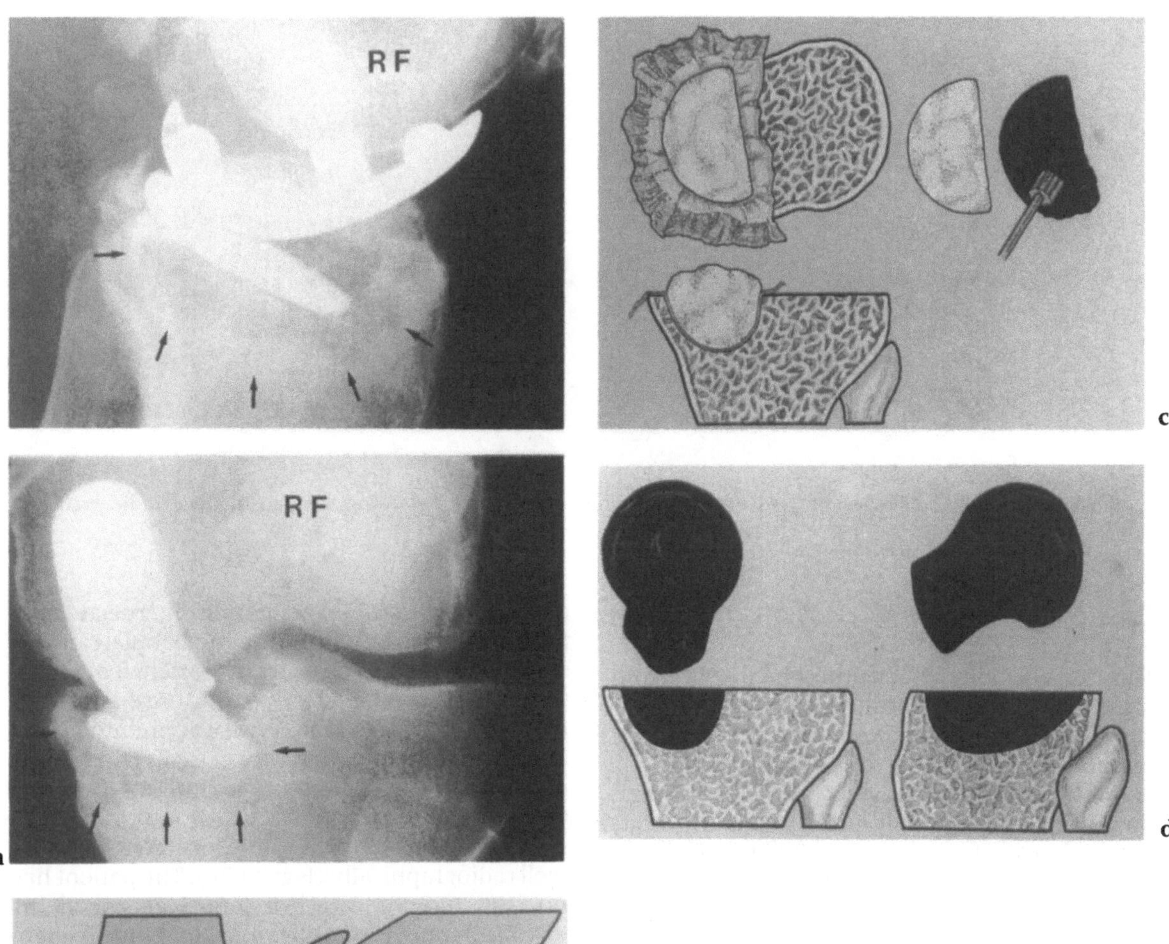

Fig. 23.1. **a** Radiographs showing a loose tibial component causing a major bone defect in the tibia. **b** Schematic representation of radiographs. **c** Sterile aluminium foil used to prevent adhesion between cement and host bone. **d** The graft shaped to the complex geometry of the bone defect using a methylmethacrylate mould.

Continued overleaf

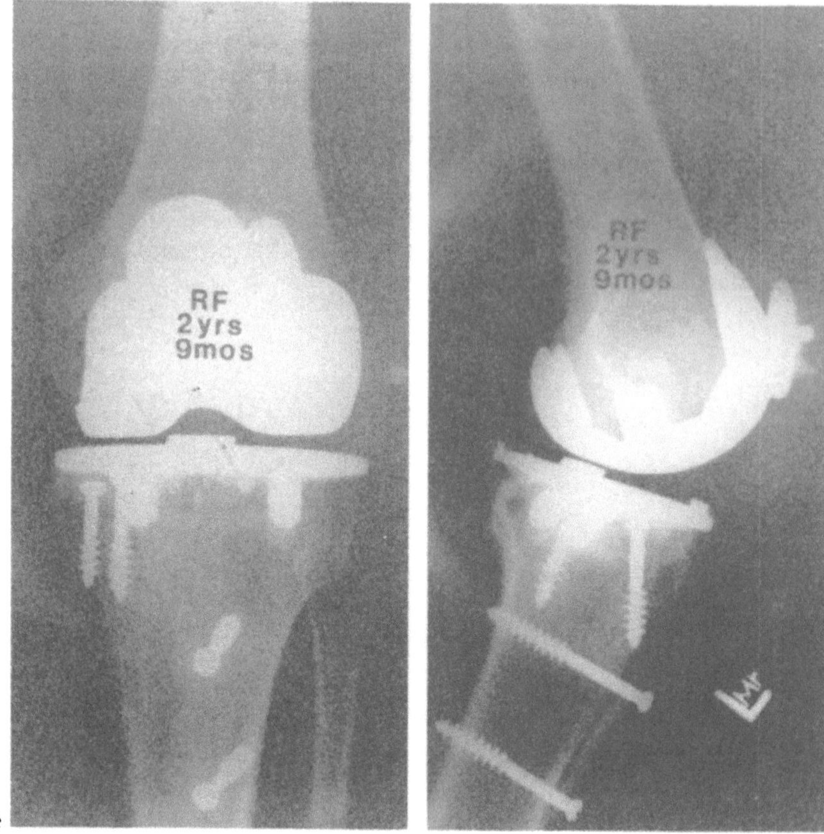

e

Fig. 23.1. (*continued*) **e** Radiographs two years, nine months post-operation showing good incorporation of the graft.

Case 2

This 62 year old female presented with complaints of pain and deformity of her left knee. Five years previously, she had a proximal tibial valgus osteotomy. Correction was excessive and the lateral tibial plateau was very thin (Fig. 23.2a). Two years later she fell and sustained a lateral tibial plateau fracture (Fig. 23.2b). This was treated non-operatively and healed in even more valgus (Fig. 23.2c). Clinically she was in 35° of valgus.

A tibial tubercle osteotomy was done to facilitate extensive exposure. A standard tibial cutting jig aligned appropriately to the shaft of the tibia was first used to remove the articular surface of the medial plateau. This jig was then moved distally in the same axis by 12mm and a disc of bone from the medial tibial plateau was removed. The jig was then moved distally 12mm further and the distal portion of the lateral plateau was cut in the same alignment. To simplify the carpentry, a vertical cut was made at the lateral portion of the tibial spine. The posterior cruciate was preserved. The disc from the medial plateau was shifted laterally to reconstruct the deficient lateral plateau and was held by two countersunk cancellous screws (Fig. 23.2d). An uncemented tibial and femoral component was used. The patient protected the leg with crutches for six weeks and then with a cane for one more month.

Three years later, the graft appeared to be doing well radiographically (Fig. 23.2e). The patient had no pain, used a cane for long walks, had full extension and 120° of flexion. Her clinical alignment was 10° of valgus. Follow-up is now four years and she continues to perform at the same level. Despite this satisfactory result, we would now use cement for fixation of the tibial component.

Case 3

This 69 year old woman with rheumatoid arthritis had her first total knee replacement 17 years previously. This failed at 16 years because of

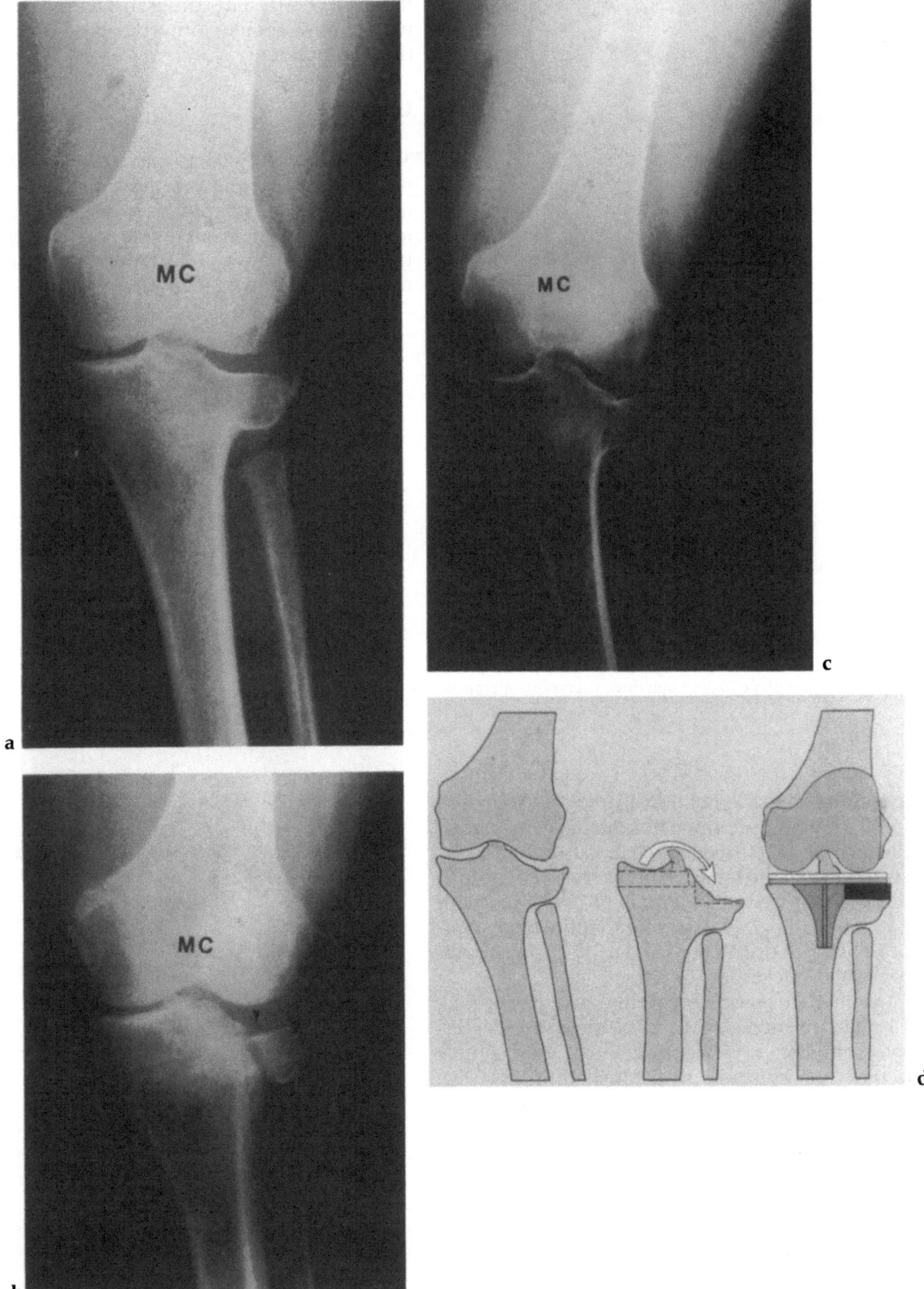

Fig. 23.2. **a** Radiograph showing a previous proximal tibial valgus osteotomy. **b** Radiograph showing a fracture of the lateral tibial plateau. **c** Radiograph showing fracture united but increased valgus deformity. **d** Schematic illustration demonstrating reconstruction of the lateral plateau.

Continued overleaf

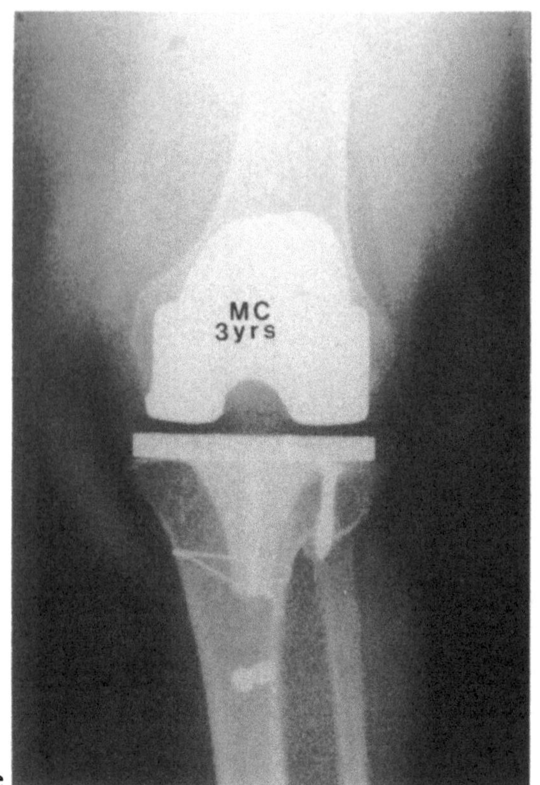

Fig. 23.2. (*continued*) **e** Radiograph three years post-operation.

loosening, and a femoral head allograft was used to fill a massive proximal tibial defect. The medial tibial cortex fractured two weeks after this reconstruction and the knee fell into varus. Ten months after this procedure, the patient presented with pain and radiographs showed that the tibial component was loose. There was an impressive proximal tibial defect (Fig. 23.3a).

A tibial tubercle osteotomy was performed. Because the medial cortex was thin and the defect was complex, it was converted to a more simple geometry by means of an oscillating saw (Fig. 23.3b). An ipsilateral allograft tibia of the same size served as the donor source. Four cancellous screws were used. The stem and base plate of a revision tibial component were cemented. The graft–host junction is hard to visualise on early post-operative X-rays (Fig. 23.3c). At follow-up of three years, three months, X-rays show the graft–host junction is still hard to visualise (Fig. 23.3d). At follow-up of four years, the patient denies pain and can walk without support or a limp around the house, but prefers to use a cane for long outside walks. She extends fully and flexes 100°. Despite this satisfactory result, we would now cement only the base plate of the tibial component.

Case 4

This 66 year old woman sustained a lateral tibial plateau fracture one year prior to admission. An open reduction was done. A deep *Staphylococcus aureus* infection ensued and despite antibiotic treatment, removal of hardware and debridement, she continued to drain copiously. At admis-

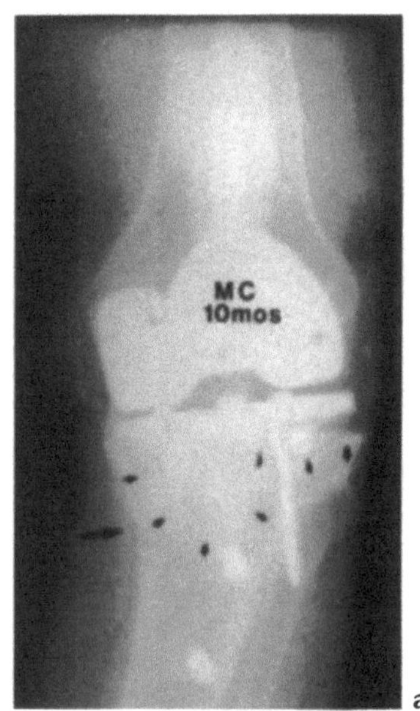

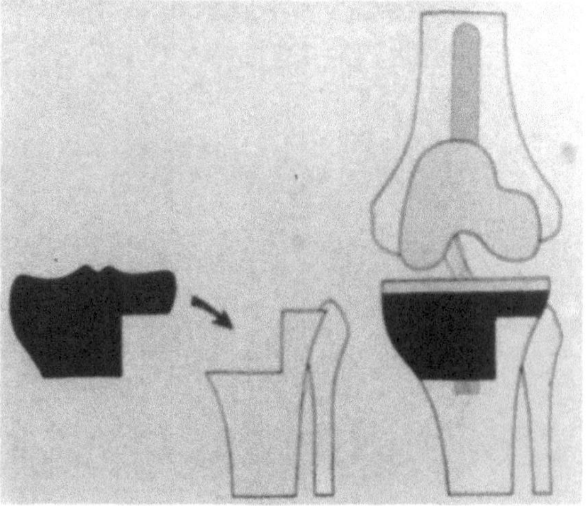

Fig. 23.3. a Radiograph showing an impressive proximal tibial defect. **b** Schematic illustration showing conversion from a complex defect to a more simple geometry.

continued

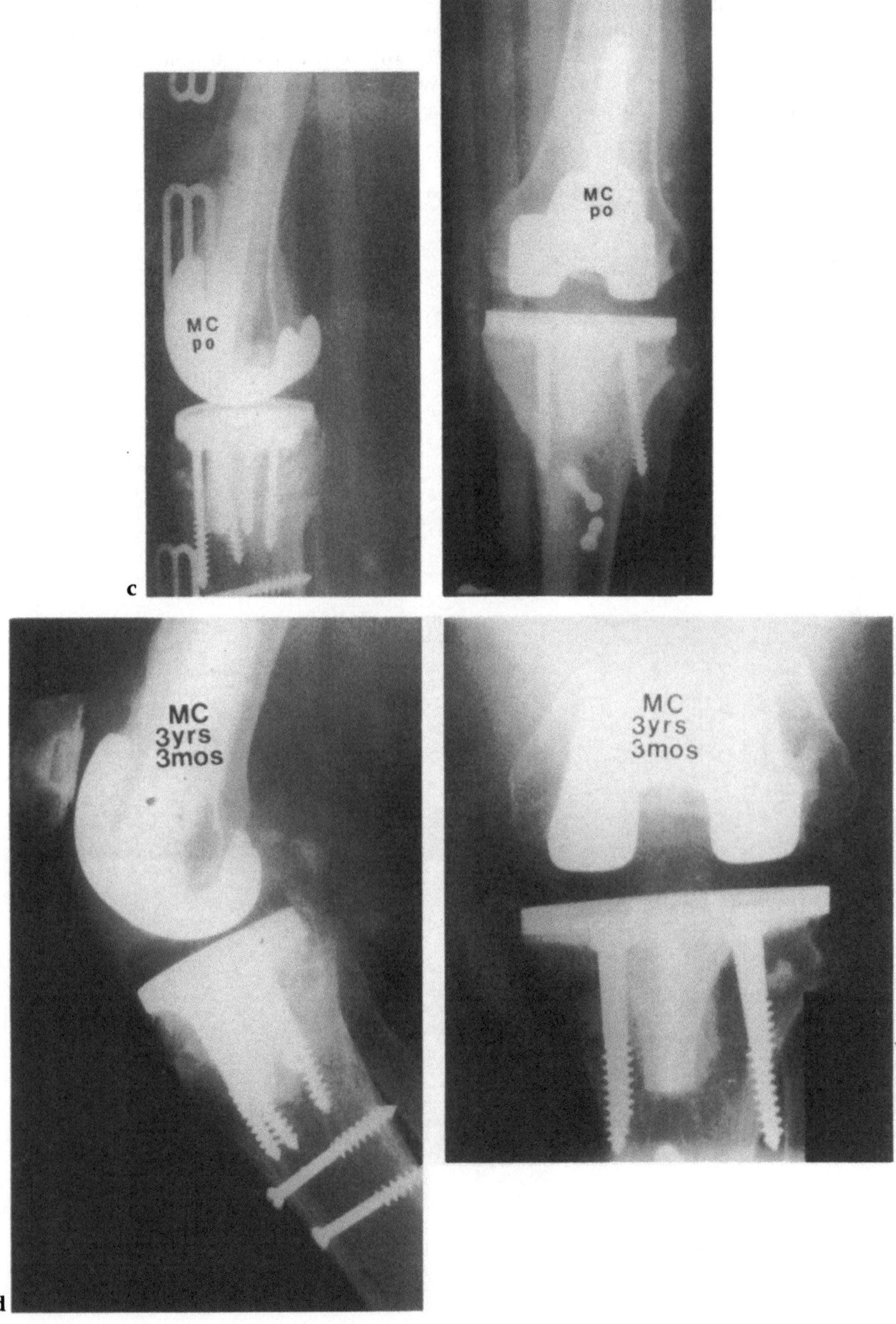

Fig. 23.3. (*continued*) **c** Early post-operative radiographs. **d** Radiographs at three years, three months post-operation.

sion, she had a fibrous ankylosis of her knee with a lateral tibial sinus. Radiographs showed a major lateral tibial plateau defect with osteomyelitis (Fig. 23.4a).

During the first procedure, a tibial tubercle osteotomy was used to gain safe exposure. An extensive debridement of the lateral plateau was performed and tobramycin-impregnated beads were used to fill the defect post-operatively (Fig. 23.4b). Appropriate antibiotics were used for six weeks and then the knee was reconstructed after a negative aspiration. The wound was benign.

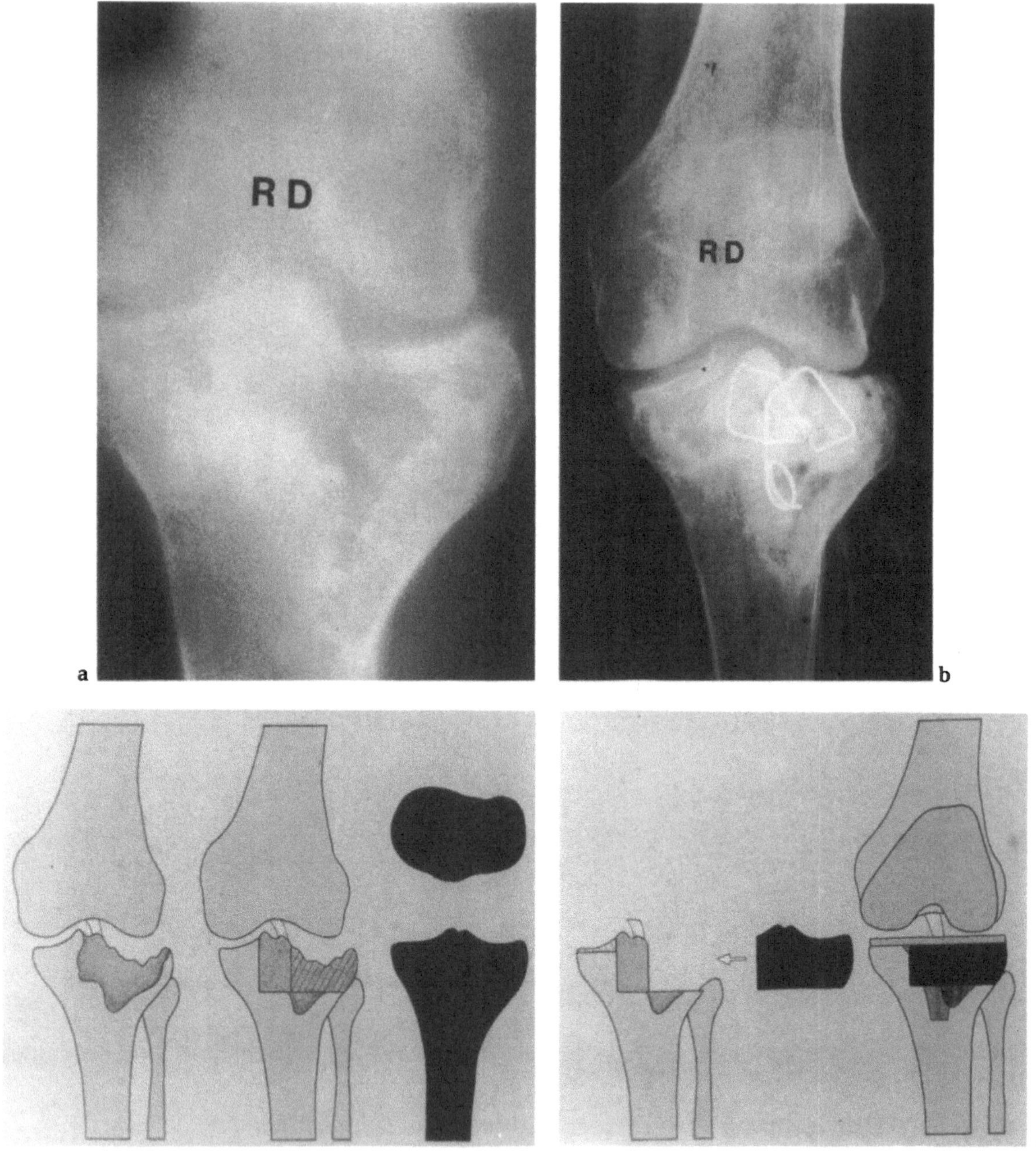

Fig. 23.4. a Radiograph showing osteomyelitis causing a major defect in the lateral tibial plateau. **b** Radiograph showing tobramycin-impregnated beads in bone defect after debridement. **c** Schematic diagram to show a complex bone defect converted to a more simple geometry. **d** Schematic diagram to show defect filled with a mixture of cancellous autograft and allograft.

continued

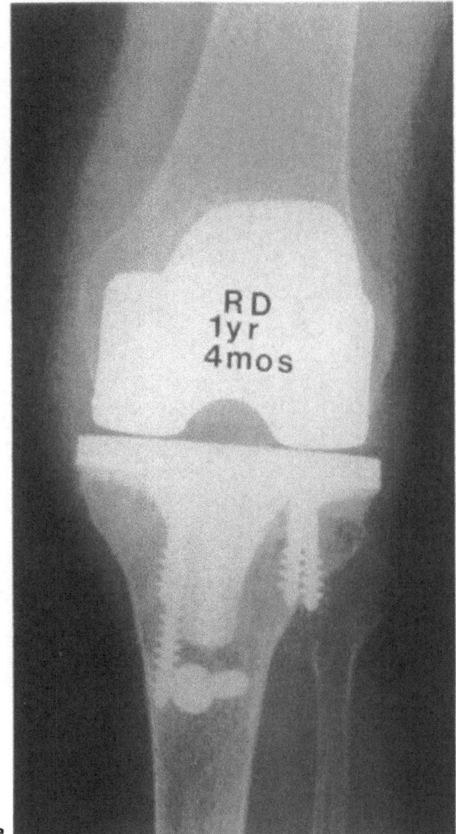

e

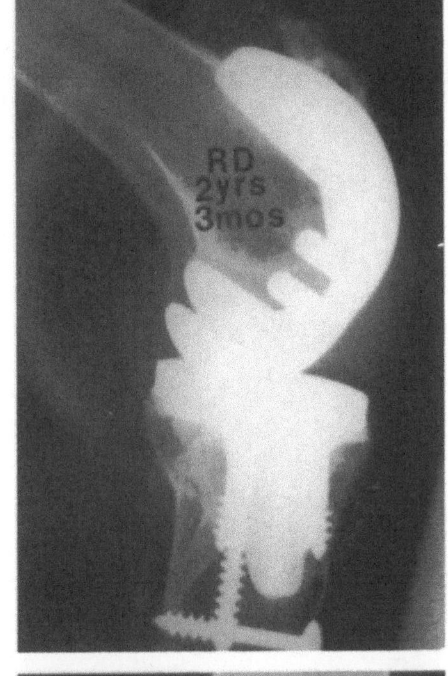

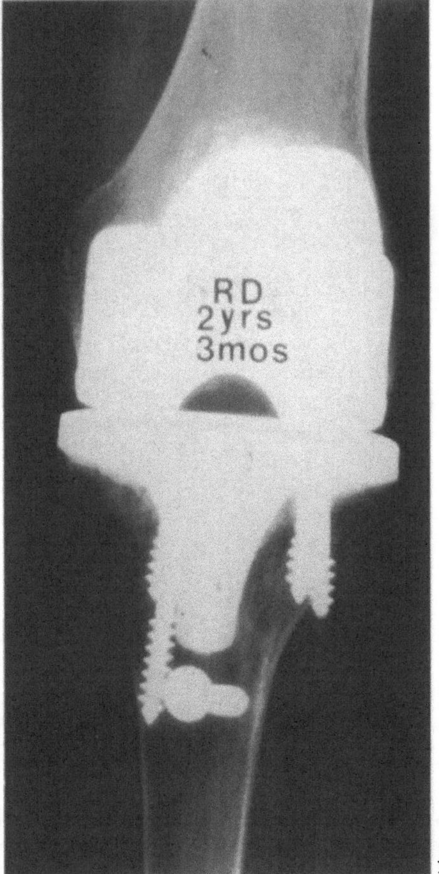

f

Fig. 23.4. (*continued*) **e** Radiograph one year; four months post-operation. **f** Radiographs two years, three months post-operation.

At the time of reconstruction, a tibial tubercle osteotomy was again used to facilitate exposure. The complex defect was converted to a more simple geometry with an oscillating saw (Fig. 23.4c). The posterior cruciate was preserved and remained attached to bone posteriorly. An ipsilateral allograft proximal tibia was used. The carpentry was quite simple because of the transverse cuts. Distally, a contained defect was left intact and was filled with a mixture of cancellous autograft and allograft (Fig. 23.4d). Four cancellous screws were used to hold the graft to the host. The tibial plate was cemented, but the stem of the component was used without cement. The femoral component was uncemented.

The patient used crutches for six weeks. On initial follow-up, she extended fully and had flexion to 95°. Radiographs looked good and she began to use a cane. Some time between that visit and her six month follow-up, she avulsed the

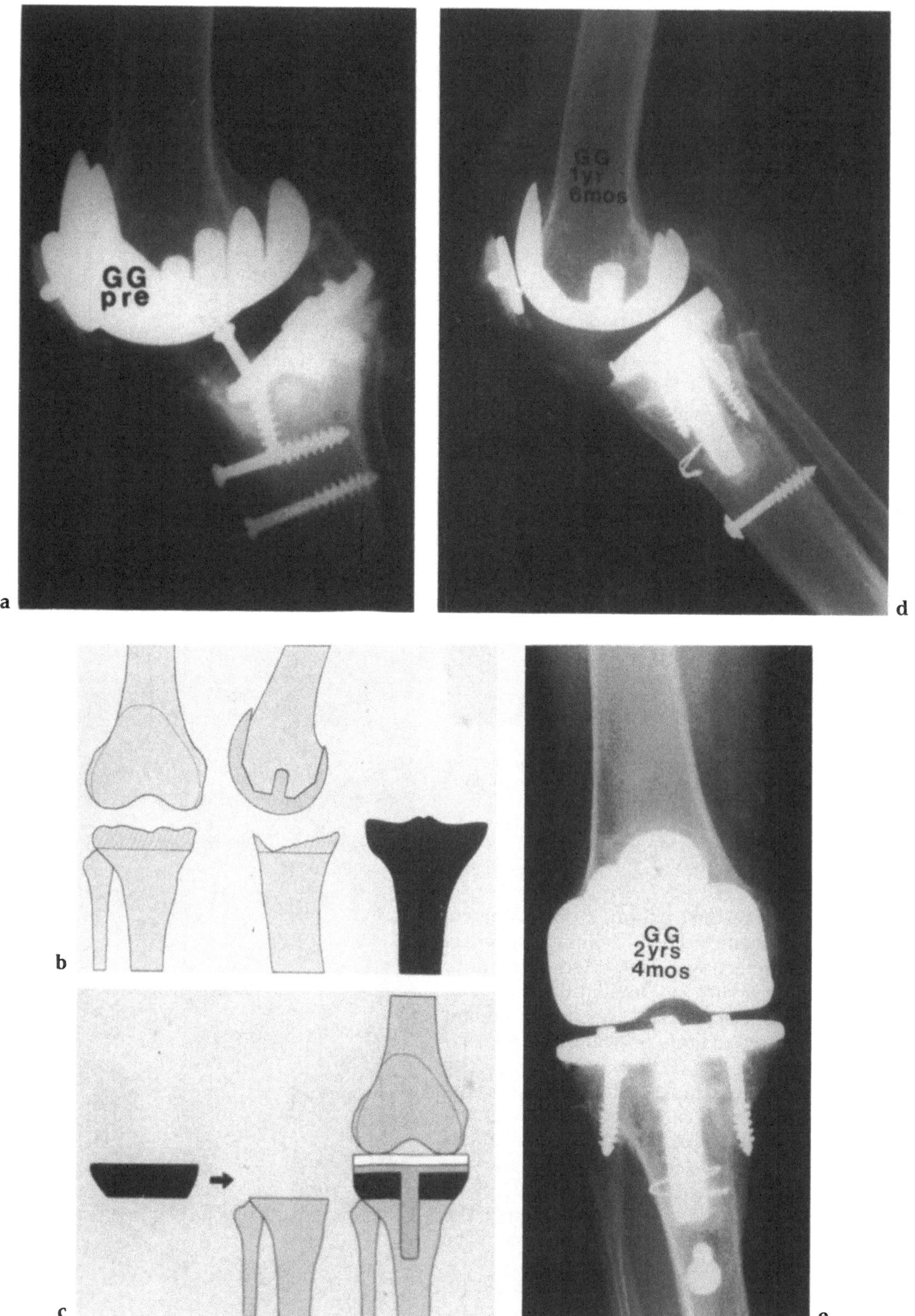

Fig. 23.5. **a** Radiograph showing a significant proximal tibial defect associated with a loose tibial component. **b** Schematic diagram to show a tibial tubercle osteotomy. **c** Schematic diagram showing an ipsilateral proximal tibia used as a donor graft. **d** Radiograph one year, six months post-operation. **e** Radiograph two years, four months post-operation.

patellar tendon from the tibial tubercle. The graft appeared to be doing well at one year, four months (Fig. 23.4e). At follow-up of two years, three months, she had full passive extension and 100° of flexion. However, she had a 40° active extension lag. She denied pain, walked without support around the house and used a cane for outside walking. She had no sign of infection and was not interested in having a procedure to reattach the patellar tendon to the tibia. Radiographs showed that the graft is doing well but the patella is high-riding (Fig. 23.4f). She continues to have the same functional level at three years, seven months.

Case 5

This 62 year old woman had the original diagnosis of traumatic arthritis. Her first total knee replacement was done in 1982. This failed because of

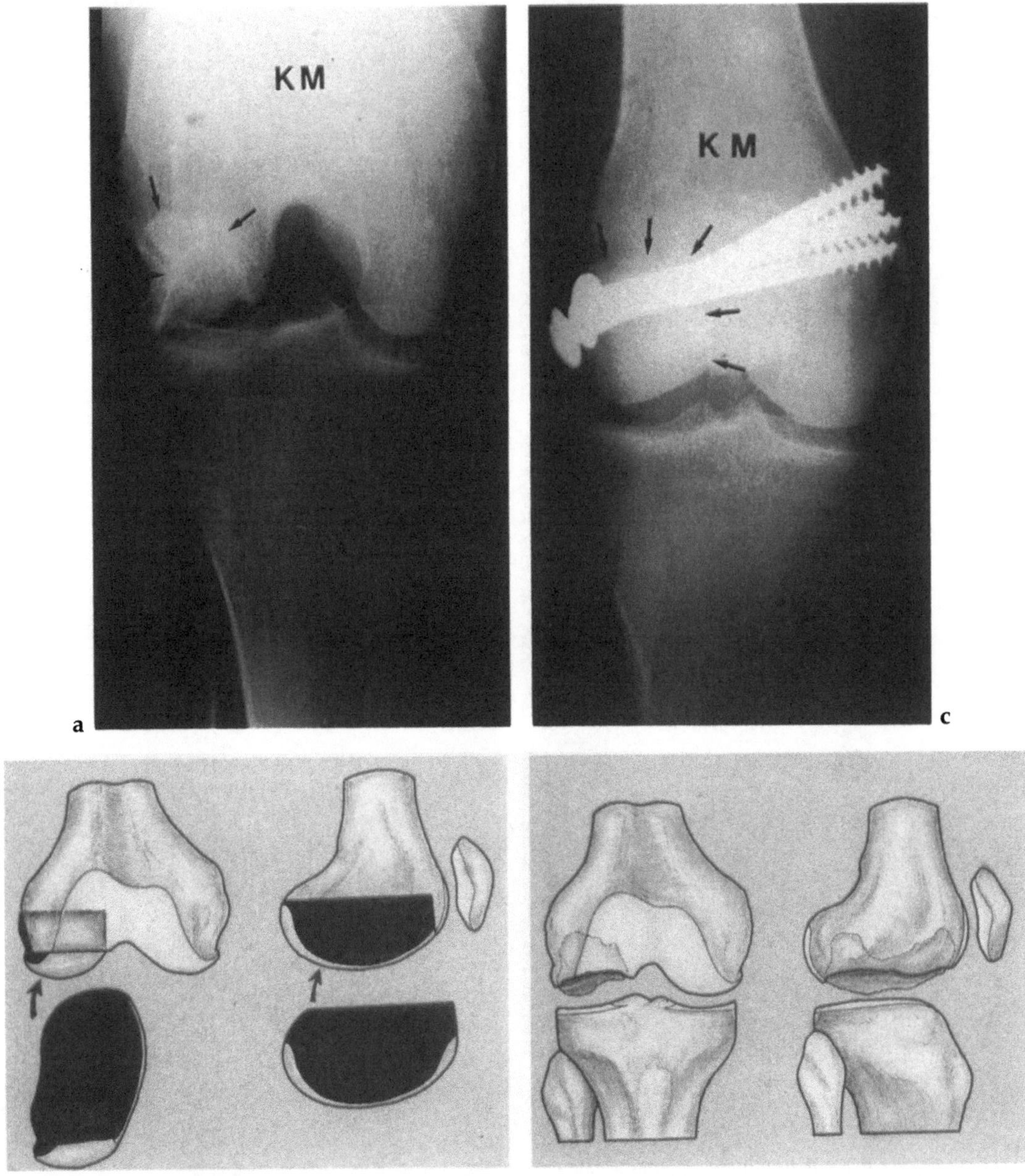

Fig. 23.6. a Radiograph showing major avascular necrosis of the lateral femoral condyle. **b** Schematic illustrations showing an ipsilateral allograft tibia used to replace bone defect. **c** Initial post-operative radiograph.

continued overleaf

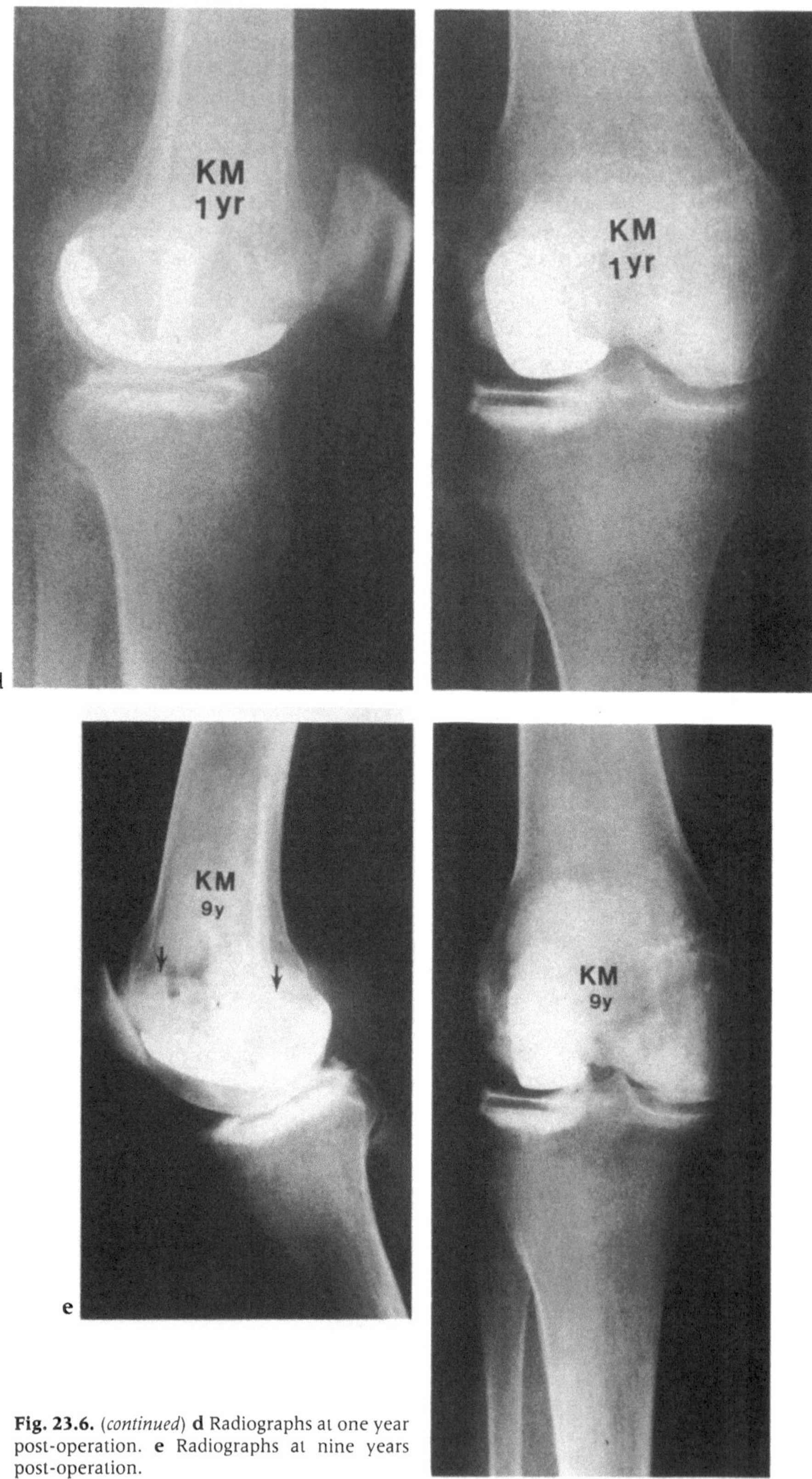

Fig. 23.6. (*continued*) **d** Radiographs at one year post-operation. **e** Radiographs at nine years post-operation.

tibial loosening and was revised one year later. Five years after that revision, she presented with pain and limited motion. Radiographs showed a loose tibial component with a significant proximal tibial defect (Fig. 23.5a). At surgery, a tibial tubercle osteotomy was used for exposure and the proximal tibia was cut back to viable bone using conventional tibial cutting jigs aligned with the shaft of the tibia (Fig. 23.5b). The defect that remained was over 3cm. The posterior cruciate was sacrificed. An ipsilateral proximal tibia of the same size was used as a donor graft. Again the carpentry was simple because of the flat surfaces (Fig. 23.5c). The femoral component was not changed and the base plate and the stem of the revision tibial component were both cemented. Radiographs at one year, six months, show the graft appears to have united (Fig. 23.5d). At follow-up of two years, four months, the tibial graft continued to do well (Fig. 23.5e). The patient had a 10° flexion deformity and further flexed to only 60°. Nevertheless, she walked without support and denied pain. Unfortunately, an alignment error was made and she is in 2° of true varus. Despite this, she remains satisfied with her knee and at last follow-up of three years, eight months, her function and X-rays were unchanged.

Case 6

This 17 year old woman was treated with massive doses of steroids for a collagen disease. She presented with complaints of lateral knee pain. Radiographs showed major avascular necrosis of the lateral femoral condyle extending almost to the epiphyseal line (Fig. 23.6a). Because of her young age, an allograft osteochondral graft from the ipsilateral allograft tibia of the same size was initially used (Fig. 23.6b). The patient did well for the first few months (Fig. 23.6c), but had pain at eight months and another physician removed the screws. At one year, she presented with increasing pain and X-rays showed complete loss of the articular cartilage but union of the graft. A cemented unicompartment replacement was used. Radiographs at one year (Fig. 23.6d) and again at nine years (Fig. 23.6e) look unchanged and she functions at a high level with close to normal motion. She does not use support.

Post-operative Management

If the trabeculae of the graft are properly orientated, and the graft–host junction has been fashioned so that it is transverse to compression weight-bearing forces, early weight-bearing is advantageous to stimulate healing with uncemented components. Twenty five kilograms of weight is encouraged for six weeks and then weight-bearing is allowed as tolerated, governed only by the patient's ability to walk without pain or limp. With stable, fully cemented components, full weight-bearing is encouraged whenever the patient can walk without discomfort. The presence of the graft and the status of its healing or incorporation does not influence the decision to bear weight.

References and Further Reading

Basset CAL, Becker RO (1962) Generation of electrical potentials by bone in response to mechanical stress. Science 137:1063

Springfield DS (1987) Massive autogenous bone grafts. Orthop Clin North Am 18 (4):249–256

24 The Use of Fresh Osteochondral Grafts for Traumatic Joint Defects

A.E. Gross

The orthopaedic transplant programme in Toronto is for traumatic defects, tumours, joint revision surgery and sports medicine. The technologies of the tumour and revision procedures closely parallel each other.

The osteochondral allografts for traumatic defects are done fresh and come from multiple organ donors. Our patients come from all over the world and are "on-call" for their operation. Those from Canada and the United States carry bleepers and are admitted within 24 hours of harvest. Patients coming from other continents wait in an apartment at the hospital or a hotel in Toronto. Most are operated on within 7–10 days and the longest wait has been three weeks. The harvest is done in the operating room after the solid organs have been taken.

Surgery is kept simple through an anterior midline approach. The defect is squared off and the graft inserted. The grafts should be at least 1cm thick so there are no stress fractures, and rigidly fixed. If the patient has a meniscus it is left; if reparable, we repair it; if they do not have one, a meniscal allograft is carried out. Survival of these grafts depends on the correct biomechanical environment.

The two most common clinical presentations are the valgus knee with an old lateral tibial type fracture and a medial femoral condyle defect. For the former, a distal femoral varus osteotomy and a lateral tibial plateau allograft are performed. The fracture bed must have healed, since operation is not possible on top of a fresh fracture (Fig. 24.1a,b).

A graft plus a valgus tibial osteotomy is performed for a medial femoral condyle defect. This is done at the same procedure because we are operating on opposite sides of the joint (Fig. 24.2a,b). If, however, we have to operate on the same side of the joint, the procedure has to be staged.

The post-operative care is very strict for these patients. They use a long leg brace for a year. Many are young patients still at school who wear baggy jeans or slacks over the brace and do not have to use crutches or canes.

Results

A series of 51 patients was reviewed at an average follow-up of 4.5 years. In the clinical and radiographic review, we looked at several parameters. The success rate was 76% in a very difficult

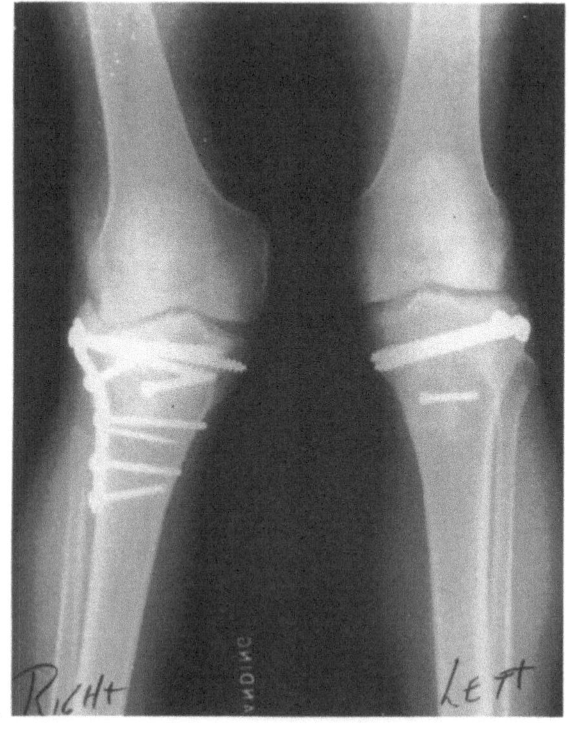

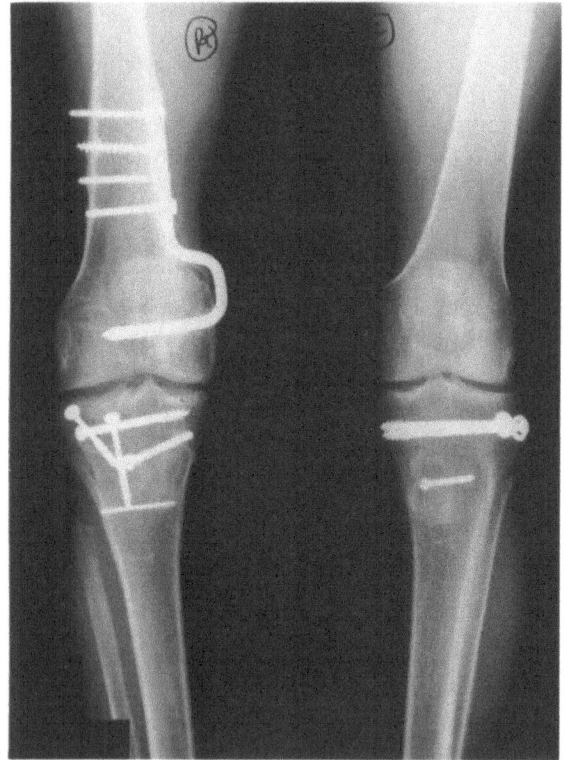

Fig. 24.1. a Radiograph of a 35 year old male with bilateral tibial plateau fractures. **b** Radiograph seven years after osteochondral fracture to replace lateral tibial plateau done in conjunction with a distal femoral varus osteotomy.

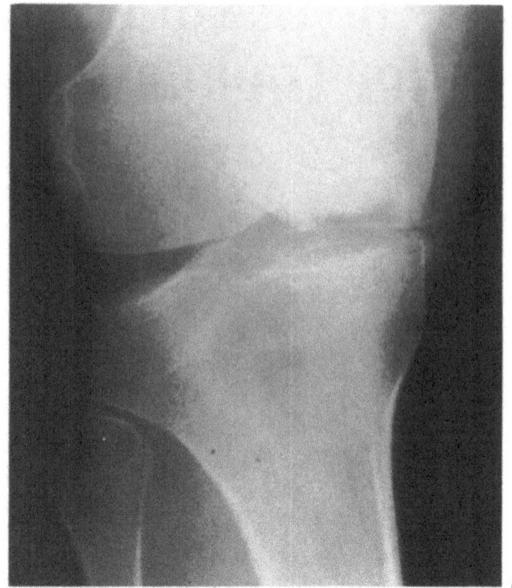

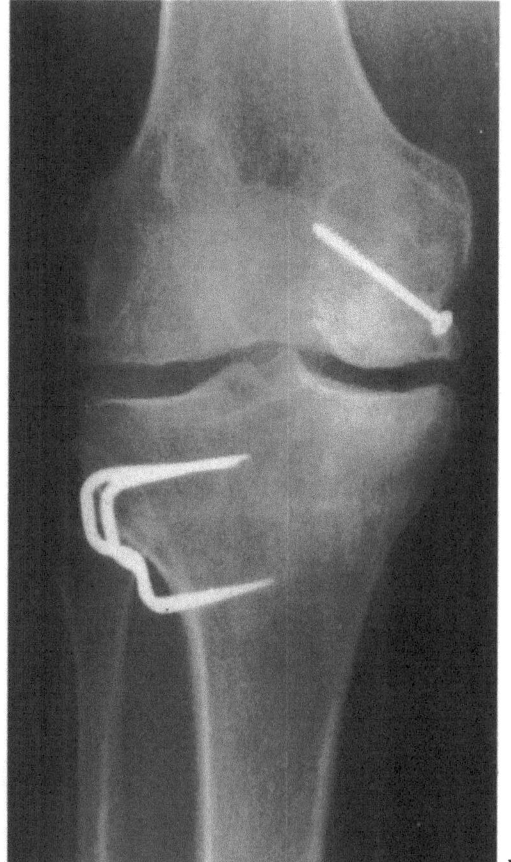

Fig. 24.2. a Pre-operative radiograph of a 30 year old female with traumatic osteonecrosis of medial femoral condyle. **b** Post-operative radiograph nine years following insertion of a medial femoral condyle allograft and a high tibial osteotomy.

patient population. There were 13 failures: four bipolar, eight lateral plateau and one medial plateau. Four of the 13 have been revised, the other nine are still functioning with their grafts in place.

All of the bipolar grafts failed, which indicates patients should be seen before they develop secondary changes in the contralateral plateau. The joint space was preserved in 30 patients. Ten had a decreased joint space and in 14 patients it was completely absent. Bony union is not a problem because cancellous bone is put against cancellous bone. Creeping substitution may take two or three years and cannot be avoided. Patients wear a brace but it is only necessary for one year.

Optimal alignment is achieved by over-correction of 2–3° so that the transplant is in a good biomechanical environment. With optimal alignment there was only one failure in 14 patients; with sub-optimal alignment there were six failures in 28 patients.

Complications in this series were minor compared to the revision and tumour series. Infection was not a problem because the patients were healthy. A prospective study of successful functioning grafts using radioactive isotope uptake techniques showed that chondrocytes survive for six years. Microscopy of failed grafts has shown chondrocyte viability up to nine and a half years.

The survivorship of 92 patients with bipolar grafts was 70% at 14 years. If patients pass the 10-year mark, survivorship is maintained even in very difficult knees which have had multiple previous operations. It is an operation for young people, who fare much better than older patients. There is no difference between medial and lateral or male and female.The diagnosis is very important for trauma patients do much better when compared with those with osteonecrosis or osteoarthritis. Osteochondritis dissecans is still an indication for this operation.

Conclusions

This study has shown that the best indication is the post-traumatic unipolar defect in young, highly motivated, compliant patients. Mechanical factors are now more important than immunology, but immunology may play a greater role the longer the life of these grafts.

Reference and Further Reading

Gross AE (1992) Use of fresh osteochondral allografts to replace traumatic joint defects. In: Czitrom AA, Gross AE (eds) Allografts in Orthopaedic Practice. Williams and Wilkins, Baltimore

25 Clinical Use of Allograft Bone in Tumour Surgery Around the Hip and Knee Joints

A.A. Czitrom

Historically the demand for allograft bone arose from the need to reconstruct skeletal defects in tumour surgery (Ottolenghi 1972, Parish 1973, Mankin et al. 1983, Gross et al. 1984). Today, with the expansion of hip and knee revision surgery, it is the tumour surgeons who learn how to do the reconstructions from the revision surgeons. The accumulated experience in revision surgery helps to elucidate the problems with the use of allograft bone in general.

The options for reconstruction in limb salvage are autogenous bone grafts, prosthetic devices, allogeneic bone grafts or a combination of all these. This chapter describes the combination of allogeneic bone grafts and prosthetic devices because this modality of reconstruction is applicable to hip and knee revision surgery.

Allograft reconstructions have been done in a variety of conditions – sarcoma, giant cell tumours and metastatic disease. Osteo-articular grafts were used in giant cell tumours, but these will not be described. Examples of reconstruction in sarcomas and metastatic bone disease using allograft–prosthetic composites will be discussed.

An important advantage of using allografts is that they can be attached to soft tissues. There is no stress shielding in allografts because these are dead pieces of material. If one has access to a bone bank, they are readily available and can be cut and custom-tailored. We know from clinical experience that allografts are very strong implants and they hold up with time if they are not revascularised or resorbed.

Meticulous fitting and rigid fixation of allografts is important. I use cement in the allograft but not the host and always autograft the junction between host and allograft (Czitrom 1992). The following case reports will illustrate the principles and techniques of allograft–implant reconstruction at the hip and knee.

Case Reports

Case 1

The patient had a supra-acetabular tumour (low grade chondrosarcoma). The pelvic side had to be reconstructed with a massive allograft (Fig. 25.1).

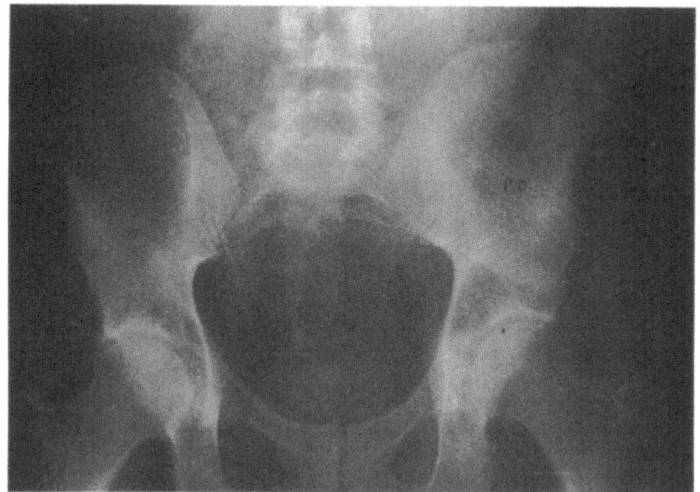

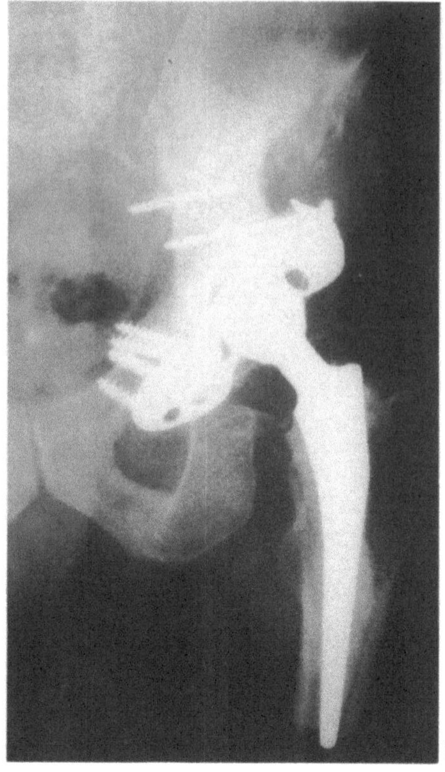

Fig. 25.1a,b. Allograft-prosthesis reconstruction for primary supra-acetabular sarcoma. **a** Pre-operative antero-posterior radiograph showing an acetabular chondrosarcoma in a 40 year old male. **b** Reconstruction at 7 year follow-up showing maintenance of the integrity of the pelvic allograft. (Reproduced with permission from Guest et al. (1990).)

The principle is exactly the same as in Type IIIC or B acetabular reconstructions in revision surgery. It is our longest follow-up to date of a massive allograft on the pelvic side (7 years). It proves that these allografts do hold up with time, but the complication rate is quite high in these acetabular reconstructions.

Case 2

A young patient had metastatic testicular carcinoma with a central dislocation of his hip. The graft was a fragment of proximal tibia which was fitted to serve as a support for the acetabulum in a transverse orientation (Fig. 25.2). In these cases the acetabulum is always supported by a reconstruction ring which goes on the allograft. The cup is cemented inside the ring.

Case 3

The patient was a 41 year old woman with a sarcoma destroying the femoral neck and head. The allograft-prosthetic composite played a major role replacing the tumour in the proximal femur of this patient. Fixation was by a long-stem femoral implant, step-cut osteotomy and cerclage wiring. The patient has done well and did not require any further surgery (Fig. 25.3).

Case 4

The patient was an elderly man with a metastatic fracture of the proximal femur (Fig. 25.4a). The reconstruction was done with plate fixation instead of cerclage wire and step-cut osteotomy (Fig. 25.4b,c). Initially he did well, but approximately six months after surgery developed a recurrence (Fig. 25.4d). The advantage of allografts is that the recurrence does not develop in the allograft but in the host bone. At re-operation the composite implant was removed and replaced with a new allograft-prosthesis (Fig. 25.4e). Rotational stability was achieved with a plate in this instance but cerclage wiring and a step-cut osteotomy is usually better to control rotation in these composite implants.

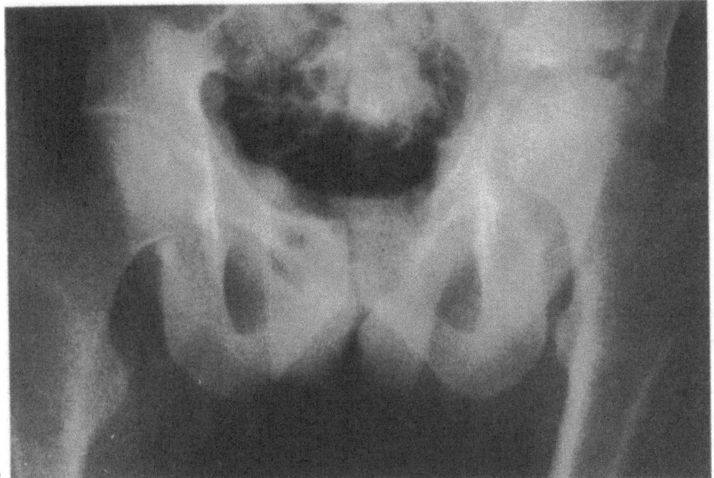

a

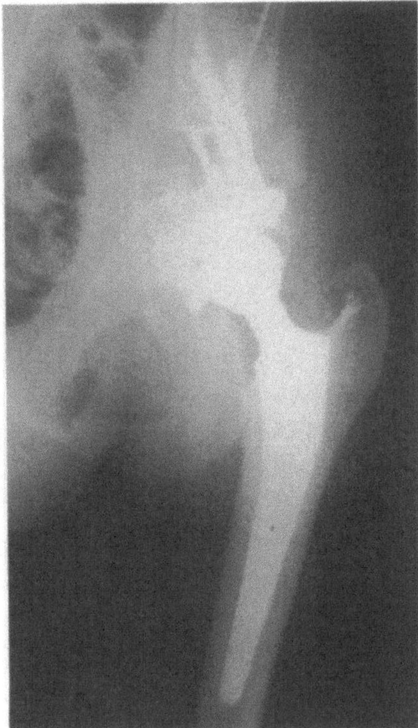

b

Fig. 25.2a,b. Allograft-prosthesis reconstruction of the pelvis in metastatic disease. **a** Antero-posterior radiograph of the pelvis showing central dislocation of the hip secondary to destruction by metastatic testicular carcinoma in a 25 year old male. **b** Post-operative radiograph showing the reconstruction with a transversely oriented tibial allograft strut, acetabular reconstruction ring and total hip arthroplasty.

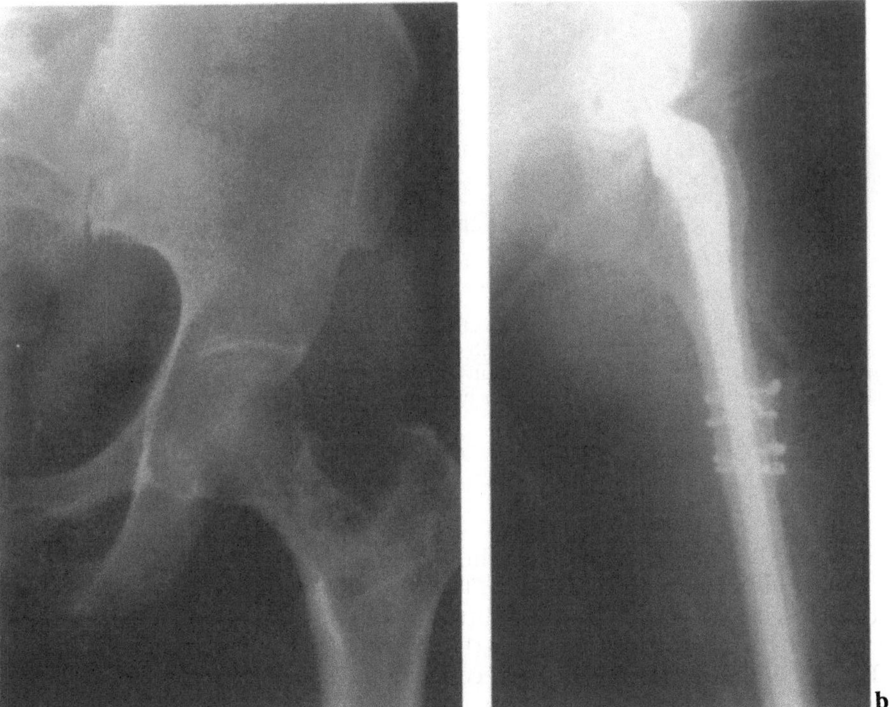

a b

Fig. 25.3a,b. Allograft-prosthesis reconstruction in the proximal femur for sarcoma. **a** Antero-posterior radiograph of the hip in a 41 year old female showing an angiosarcoma of the femoral neck. **b** Early post-operative radiograph after proximal femoral resection and reconstruction with allograft-prosthesis showing step-cut osteosynthesis fixed with long-stem femoral component and cerclage wires. (Reproduced with permission from: Czitrom (1992).)

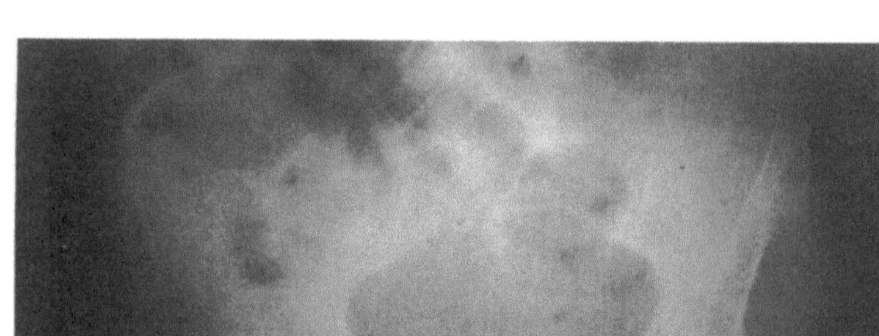

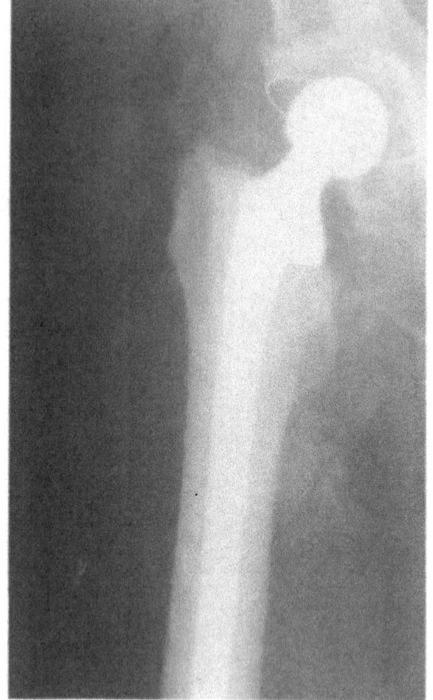

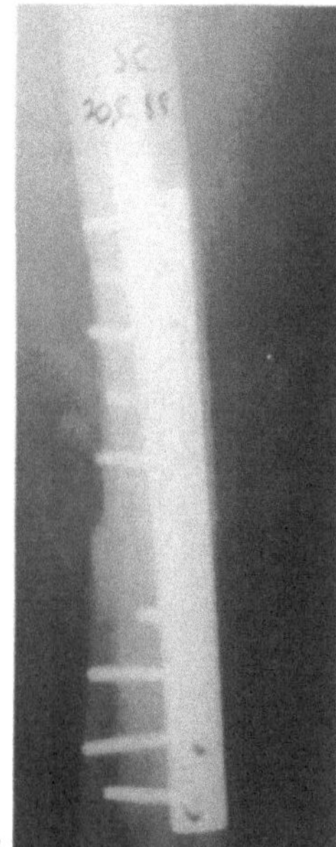

Fig. 25.4a–e. Allograft-prosthesis reconstruction in the proximal femur in metastatic disease. **a** Antero-posterior radiograph of pelvis showing pathological fracture of right proximal femur in an elderly man. **b** Post-operative radiograph showing reconstruction with allograft and total hip replacement. **c** A plate was used for osteosynthesis.

Case 5

The patient was an 18 year old girl with a chondrosarcoma of the distal femur (Fig. 25.5a,b). The reconstruction used a composite implant fitted exactly in order to accomplish some stability (Fig. 25.5c,d). The posterior capsule can only be reconstructed before the allograft is put in. The stem is the difficult technical part which in this case consisted of an intramedullary nail attached to the femoral component. Inserting this prosthesis is not an easy procedure and the composite implant in this case is not as stable as one would like. A better solution may be a custom prosthesis to replace the distal femur.

The distal femur is a controversial area for allografts at present. Although it can be done, I am uncertain that allografts in tumours of the distal femur are the ideal solution. The major deficiencies require massive grafts. An exact measurement of the distance being removed is vital to achieve the necessary tightness at the knee. Usually all the ligaments are gone. Either the ligaments attached to the allograft have to be used

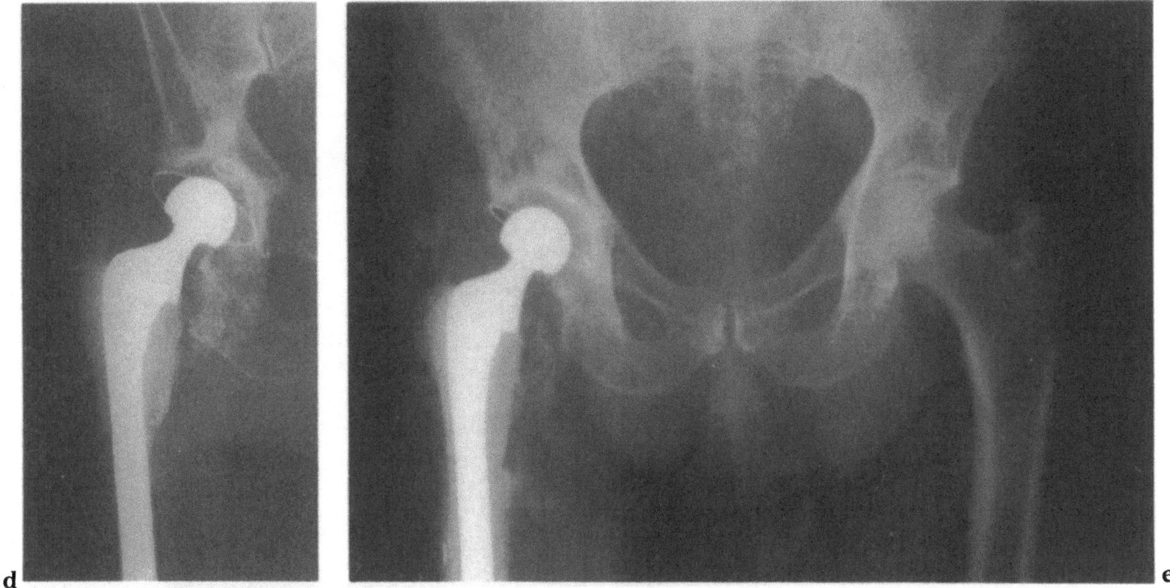

Fig. 25.4a–e. (*continued*) **d** Radiograph at six months showing recurrence with massive destruction of host bone and partial resorption of allograft. **e**. Post-operative radiograph of repeat reconstruction with new allograft-prosthesis.

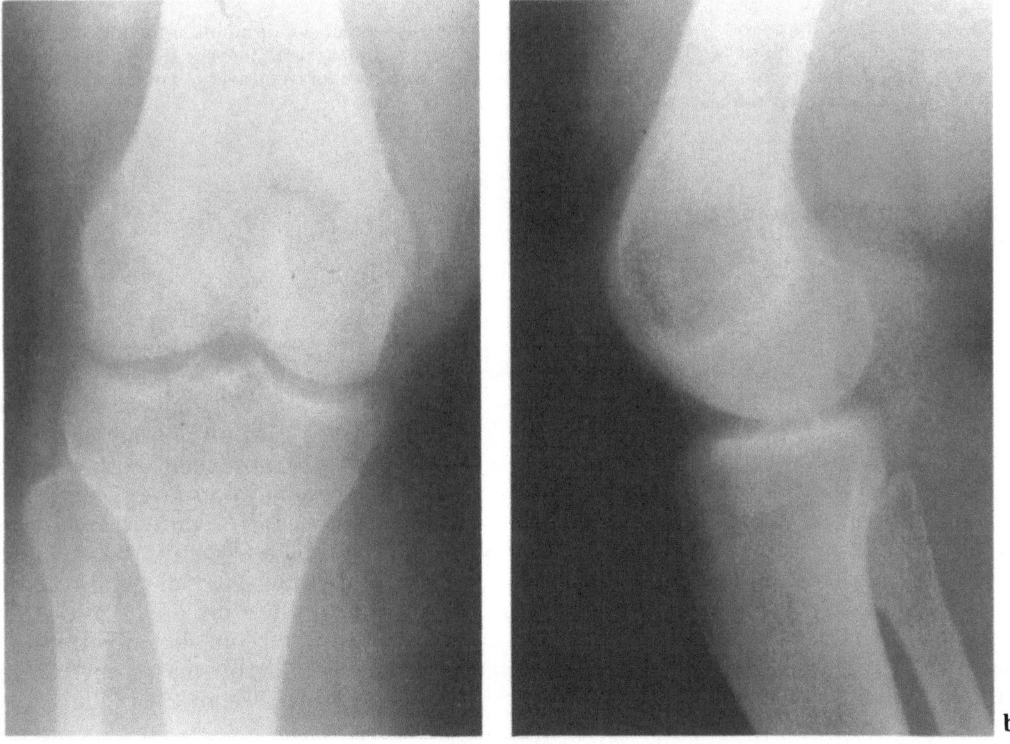

Fig. 25.5a–d. Allograft-prosthesis reconstruction in the distal femur. **a,b** Antero-posterior and lateral radiographs of the knee showing a chondrosarcoma of the distal femur in an 18 year old female.

continued overleaf

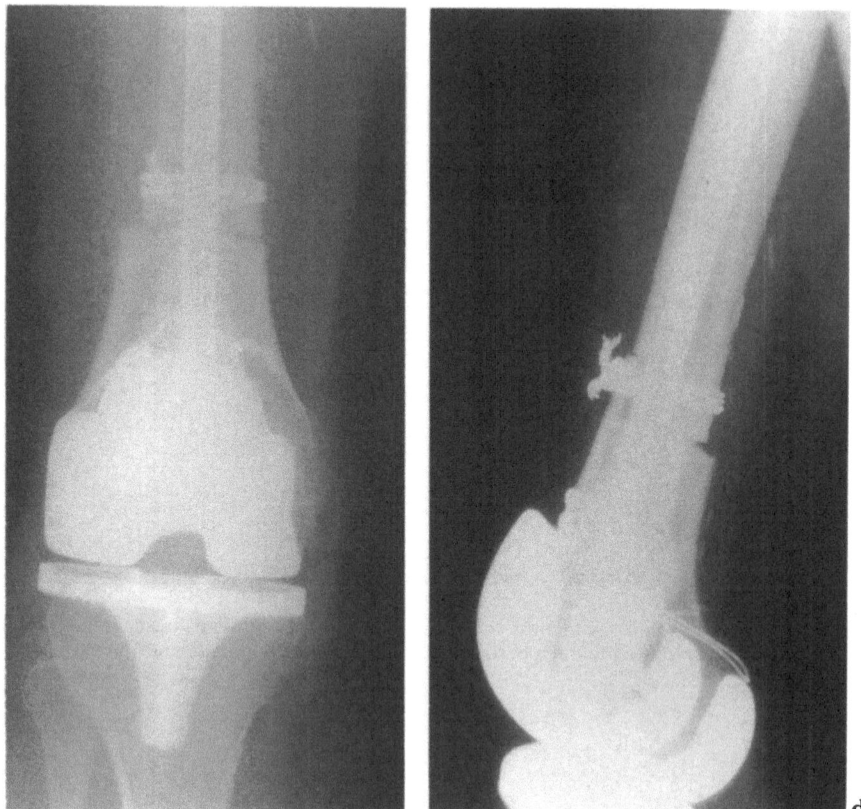

Fig. 25.5a–d. (*continued*) **c,d** Antero-posterior and lateral views of the distal femur at nine months showing reconstruction with unconstrained prosthesis combined with allograft and retrograde intramedullary nail (note the step-cut osteosynthesis with cerclage wires, the use of methylmethacrylate in the allograft and the wires used for reconstruction of the posterior capsule and ligaments). (Reproduced with permission from Czitrom (1992).)

to reattach them, or else some kind of constrained prosthesis.

Case 6

The patient was a 47 year-old man with a fibrosarcoma in the proximal tibia. There was no leg length discrepancy and the patient had a flexion deformity because of pain (Fig. 25.6a,b). The MR scan showed extensive tumour in the proximal tibia (Fig. 25.6c,d). This required a wide excision and reconstruction of the knee. A constrained knee with a long-stem combined with allograft was used. Fixation was by step-cut osteotomy and cerclage wiring (Fig. 25.6e,f).

In the proximal tibia, the allograft has a major advantage because you can reattach the patella tendon and the collateral ligaments. The attachment of the patella tendon is a "major plus"; it cannot be attached to a custom prosthesis.

Results

The Toronto experience includes 80 sarcoma patients until 1988 with a follow-up of two to ten years (Gross et al. 1984, Czitrom 1992, Bell et al. 1992 a,b). The infection rate was 9%; the fracture rate 12%; the non-union rate 7%. These are similar figures to those seen in other large tumour series. The complication rate is high since these are operations which require a great deal of dissection and time and many of these patients are on chemotherapy. Overall good and excellent results run at the 70% level using the rating system of the Musculo-skeletal Tumour Society (Enneking 1987). This makes these reconstructions worthwhile, despite the high risk of complication.

An early report includes ten patients with allograft implants of the pelvis for sarcomas with a follow-up of 25 months (Guest et al. 1990). One

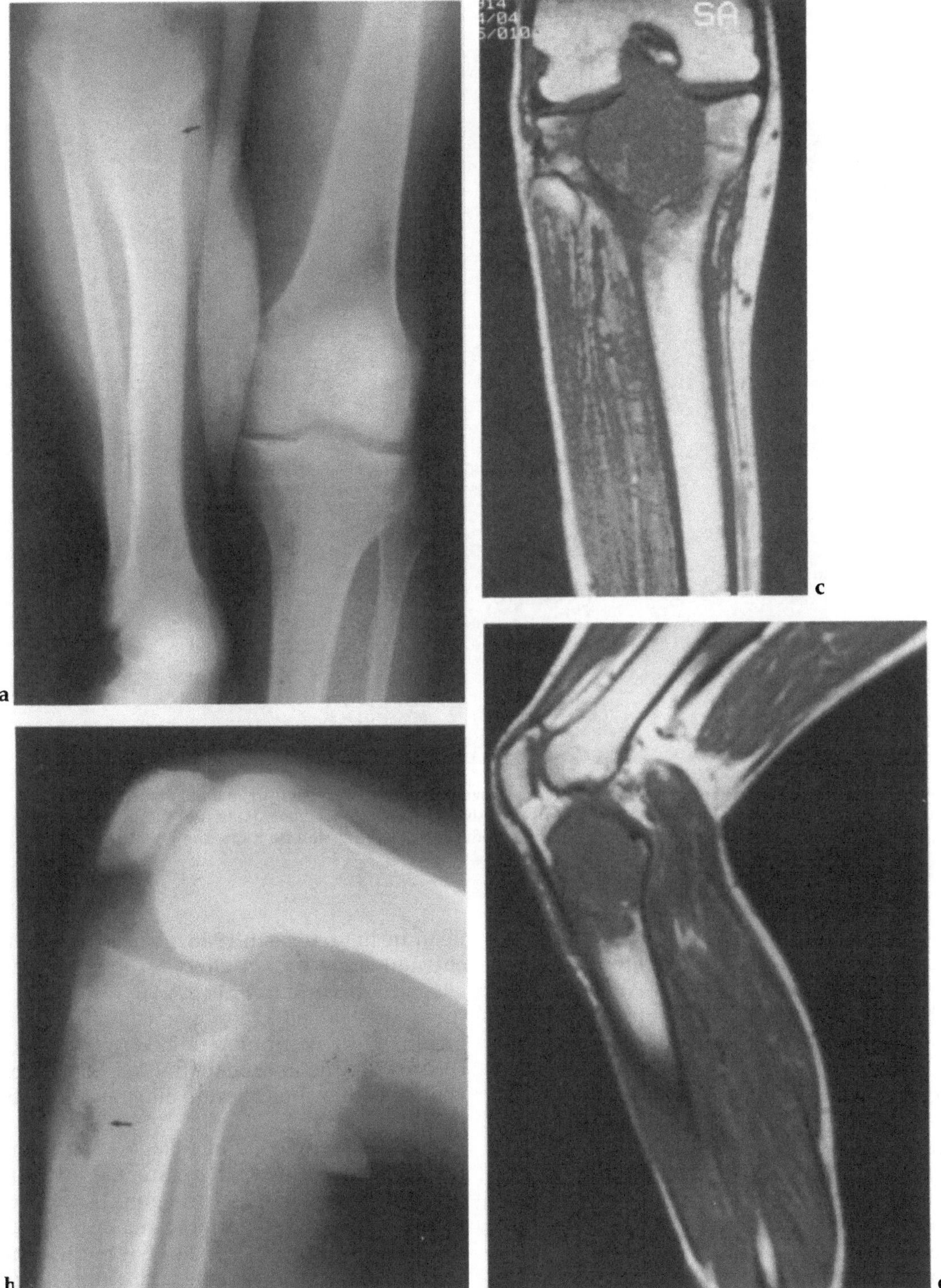

Fig. 25.6a–f. Allograft-prosthesis reconstruction of the proximal tibia. **a,b** Antero-posterior and lateral radiographs showing a fibrosarcoma of the proximal tibia (arrow) in a 47-year-old male (note the extreme flexion deformity of the right knee). **c,d** Magnetic resonance images in coronal and sagittal planes demonstrating intra-osseous extent of the tumour.

continued overleaf

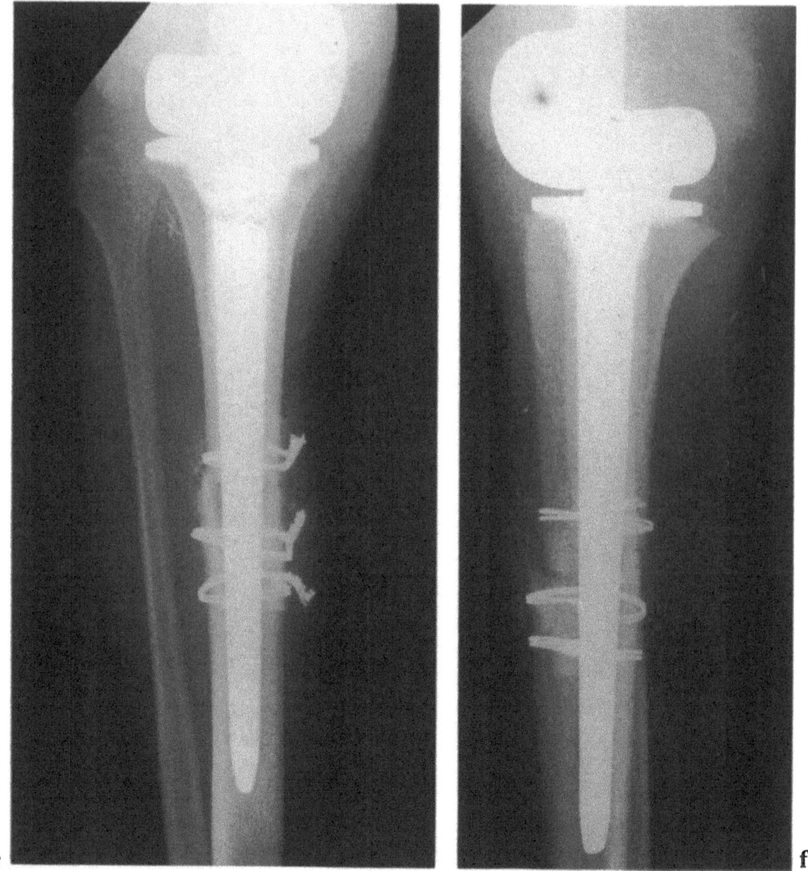

Fig. 25.6a–f. (*continued*) **e,f** Antero-posterior and lateral radiographs of the knee and proximal tibia taken at six weeks after reconstruction with allograft-prosthesis (note the osteosynthesis by methylmethacrylate fixation of the stem in the allograft, press fit in the host bone, step-cuts and cerclage wires). (Reproduced with permission from Czitrom (1992).)

patient died, three out of nine had poor function and seven out of nine had complications. Most were dislocations which could be corrected. This is still considered to be experimental surgery and it should be done with caution.

In metastatic disease the experience has been surprisingly good (Czitrom et al. 1991). In 52 patients treated between 1982 and 1989 the infection, fracture and non-union rates were all 2%. The follow-up is short because patients die early and do not have time to develop complications. The infection rate is low because the surgery is not as extensive as in a wide or radical tumour excision. These are generally intralesional curettages and marginal excisions. The aim is to achieve increased mobility, and the relief of pain. The results were 87% excellent or good for increased mobility and the relief of pain was accomplished in 94% of patients.

A review done in collaboration with the Rizzoli Institute in Bologna in 1988 investigated whether chemotherapy has an effect on the outcome of allografts (Czitrom et al. 1991). We found that there was no effect on the complication rate including the non-union rate. However, chemotherapy may delay time to union. Chemotherapy is not a contra-indication to allograft reconstruction.

Another study done in collaboration with the Rizzoli Institute compared autografts with allografts in segmental reconstructions (Huckell et al. 1990). It showed the same complication rate in the different procedures. Moreover the time to union was the same in large segmental autografts and allografts indicating that there is no clinical advantage in the use of autografts.

Future challenges in allograft implant surgery include better methods of fixation in the pelvis, early distal fixation of the stem of the proximal femur and the design of special prosthetic implants to be used in conjunction with the allografts.

References and Further Reading

Bell RS, Davis A, Allan DG, Langer F, Czitrom AA, Gross AE (1992a) Fresh osteochondral allografts for advanced giant cell tumours at the knee. J Arthroplasty

Bell RS, Davis A, Langer F, Czitrom AA, Gross AE (1992b) Reconstruction of primary malignant knee tumours using irradiated allograft bone. J Bone Joint Surg (Br) (in press)

Czitrom AA (1992) Allograft reconstruction after tumour surgery in the appendicular skeleton. In: Czitrom AA, Gross AE (eds) Allografts in Orthopaedic Practice. Williams and Wilkins, Baltimore

Czitrom AA, Langer F, Bell RS, Shahin AM (1991) Allograft reconstruction for bone metastases. In: Langlais F, Tomeno B (eds) Limb Salvage – Major Reconstructions in Oncologic and Nontumoral Conditions. Springer-Verlag, Berlin Heidelberg New York p733

Czitrom A, Capanna R, Donati D, Bacci G, Campanacci M (1991) Segmental allograft reconstruction concomitant with neoadjuvant chemotherapy. In: Langlais F, Tomeno B (eds) Limb Salvage – Major Reconstructions in Oncologic and Nontumoral Conditions. Springer-Verlag, Berlin Heidelberg New York, p95

Enneking WF (1987) Modification of the system for functional evaluation of surgical management of musculoskeletal tumours. In: Enneking WF (ed) Limb Salvage in Musculoskeletal Oncology. Churchill Livingstone, New York, p626

Gross AE, McKee N, Farine I, Czitrom A, Langer F (1984) Reconstruction of skeletal defects following en-bloc excision of bone tumours. In: Uhtoff HK (ed) Current Concepts of Diagnosis and Treatment of Bone and Soft Tissue Tumours. Springer-Verlag, Berlin Heidelberg New York, p163

Guest CB, Bell RS, Davis A, Langer F, Ling H, Gross AE, Czitrom A (1990) Allograft–implant composite reconstruction following periacetabular sarcoma resection. J Arthroplasty 5S:25–34

Huckell C, Czitrom AA, Capanna R, Langer F, Gross AE, Campanacci M (1990) Segmental reconstruction of bone defects after tumour resection – a comparative analysis of complications using autografts and allografts. J Bone Joint Surg (Br) 72:540

Mankin HJ, Doppelt S, Tomford W (1983) Clinical experience with allograft implantation. The first ten years. Clin Orthop 174:69–86

Ottolenghi CE (1972) Massive osteo and osteo-articular bone grafts. Technique and results of 62 cases. Clin Orthop 87:156–164

Parish FF (1973) Allograft replacement of all or part of the end of a long bone following excision of a tumour: Report of twenty-one cases. J Bone Joint Surg (Am) 55:1–22

Discussion: Knee Problems

Chairman: **Mr Older**

The Panel: **Dr Chandler**
Dr Czitrom
Dr Hedley

Mr Older: Do we use standard components in our revision knees, Dr Hedley? What do you see as the place of a hinged prosthesis? Do you think that modular prostheses should be used, or is there benefit to be gained by any form of custom prosthesis?

Dr Hedley: I think that as a general principle we would all agree that we should use as unconstrained a device as possible for the particular circumstances. In a lot of revisions it is possible to use the prosthesis that you are accustomed to using. I found that the advent of the PCA revision components helped enormously, because we were able to solve many of the easier revision problems and the result was a stable knee. That was the first prosthesis available that built up the posterior condyle and the distal femur to whatever extent was needed. That prosthesis enabled us to maintain the joint line, and put it back where it belonged, without adding any further constraint.

The next stage is stabilisers and a number of them are available. The Total Condylar III was probably one of the first available and it solved many problems. Where we might in the past have used a hinged prosthesis, we could now get by with an unlinked prosthesis. The hinged or linked prosthesis should really be used as a last resort. If you use a hinge it should be a rotating hinge, because its performance is a little better than some of the earlier ones. I think we all like to use as little constraint as possible and work towards a rotating hinge.

I think that custom components, by and large, will become less and less profitable. One of the reasons is that there are such a large number of prostheses available in hips and knees that one really should not have to look at custom components any more. It takes time to get custom components and there are many drawbacks. The price has become a major concern. With any custom component you have only one chance. If you are lucky, the custom component will come with an appropriate instrument but sometimes you are caught unawares and cannot use it. I think modularity will be on trial for the next five or ten years, allowing us to move from a standard prosthesis to a more constrained device by adding bits to the component.

Dr Chandler: It is interesting that the designs of knee implants are all converging and there is not a great deal of difference. They all have two pegs and the tibial components look very similar. Standard knee design is much better. The PCA was the first revision component that really solved the problem allowing adjustment to be made at the centre of rotation. It is better to add those changes on to another knee. I like the idea of modular knee components by which one can add wedges to a prosthesis and use it for revision surgery.

Modularity can keep the cost down and this is becoming a very important factor. I can see us in the USA very soon being told by the government that this generic knee or hip will be what people are going to get. It is so expensive to have such a wide variety. With big defects such as condyles however, I would rather use graft.

I am not sure that there is a place for the hinge any more but a stronger constraint than a rotating hinge may be possible.

Mr Older: Dr Czitrom, from your knowledge and experience, do you think that we should be using bone allograft to complement massive joint prostheses to replace tumours involving both the knee and the hip, rather than use enormous prostheses, sometimes virtually a whole femur?

Dr Czitrom: I think that massive prostheses are to be avoided whenever possible. In young people it is better to have a budget type of replacement, but that is not possible all the time. I have had experience with a modular system from Vienna. Its advantage is that it is cementless and the surface allows ingrowth that is so strong that even though there is breakage of screws and resorption or infection, it is very difficult to remove them because they are solidly fixed.

I think that some form of cementless fixation of the prosthetic component is most likely in the future with a porous surface or press fit. This applies especially in the distal femur. In the knee, it is a hinged prosthesis which allows some toggle. That has not been a problem. The axle has been broken a few times and has now been redesigned.

Dr Chandler: I think that we ought to also plan for its removal whenever we put in a component. It is very frightening to have a component that cannot be removed.

Dr Hedley: The next decade will, I think, be the decade of the granuloma. We shall pay the price for misadventures in design. We already see granuloma on the tibia and shall certainly see it in the acetabulum. There is little doubt that the main culprit will turn out to be plastic. There seems to be a synergistic reaction between plastic debris and certain kinds of metal particles. The synergist, of course, is massive osteolysis. The development of granuloma is making us re-think thin plastics.

Fortunately, diagnosis of knee problems is easier than the hip, because when a patient with a well performing knee starts to develop recurrent effusions, the polyethylene in the fluid can be aspirated. The arthroscope also allows us to look inside the knee. It is easy to identify when plastics wear through. Patients have pain, there is metal to metal contact and the knee makes a noise. With the hip, you have to rely almost exclusively on X-rays which may show metal erosions on the calcar. We already see erosions at the margin of acetabular components. In some devices with an incomplete porous coating that allows the passage of plastic particles to the distal parts of the stem, erosion is occurring at the tip of the stem.

Excessive plastic wear is significant. The patient should be made aware of it, and the frequency of follow-up increased.

A patient of mine had a granuloma that occurred with great rapidity six years post-operatively, which is very early. There was some erosion around the stem and it "galloped" over a period of 12 to 18 months. The cemented tie-back in which the polyethylene is held in position by moulding it into a sintered surface inside the cup had broken loose. The plastic was rotating on a roughened sintered surface which acted like a cheese grater and produced an enormous amount of debris. In this case, the femoral head had subsided into the margin of the acetabulum. This was difficult to discern on X-ray, and it is becoming increasingly important to tune in to the finer points on X-rays. It also emphasises the need for periodic X-ray review.

Mr Older: There was a time when we talked about "cement disease". Many of us now feel that we should be talking about particulate disease which is very real.

Dr Hedley: Charnley said that if you use a biological surface for an ingrowth fixation and there is complete ingrowth, then the passage of particles in and around that interface will be prevented and dissolution of bone will occur on the surface, that is, the calcar. But if ingrowth into the porous surface is incomplete, there is a passage for the ingress of particles and bone destruction will be worse than with cement.

Dr Chandler: Harris has shown lysis around a hip. In failed revision total knee replacements, polyethylene particles were wearing at an astronomic rate. The pieces are big because of the shear forces of the knee as opposed to the grinding forces of the hip. Big flakes of polyethylene come off the knee. You can go through a plastic tray very quickly and get down to the metal and then fine dust. You do not see as many massive destructions of bone around the cemented knee as you do with the hip. One reason may be because the particles are different. I do not think it makes any difference what the particles are; it is their size that matters.

There is a great deal of discussion now concerning the McKee-Farrar concept, an accurate metal-to-metal fitting. We are going to see a great many of them. The aero-engine and automobile industry have metal-to-metal parts which are put under unbelievable stresses for an incredible number of cycles. I took out a prosthesis two years ago that had been implanted for 20 years. I was very surprised at the small amount of metal. Maybe a perfectly fitting metal-to-metal prosthesis would be the solution.

Mr Older: The problem of metal upon high density polyethylene and the subject of particulate disease are very important issues. There is no doubt that, in Europe, people are looking at metal-to-metal problems very seriously. They are also looking at the whole subject of ceramic heads in polyethylene sockets to overcome the problem of particulate disease.

Dr Hedley: I think you need to differentiate stress-related osteopenias from true erosion. There is a lot of stress shielding in and around knees. The antero-distal part of the femur is typically stress shielding. I do not think it can be avoided. Granuloma has much in common with a bone tumour. It is not symmetrical; it has an endosteal margin depending on how rapidly the bone has been destroyed. The key is the location; you need to be suspicious if it is somewhere you do not usually see stress shielding. It is the rapidity with which it grows that is important. Whether it is associated with effusions, I do not know.

One option is to examine the synovial fluid, obtain cell counts and test for metal content. The problem is that the granulomas do not occur in the first year, but in patients with well functioning knees five or six years post-operatively.

Dr Dall: If you do not see a definite endosteal lesion on the lateral radiograph, can you reasonably assume that the cortical disappearance on the sides is stress yielding?

Dr Hedley: If it is symmetrical, uniform and does not have a margin, that is reasonable. There is one small caveat. If you are looking for polyethylene in synovial fluid, the temptation is to take some of the fluid and spin it in the centrifuge and look at the sediment. You need to look at the top end, not the bottom!

Dr Learmonth: Do you think that the interface contact congruence and the line or point contact that you get in the unconstrained posterior cruciate-retaining prosthesis in the medium term will be a significant factor in the formation of granuloma?

Dr Hedley: I do. The more anatomical the prosthesis becomes, the more point contact. All we have done in trying to be anatomical is to create a new monster. Every solution has a new problem! The theory behind the meniscal knee is great and you must try to make that work.

Mr Tasker: If you look at some of the hip prostheses that are taken out and examine the head, you may find a tiny little scratch which can only be picked up with a microscope. That acts like a cheese grater. When we are putting in the prosthesis, we may damage the head of the femur or the femoral component of the knee prosthesis. We may put in a couple of scratches and eventually it starts to wear, and then it accelerates. I agree that it can produce very fine particulate matter, and once it starts to go, the granulomas form. These are giant cells and act like a tumour which eats the bone.

Dr Hedley: There can be a great deal of osteolysis with mononuclear cells without having giant cells. There can also be considerable osteolytic enzyme production. It is all provoked by particle size and the wear surface is important. I never touch the head, I put a cloth over it and tap it on. It is an interesting point because sometimes in revision surgery, you look at an X-ray and make a decision as to whether to revise an acetabulum. You often see a well fixed femoral component and decide to leave it and just deal with the socket. You need to look very hard, because if the socket is loose enough and some acrylic has fallen into the acetabulum and the head is scratched, you are obliged to revise that.

Mr Tasker: Maybe it is caused by tiny bits of cement getting in between and producing the first small scratch?

Dr Loty: I agree absolutely, and I would like to ask your opinion about the hinged prosthesis of the knee. We do not use them much now. Do you really believe that the rotation would be beneficial or will it not be a reason for more wear? I have often been questioned about the fact that our patients who have a rotational knee always walk in only the external position. What is your opinion about this?

Dr Hedley: I would agree. The potential for wear is obviously high. If you have a knee such as the Total Condylar III which has a big central eminence held in position by the housing, there will be considerable wear on that, too. I try always to restrict the use of hinges to elderly people whose activity is low. If you put that type of device in and the patient is active on a cycle, you will see problems. I think you will see problems with the other kind, too. Fortunately, the majority of the patients are elderly, and it is really end-stage salvage situation in old patients.

Dr Chandler: I agree that often when you solve a problem you create another one. If you use a true hinge you will probably have less rotation and less polyethylene wear, but you will surely loosen the stem because it is too constrained.

Mr McLardy-Smith: I have some interest in meniscal bearings. It certainly seems to be one potential solution for the wear problems in knee replacements. The wear in a truly congruous meniscal bearing is very low indeed. In clinical studies it has been shown to be about 0.025mm/year in normal usage. I am very happy to put in a bearing which is 3.5mm thick at its thinnest point, and I am confident that because it is congruous, we will not have any problems with it wearing through to expose metal on metal. There are other problems with meniscal bearings.

Dr Delloye: Have you any experience of treating the granuloma with indomethacin?

Dr Hedley: No, I think I would rather get to the root of the problem and change the plastic, but I understand the problem.

Mr Older: Do you use indomethacin, Dr Chandler?

Dr Chandler: I have no experience of its use for this purpose.

Mr Older: Dr Czitrom, do you have any comments on the use of Indomethacin?

Dr Czitrom: There was a paper from Finland on the histology of granuloma versus normal aseptic loosening in which several cases were compared and it clearly showed that there are different cell types involved.

Dr Dall: If you are putting in a large allograft, how do you size it pre-operatively? Secondly, could we have some discussion on the preference for the kind of insertional graft to which Dr Hedley referred – the femur in the femur or the tibia in the tibia – versus the end-on graft that Dr Chandler seems to favour?

Dr Chandler: Our bone bank now X-rays all grafts to determine the size. In reply to the second question, I think that both have an application.

Dr Learmouth: Is the bone defect involved in the unilateral ligamentous disruption? Would you make an effort to reconstruct the ligamentous deficiency or would you go straight to a more constrained type of procedure?

Dr Chandler: Very often it can be done surgically if the ligament is there. If you have a collapsed knee in varus, once you reverse that and build the bone up, often you find a ligament that was not obviously there. I have not tried specifically to reconstruct the ligament. You can take the sartorius and try to build it up although I do not have personal experience of this.

Dr Hedley: It usually occurs when you have massive femoral destruction. Tibial destruction does not usually upset the ligaments. The medial ligament attaches so far down the tibia that you must have a tremendous amount of tibial bone destruction to lose it. If you lose both condyles in the femur, you will have real trouble. If it is just one condyle, the ligament is there and you have an eggshell of bone; even if you have a femur inside a femur you can reconstruct and it would probably be alright. So the answer is the tibia seldom, and the femur only when the whole condyle has gone.

Dr Czitrom: Even if it occurs very rarely, I would urge that the ligaments be attached to allograft because there is good evidence from clinical studies that ligaments heal solidly into allografts by forming bone just outside the allograft. If one of these ligaments that has been attached to an allograft is retrieved, you find that whilst the allograft may be dead, right outside the cortex there is new bone formation inside that ligament which attaches it firmly to the surface, so the rationale would be there to reattach.

Dr Hedley: In many of these knees that have had multiple surgery there are problems with the patella. Those factors have to be taken into account, but my decision on whether to use plastic or not is based entirely on whether I can obtain flexion and extension stability. If you cannot, you must go to the femur. In fact, the basic philosophy of the PCA revision was first to get stability in flexion because that is the big variable. You can get that stability in flexion by adding plastic until it is stable and then adjusting the distal femur up or down. By being able to add on the distal femur you often end up with the joint close to where you want it.

Dr Paprosky: If you have a gap of more than 3cm, you will have some degree of instability. That is used as a rule for adding on the femur.

Dr Chandler: If you have a very distorted knee with very little anatomy to tell you where the rotation is, the other knee will sometimes give you that answer, and you can measure down. Or, if the other knee has been equally destroyed, you can measure down a normal X-ray of another person who is roughly that size to get at least a rough idea of the centre of rotation. If the knee is unstable in extension and stable in flexion, you need it both ways.

Part IV

Bone Banks

26 Bone Banking

A.A. Czitrom

The increased demand for bone and tissue banking is related to the need for tissue grafts for the revision of joints, limb salvage surgery and procedures of knee reconstruction. This large demand is limited by the quality standards which have to be met, the complex methodology and currently, in the USA, by the risk and liability involved in bone banking.

At the present time in the United States and Canada there are both regional tissue banks and surgical bone banks. The former are large banks which process bone and other tissues whilst the latter process bones from living donors, usually femoral heads.

The American Association of Tissue Banks is the accrediting organisation which sets standards for tissue banking in the United States and Canada (American Association of Tissue Banks 1989).

As of December 1990, there were 33 banks accredited by the AATB in 19 states in the USA and one in Toronto, Canada. This bank, the first in Canada, received accreditation by the AATB in November 1990. A network exists amongst these banks and tissue is being transferred from one bank to another. There are 300 surgical bone banks in the USA today and they follow AATB guidelines. Technical manuals on surgical bone banking have been published.

Tissues are banked in a deep-frozen, freeze-dried, cryo-preserved or demineralised state. Deep-frozen bone is used either morsellised or as cortical segments. Cryo-preserved joints are being used mostly in tumour surgery and for post-traumatic defects as osteochondral grafts. Demineralised bone is used for osteoinduction but this is still in the research phase. Ligaments and menisci are used deep-frozen, freeze-dried or cryo-preserved.

The ideal age for donor selection is between 16 and 65; fresh grafts, ligaments and menisci are best derived from younger donors from 16 to 35 years. Age limits for frozen morsellised bone are 16 to 80 years. When the body is kept at 4 °C collection is done within 24 hours; if the body is at 20 °C it must be done within 12 hours.

Regional tissue banks obtain all major bones, ligaments and menisci from cadaver donors. Surgical bone banks only harvest live donor bone.

The most important aspect of screening is exclusions. Every major disease, including AIDS, is a contraindication to the harvesting of tissues. Serological screening for syphilis, hepatitis B, AIDS and hepatitis C is currently a required standard of the AATB. Human T-cell leukaemia or anti-HTLV-I screening is also done by many tissue banks. Blood culture, blood group and Rh testing is also done by most of the banks. The histology of lymph nodes to exclude disease is performed by some banks. The Miami Tissue Bank does this routinely and it should probably be done universally. Donor bone microbiology is tested by most of the banks but only some test the histo-pathology of the bone graft.

Preservation and storage in the regional banks is done either at 4 °C or by deep-freezing at −70 °C −196 °C with liquid oxygen. Cryo-preservation is carried out at −196 °C. After freeze-drying, tissues can be stored at room temperature.

Surgical bone banks only use the deep freezing method and the shelf life of this bone is generally believed to be five years. The quarantine required for this type of bone is now 180 days, and HIV re-testing has to be done after this period prior to use. The bones are triple wrapped, and stored in glassware or polyethylene. Storage is preferably done in upright freezers because the various bones can be better organised and more readily located.

The setting up of a surgical bone bank in a small hospital requires a local bone bank committee, simple record keeping and a bone bank coordinator (Czitrom et al. 1988). Femoral heads are stored and an inventory of bones purchased from regional banks. The processing cost of these bones is high and in the USA the expense is covered by the patient's insurance company.

Retrieval is done in the operating room. Aerobic and anaerobic cultures are taken. The grafts are soaked in antibiotic solution, and thawed for one to two hours. HLA matching is optional; the AOB blood group is taken as a record. No attempts are made to match grafts. Rh matching should be done for Rh females at risk, because there can be Rh sensitisation to a bone graft.

Because HIV is 100% fatal, its possible transmission is the most important topic in bone banking at the present time. The prevalence of HIV in donor blood is 1 in 40,000; but it is unknown in donor bone. There has been more than one case of HIV transmission by bone. The risk estimates vary between 1 in 10,533 and 1 in 1,667,600 (Buck 1989). These are mathematical estimates based on calculations which are weakened by uncertainties related to current testing procedures. The negative window may be as long as three years and all current screening tests have serious limitations.

The screening procedure currently done for HIV is the antibody test. The Western Blot and RIPA both confirm a positive antibody test. The PCR (polymerase chain reaction) is a new assay which is emerging as more sensitive than the antibody test. It relies on RNA translated into DNA, amplified so that trace amounts of HIV can be detected from the cells. The AATB guidelines regarding quarantine changed on 11 April 1991. Femoral heads must now be stored for 180 days before use. It should be emphasised that there has been no case of HIV transmission from live donor bone under the circumstances of quarantine before or after 11 April 1991.

The Public Health Service has stated that bone from donors not available for re-testing, including cadaveric donors, should be used only when bone from re-tested living donors is unavailable or inappropriate for use in the anticipated surgical procedure. This poses a small dilemma and stimulates the search for methods by which cadaver bones can be made safe. Should cadaveric bone be sterilised, for example? This is not a substitute for screening however, and the methods that have been used are ethylene oxide or radiation. Data exists in the literature which should be examined critically as to whether radiation is effective in inhibiting HIV.

Data on radiation was first reported in France by Montagnier's group, the co-discoverer of HIV virus (Spire et al. 1985). Various doses of radiation were administered to a viral culture. There was total inhibition with the dose of 0.25Mrad (2.5kGy). The virus dose measured was in counts per minute, reverse transcriptase equivalent. It is unfortunate that each of the subsequent experiments looked at a different dose of virus, with a different method of assay so that no valid comparisons can be made.

Experiments by us with 2.5Mrad (25kGy) caused complete inhibition of reverse transcriptase activity of HIV virus placed in the medullary cavity of bones (unpublished results). Only one dose was tested.

Experiments at the FBI Laboratory at Quantico showed that 75krad (750Gy) inhibit 10^{11} viral particles per ml; 50krad (500Gy) inhibit 5×10^{8} and 25 krad (250Gy) inhibit 5×10^{7}. Viral load was measured in particles/ml (Bigbee et al. 1989). The level in human blood infected with HIV is approximately 100–200 particles/ml. If 5×10^{7} particles/ml are inactivated by 25krad (250Gy), current routine doses of 1.5–2.5 Mrad (15–25kGy) used by tissue banks would be expected to give a good safety margin. Studies by Conway and associates used TCID 50/ml as a method of measuring the viral load. Reduction of viral bioburden but no inactivation by 0.4Mrad (4kGy) was demonstrated (Conway et al. 1990). Another study used the feline virus and heavy infection in cats. The cat bones were assayed and no inactivation was found in 2.9Mrad (29kGy) (Withrow et al. 1990).

Recent studies by Conway and associates studied the decrease of HIV bioburden in relation to the increase in doses of radiation. A decrease of viral bioburden by 1 log was shown to require

0.4Mrad (4kGy) assuming an initial bioburden of 10^2 and 10^3 TCID 50/ml (unpublished results).

Using this data, a decrease in the viral concentration from a bioburden of 10^2 TCID 50/ml (the expected concentration in the tissues of a heavily infected HIV victim) by six logs to 10^5 would require 2.4Mrad (24kGy).

The problems with all the above studies are the many unknown factors and the variability in the methods. The variables are the strain of virus, the method used for establishing the titre, the recipient cell type and the assay used for detection, all of which may give a different reading on the extent of infection. So the data is not entirely reliable, because in each assay system the result may be slightly different. Moreover, the exact bioburden of HIV in tissues has not been established.

Radiation up to 3.0Mrad (30kGy) has no effect on the mechanical properties of bone in bending, compression or torsion (Kommender et al. 1976).

Knowing that irradiation does inactivate the HIV virus but the dose required depends on the bioburden in the tissue, specifically in the bone of an HIV infected victim, future experiments must be aimed at measuring and standardising the HIV bioburden in musculo-skeletal tissues. In the meantime, the use of 2.5Mrad (25kGy) of radiation is considered a safe sterilisation method which reduces HIV bioburden.

References and Further Reading

American Association of Tissue Banks (1979) Guidelines for the banking of musculoskeletal tissues. Am Assoc Tissue Banks Newslett 3:2

Bigbee PD, Sarin PS, Humphreys JC, Eubanks WG, Sun D, Hocken DG, Thornton A, Adams DE, Simic MG (1989) Inactivation of human immunodeficiency virus (HIV) by ionizing radiation in body fluids and serological evidence. J Forensic Sciences 34:1303–1310

Buck BE, Malinin TI, Brown MD (1989) Bone transplantation and human immunodeficiency virus. Clin Orthop 240:129–136

Conway B, Tomford WW, Hirsch MS, Schooley RT, Mankin HJ (1990) Effects of gamma irradiation on HIV-1 in a bone allograft model. Transact Orth Res Soc 15:225

Czitrom AA, Gross AE, Langer F, Sim FH (1988) Bone banks and allografts in community practice. In: Bassett FH (ed) Instructional Course Lectures. The American Academy of Orthopaedic Surgeons, Park Ridge, Illinois, Vol. 37, pp13–24

Komender J, Komender A, Dziedzic-Goclawska A, Ostrowski K (1976) Radiation-sterilised bone grafts evaluated by electron spin resonance technique and mechanical tests. Transplantation Proceedings 8 Suppl. 1:25–37

Spire B, Dormont D, Barre-Sinoussi F, Montagnier L, Chermann JC (1985) Inactivation of lymphadenopathy-associated virus by heat, gamma rays, and ultraviolet light. Lancet 1:188–189

Withrow SJ, Oulton SA, Suto TL, Wilkins RM, Straw RC, Rose BJ, Gasper PW (1990) Evaluation of the antiretroviral effect of various methods of sterilizing/preserving cortico-cancellous bone. Transact Orth Res Soc 15:226

27 Allogeneic Bone Bank: The Nottingham Experience

H.G. Prince

In the UK we are far behind the USA in the organisation and running of an allogenic bone graft service. In 1988 I was delegated to go and organise a bone bank. I formally approached one of the Trust funds and we purchased a −80 °C freezer, and set about investigating the problems of bone banking. The easy harvest is of the femoral heads. The massive bone harvest has to be done by special teams.

We started with the fractured necks of femur. The basic screening is as described in Chapter 28. We use a separate informed consent form to harvest bone for donation. Patients are tested for HIV and have to be visited by the AIDS counsellor before we can operate on them. The guideline from the UK Department of Health is still that we test at 90 days. We have not yet gone to 180 days, but I can see that happening in the future.

All our bone, whether it is cadaver bone or heads of femur, is harvested in a sterile laminar flow theatre. Deep cultures are taken and we soak and wrap the bone in povidone-iodine; it is then placed in double-layer self-sealable polyethylene bags which are stored at −80 °C. We like our surgeons to give us advance warning as to when they need them.

One of the big problems with harvesting massive allografts is obtaining the support of the medical and nursing staff in theatre, for it is an unpleasant procedure. In a major harvest it is possible to take the long bones from the upper arm, forearms, femurs, and some foot bones. It is a long procedure and can be emotionally traumatic for the staff. There has to be extremely good liaison with them.

Unlike some workers, we have found that when using autograft bone anteriorly from the iliac crest, there is an 85% incidence of discomfort on the donor side, and deformity. Femoral heads in total hip replacement procedures were found to give a much better sized graft and good bone strength. The femoral heads are thawed out and irrigated with a Hartmann solution before use. There is obviously a certain amount of damage caused by the aqueous povidone-iodine.

The conclusions we have reached are similar to those of others; there is not a substantial revascularisation of massive allografts. In fact, we would rather it did not happen. You need to have firm fixation with cement in the allograft but not in the host bone. Ideally in the long grafts you should be using autogenous bone graft along the host allograft junction.

28 Setting Up a Femoral Head and Massive Allograft Bone Bank

H. Stafford

Excellent results can be obtained by using allograft material. The problem, however, is where do you get the allografts from, especially in the UK? It is obvious that banks are needed. I shall describe our experience in Leicester in setting up a femoral head and a massive allograft bone bank to meet this need. The Leicester Bone Bank is part of the University of Leicester Department of Orthopaedic Surgery based at Glenfield General Hospital.

Femoral Head Bank

The femoral head bank was started in January 1989 and expanded to include massive osteoarticular and intercalary grafts a year later. It should be possible to set up a femoral head bank in most district general hospitals, but clearly the decision to do so will depend on the anticipated number of femoral heads likely to be used each year. Smaller units may find it more cost effective to obtain their femoral heads from other banks.

The initial cost covered purchase of a −20 °C freezer, storage containers and the necessary stationery required in order to run the bank. In Leicester we have estimated that, initially, at least one session a week is required to get the bank running. In addition, it is highly desirable to have a member of either the technical or nursing staff who will be responsible for the day-to-day running of the bank.

Several steps have to be followed in the planning and running of such a bank. The first is to obtain consent to store femoral heads removed at total hip replacement. This is done prior to admission, by sending the patients an information sheet and identifying the sorts of conditions that would preclude them from becoming a donor. These include a past history of jaundice, malignancy, tuberculosis and at-risk factors for HIV infection. The form goes out with the routine instructions for admission from the waiting list office, and patients are asked to return the tear-off consent slip.

We also follow this up by checking on admission that the patients understand exactly what will be involved. All potential donors are screened by checking their VDRL, the hepatitis B and C antibodies and HIV status. The screening process is completed in the operating theatre when the femoral head is swabbed and subsequently examined for contamination with micro-organisms. Finally, the patient's notes are tagged and at the three-month outpatient follow-up appointment the patient's HIV status is checked again. The expert advisory group on HIV infection at the Department of Health has stated that this is currently considered to be the best practice.

The femoral heads are retrieved during normal total hip replacement operations. When the femoral head is removed it is passed to the scrub nurse who trims off any remaining soft tissue attached to the femoral neck. It is then put into a sterile glass jar which is itself placed in a second outer polyethylene jar and then handed off the instrument table.

Accurate documentation is essential and it should be the surgeon's responsibility to make sure that the first part of the data sheet which confirms that there are no at-risk factors for transmission of disease is completed. The time of removal is noted and the head categorised roughly into size – small, medium or large.

The second part of the data sheet records that all the pre-operative blood tests are normal, including the 90-day HIV. Once these are completed, the femoral head can then be released for implantation.

Until recently we had been storing our femoral heads at −20 °C for a maximum period of one year. In practice, we have found that the four surgeons at Glenfield General Hospital have been able to provide enough femoral heads to service the needs of 12 orthopaedic surgeons practising in Leicester.

The vast majority of the femoral heads are used between three and six months after retrieval. The advantage of storing at −80 °C is that it is now possible to store the femoral heads for five years.

Occasionally, a three-month HIV test is missed during the normal patient follow-up. Extended storage allows us time to follow up these stragglers and to pick them up later.

The steps involved in the implantation process are clearly described on an instruction sheet which accompanies every femoral head sent out from the bank to the operating theatres. We consider it essential that the femoral head is swabbed and the specimen sent for culture and sensitivity prior to implantation. Should the recipient develop problems with sepsis, early identification of the organism is made possible. In practice, we have found that the positive results prior to implantation are extremely low.

We do not match for ABO blood groups but we do try wherever possible to provide Rhesus-negative bone for Rhesus-negative recipients. If this is not possible then the risk of producing Rhesus antibodies can, for practical purposes, be prevented by giving the recipient an obstetric-type dose of anti-D.

Lastly, it is mandatory to audit the bank's performance, not only to monitor the bacteriology aspects of the process but also to identify clearly where wastage is occurring and, hopefully, to educate your colleagues in order to reduce this, which is not easy.

Massive Allograft Bank

If you decide to establish a massive allograft bank, it is advisable to first cut your teeth on a femoral head bank! Increasing the scope of your bank to include retrieving and restoring massive allografts is a great deal more costly and complex to establish than a femoral head bank. Present and anticipated future demand clearly lies in the field of revision surgery, for which intercalcary rather than osteoarticular grafts are required. Our allografts are used locally for both revision and tumour surgery.

Educating people in your area should start before you set up the bank. The most critical personnel are the local transplant coordinators. You have to be able to convince them that retrieving bones is an important part of the whole process of organ retrieval. We have been very fortunate in the amount of support that our transplant coordinators have given us in Leicester. In the early stages we were very surprised at the degree of antipathy expressed concerning the process of bone retrieval. This largely stems from the fact that others involved in organ transplantation immediately have an image of an amoeba-like corpse following the retrieval of bone allografts.

The reconstruction of the donor's limb must be fully explained to all concerned. The donor's body must be seen to appear as near normal as possible following a bone donation. The limb with no bones in it has to be reconstructed.

The process of education includes medical colleagues, in particular those who work in the intensive care units, the theatre staff who will help you with the retrievals, and the potential members of the retrieval teams. It is absolutely essential that everyone behaves in a seemly manner because, unfortunately, we know of examples where retrieval teams have badly upset the theatre staff by their callous and crude approach to an organ donation.

Finally, the education process continues into the community to make people aware of the value of bone transplantation. This does not merely include femoral head donors but everyone in the community.

Setting up a massive allograft bank is expensive and includes purchasing all the necessary hardware and the salary of a full-time technical manager.

One way to solve this problem is to persuade a local fund-raising charity to take your bank "on board". We have been very fortunate in that the Rotary Club of Leicester decided that the bone bank would be their major project during their 75th anniversary year. They considerably exceeded their target and this has allowed us to use some of the money to pay for research.

You may be able to persuade the hospital administrators that having a bone bank in the hospital is a worthwhile enterprise, and to fund the technical manager for an initial two-year period. This was achieved at our hospital and by the end of two years we hope that the bank will be completely self-financing.

Manpower can become a real problem. Ideally the bank should be headed by an orthopaedic surgeon who acts as the medical director, supported by a technical manager who is responsible for the day-to-day running of the bank. In addition, secretarial support is vital. You would be very surprised at the length of time it takes to climb the learning curve before your bank is running smoothly. You will find many problems peculiar to your own particular area and local situation.

We found it necessary to organise two retrieval teams; each is led by a consultant or senior registrar supported by a registrar, a technical supervisor and a group of medical students who we have recently recruited to help with the routine trimming of the grafts in the operating theatre.

All the retrievals are from beating-heart multiple organ donors. These occur at inconvenient hours and it is practically impossible to get a non-beating-heart donor into the operating theatre during normal working hours. We have recently been questioned about the necessity of retrieving bone allografts in aseptic conditions in an operating theatre. Since this is currently the accepted practice in North America and in Europe, we feel that it is prudent to continue. Not enough is known about the usefulness of allografts retrieved in dirty conditions sterilised by large doses of gamma irradiation. At the moment we would not be happy to obtain an allograft for one of our surgeons that had not been retrieved in an operating theatre.

Open lines of communication have to be established between the bank and several important key personnel. The advantages of being able to communicate rapidly with transplant coordinators is self-evident. It is also very useful to be in close contact with colleagues who run intensive therapy units, because they are the first to identify potential donors. It also applies to physician colleagues who may be looking after potential non-beating-heart donors. Theatre staff need to be kept informed. If there is a potential donor, they need to know because this may have significant effects on staffing levels, particularly since the work is usually done out of hours.

The golden rule when setting up a bank is "never say no to the offer of an allograft". When a donor is offered, it means that the retrieval teams have to be mobilised, but fortunately, because bone donation is usually the last tissue donation to be performed, the norm is for the bone bank to have several hours' warning.

We found it important and useful to be able to communicate with other bone banks. This is where the fax machine comes into its own, particularly when a local surgeon wants a particular allograft that is not in our own store. It also allows surgeons from farther afield to obtain a rapid reply as to whether or not we can help them. Whenever an allograft is requested, we do our utmost to meet that request, even if it means arranging to import bone from abroad.

All the data concerning a donor must be obtained right at the beginning. Very often you may miss an important laboratory result because this was not required by the soft tissue transplant teams, or an inadequate blood sample was taken prior to the bone retrieval. Our transplant coordinators very kindly ask us how much blood we want before any multiple organ donations start. They take it for us before the donor is ex-sanguinated.

All the details pertaining to the recipient should also be stored, and this includes the necessary sizing X-rays, to act as a basis for auditing the function of the bank.

The administration is largely concerned with the day-to-day running of the bank, performed by the technical manager and the secretary. We include audit as part of the research activities of the bank. We think there is still a good deal to be learned in two other main areas which concern preservation, in particular preservation of articular cartilage, and the problem of what to do with a contaminated graft. The currently acceptable mechanism is gamma irradiation.

Several important ethical issues arise. The first concerns obtaining consent from the relatives of the cadaveric donor. It is rare to be given a blanket consent to remove whatever tissues are deemed usable. It is more often the case that each tissue or organ has to be specified on the consent form. With live femoral head donors it is very important that the donor gives consent to the process because of the implications of some of the necessary screening tests carried out.

You will be aware of scandals in recent years concerning the sale of organs for transplantation. Nevertheless, it is permissible to pass on any costs incurred in retrieving, preserving or storing any tissue for transplantation to the unit in which the transplantation itself will take place. It should be made very clear that any such charges are purely reimbursement to cover their costs and that the allograft itself is free; it is a gift from the donor. It is illegal to buy or sell any human tissue for transplantation.

The relevant Act of Parliament concerning the transplantation of organs and tissues in the UK does not specify bone. In discussions with the Department of Health, it is very clear that they do not officially recognise bone as a tissue which can be transplanted. In this respect, no bone transplant has to be registered with the national registry. One bonus of the Act is that it makes it clear that it is entirely acceptable to pass on any processing costs to the user.

Discussion: Bone Banks

Chairman: **Mr Elson**

The Panel: **Dr Czitrom**
Dr Delloye
Miss Prince
Mrs Stafford

Mr Elson: Mrs Stafford said very definitely that the team in Leicester would not be prepared to accept bone which was not collected under ideal operating theatre circumstances and then irradiated. This is a very important matter. I would like to ask Dr Czitrom to comment on that.

Dr Czitrom: It would depend what it is contaminated with. In many banks, bones that are contaminated with *Staphylococcus epidermidis*, for example, will be irradiated before use. In principle, however, if it is any kind of pathogen that is considered important, one would not use a graft that has subsequently been irradiated.

Mr Elson: I do not quite understand what you mean. The extreme in this country would be that somebody would go to a post-mortem room within an acceptable period after death and remove a bone cleanly, but not in a sterile atmosphere and not in an operating theatre. The only difference between an operating theatre and a post-mortem room, provided there has not been something splashed about, is a question of air. Obviously the air will contain *epidermidis*, and perhaps other things as well in the moist air of this place. Those are the two things I am contrasting.

Dr Czitrom: You do your screening and it is all negative. You harvest in the post-mortem room and a culture comes back "Staph epi". You irradiate the bone and you can use it – that is fine. But if one of your blood cultures shows a positive bacteria or your culture shows that there is an infection with *aureus* or some other organism, then I would not use irradiated bone.

Mr Elson: But it is perfectly true, is it not, that a gamma irradiation at 3.0Mrad (30kGy) will kill *aureus*?

Dr Czitrom: Yes, it will.

Mr Elson: There is no question about that.

Dr Czitrom: The legal implication is just too strong. You are relying entirely on irradiation to sterilise known, highly virulent bacteria. I do not think you would do it with a surgical glove. If the surgical glove was infected with *Staph. aureus* and then irradiated, you would not give it to the surgeon to operate with; you would take another glove. Logically, however, you are right.

Mr Elson: Would you agree with me that *Staph. epidermidis* nowadays is a serious organism to contend with, especially in arthroplasty revision where we might be using a graft? I would be very worried if I thought there was any chance of *Staph. epidermidis* contaminating a graft. Yet you still seem to feel that this is an organism of no importance.

Dr Czitrom: I simply chose that one. There are other incidental surface organisms that are encountered. I do not personally concur with using bone that is procured under non-sterile conditions. I would not do it although I know some bone banks do. You need the clean air; you need the operating room; you need a sterile environment. There are banks which rely on irradiation to sterilise cleanly-procured bone. That is a minority of their bones, however. I do not support this concept. I think it would be better to set up clean air rooms to harvest bones.

Mr Elson: If you had a facility for a portable clean air canopy in somewhere like a mortuary and you took bone, would that suffice? The only difference between an operating theatre and a post-mortem room, assuming that you have all the defences against contagious infection and nothing is being splashed about, is the air.

Dr Czitrom: I think a great many bone banks use clean rooms to process their bones. That is exactly the same as the mortuary perhaps, except that, from time to time, they monitor bacterial contamination everywhere in the room, including the walls. Some banks have a very modern facility set up with clean air rooms where people are gowned with exhaust suits, and they process their bones under those conditions.

I do not think the question is whether you get your bone in that circumstance but what do you do if you get positive cultures from your bone? There the opinions are divided. Some banks will use bad bone, irradiating it before use. They say "I have sterilised it." Others will say "We will not use it. Throw it away."

Mr Elson: Does the bank have to declare its method? Supposing a bank were doing something that you would obviously regard as being slipshod or even dangerous, would you expect them to declare that this is how they collect bone, or would you just not know?

Dr Czitrom: You would not know. You would have to check what the bank does and how they procure most of the bone.

Mrs Stafford: This is one of the advantages of having organisations like the AATB. Hopefully, accredited banks work to minimum standards. If you acquire a graft from an accredited bank, you know what has happened to that bone and how it has been prepared by reading their manual. If you acquire bone from non-accredited banks, then you have to accept that you are taking their word for what they have done, and that is your decision.

Mr Elson: That is very fair comment.

Dr Delloye: In Belgium we always send an information sheet to the surgeon who is implanting the allograft. We rely on sterile procurement which is done in operating rooms. If there is contamination, we discard the bone and use another method. We will cut the bone, freeze-dry it and then rely on irradiation.

Dr Loty: I do not agree at all with what you have said about irradiation. First, coagulase negative *Staph. epidermidis* has the same reduced sensibility as *Staph. aureus*. There is a very well known standard in sterilisation by irradiation called *Bacillus pumilus*, which is the reference regarding radiosterilisation.

Using the example of the gloves; the gloves do not have to be sterile before sterilising, they have to be sterile after sterilising. That is the way you must see radiation.

Little was known about radiosterilisation when we started our bone bank. We had a study carried out. One hundred bones were examined before irradiation to establish which bacteria were encountered in our procedure. We then had controls after irradiation. After 2.5Mrad (25kGy), all controls were negative. We then took the *Bacillus pumilus* and put it in the middle of big bones. Further controls were carried out. They were negative.

The only problem with radiosterilisation is that it is more or less efficient given the temperature which is used. This is why at the beginning of our study we wanted to control the temperature efficiency relating to the bones we used. It is certainly completely efficient with regard to bacteria.

Given this, I do not see why we should be obliged to go into an operating theatre to obtain the bone. We do not retrieve in sterile operating rooms. We have now performed almost 400 massive allografts and we have never had any problems.

Regarding viral sterilisation, no one can guarantee that HIV will be sterilised by irradiation. We must therefore use clinical research and the routine serology just as if we were not irradiating. Irradiation for the virus is one more precaution. I believe that it is efficient, having read some of the papers on the subject. The problem is that there has been no mention of temperature in most of the papers. Some workers have made an assay on plasma at −40 °C and, after 3Mrad (30kGy), had 10.7 logarithm degrees. It appears to be efficient, but we must continue to take all the usual precautions against viral infection.

There is one precaution which is taken systematically and that is the autopsy. I am surprised that the AATB do not insist on an autopsy being performed each time. Some disease is not sterilised by irradiation or anything else; the only way to detect is by autopsy. We always have an autopsy performed in our bone bank.

Dr Czitrom: You disagree with me but I do not disagree with you at all! What you are saying is right. If there is bacteria present and you give the rads, it will be dead. There is no question about that. But in America you cannot take infected bones, sterilise them and put them into other people. It will not stand up legally.

Dr Loty: When you buy surgical gloves you do not insist that the factory is sterile. The gloves are sterile after sterilisation.

Dr Czitrom: You do not take a glove full of pus and sterilise it either!

Dr Loty: But if you sterilise them, there is no trouble.

Mr Elson: Either something is sterile or it is not. You have to put aside all sentiment in this. If it is sterile, it does not matter whether you have spat on it beforehand!

Dr Delloye: To make a reconstruction we sometimes take not only the bone but also the soft tissue including tendons around the joint. Warming up the graft at 60 °C for two or three hours is probably effective against the virus, but I have my doubts about the mechanical resistance of the collagen and the cartilage.

Mr Elson: I was not referring to the massive allograft but to methods of actually housing bones or preparing them for producing cancellous chips, and so on.

Dr Delloye: We have another method for the small pieces of bone. We go to the mortuary room, procure the epiphysis around the knee and then make an acellular implant. The bone marrow is washed with a water jet and it is put in chloroform methanol for three days. You will get a rather dry implant which is freeze-dried afterwards and then irradiated. It is very convenient, but only for filling and not for any mechanical purpose. We thereby have two or three barriers against the HIV virus.

Mr Cheah: At the present time we do two separate tests: one at the beginning and one in 180 days' time. Why do you have to do the first one? Why not just the second one at 180 days or 90 days?

Dr Czitrom: It is only for storage purposes. The test could be done at 180 days only. Very few will convert anyway, so there will be no unnecessary storage. The only reason for doing it is because of the guidelines laid down by the AATB.

Mr McLardy-Smith: Turning to storage temperature, how important is it to store at −70 °C or −80 °C, which is the most often quoted? Is it acceptable to store at −20 °C for a femoral head, as you are doing at Leicester?

Mrs Stafford: We did start storing at −20 °C, but there is a Department of Health recommendation that tissue which is going to be stored for transplantation should be at a minimum of −70 °C. We try to follow the recommendation.

Miss Prince: Professor Burwell did some work in the late 1960s on bone banking. He found that there is a slow deterioration at −20 °C. His recommended storage time at −20 °C is only three months. That does not give us the safety margin for our second testing.

Dr Czitrom: It is the same reason why one has a deep freezer to store beef. It will go bad if stored at −20 °C.

Mr McLardy-Smith: Does the lower temperature also reduce its immunogenicity?

Dr Czitrom: It will kill the cells. But −20 °C will also kill cells.

Dr Delloye: So far as immunogenicity is concerned, there is no difference in the storage temperature. Friedlander has carried out work in that area.

29 Conclusion

R.A. Elson

My personal experience of allografting is from twenty-three years ago when I visited Professor Volkov in Moscow and Professor Icomonov in Sofia. There, as I understood matters, the body of the deceased was the property of the state and the harvesting of allografts of any type presented no ethical problem.

While I had no access to really long term results, I saw remarkable examples of structural allografts with useful function. Methods of preparation included freezing and boiling. For example, one case was a sarcoma of the lower femur; it was resected with a coating of soft tissue, stripped of this, drilled and boiled (tumour-sterilised) and then replaced. Certainly it had united and two years post-operation the patient still had his leg and a useful knee, albeit with a modest range of movement and some instability.

I could not fail to have been impressed. But when I returned to England and suggested that we should commence allografting, thinking more of these massive replacements, the idea was met with severe ethical objections. Having other time-absorbing orthopaedic interests, I quickly lost heart – I wish I had had the foresight to have persisted because, while we do not know longer term outcomes, the results we now see are encouraging, to say the least. But is fatigue mechanical failure of massive grafts likely? There has been a change of ethical emphasis in the matter of harvesting from brain-dead or from totally dead donors. The influences of organ transplant generally, religious amelioration, social acceptance and understanding have all contributed. We still have progress to make, however. AIDS has seriously impeded us and conflicting reports concerning structural failure are a nuisance. The startling concepts of cementing on morsellised allograft and the incorporation of antibiotics in graft material have opened new avenues of particular interest to me personally.

Our North American colleagues have described examples of gross loss of bone stock about knee and hip replacements. They may seem to be "ahead" of us in numbers, but I can assure them that, hidden away in little pockets all over the UK, we have the same cases – perhaps proportionally as many – we don't know. In this country we tend not to follow up our patients with joint replacements, a very inadequate organisation. No news is not necessarily good news. We have established in Trent an arthroplasty collecting unit. Theoretically, all hip and knee replacements done in the region are forwarded to Professor Gregg at Leicester University who keeps a list of them. These patients will be sent for at five years. I do not know whether it is going to work. It should not be difficult to organise such a thing in a tiny country like ours. I mention this plan in order to emphasise the fact that in this, the largest field demanding bone allografting, we should strive to prevent the progressive loss of bone which will occur unless we follow up properly.

I congratulate John Older and his supporters on the publication of an important contribution to our knowledge of allografting in hip and knee revision surgery.

28 Setting Up a Femoral Head and Massive Allograft Bone Bank

H. Stafford

Excellent results can be obtained by using allograft material. The problem, however, is where do you get the allografts from, especially in the UK? It is obvious that banks are needed. I shall describe our experience in Leicester in setting up a femoral head and a massive allograft bone bank to meet this need. The Leicester Bone Bank is part of the University of Leicester Department of Orthopaedic Surgery based at Glenfield General Hospital.

Femoral Head Bank

The femoral head bank was started in January 1989 and expanded to include massive osteoarticular and intercalary grafts a year later. It should be possible to set up a femoral head bank in most district general hospitals, but clearly the decision to do so will depend on the anticipated number of femoral heads likely to be used each year. Smaller units may find it more cost effective to obtain their femoral heads from other banks.

The initial cost covered purchase of a −20 °C freezer, storage containers and the necessary stationery required in order to run the bank. In Leicester we have estimated that, initially, at least one session a week is required to get the bank running. In addition, it is highly desirable to have a member of either the technical or nursing staff who will be responsible for the day-to-day running of the bank.

Several steps have to be followed in the planning and running of such a bank. The first is to obtain consent to store femoral heads removed at total hip replacement. This is done prior to admission, by sending the patients an information sheet and identifying the sorts of conditions that would preclude them from becoming a donor. These include a past history of jaundice, malignancy, tuberculosis and at-risk factors for HIV infection. The form goes out with the routine instructions for admission from the waiting list office, and patients are asked to return the tear-off consent slip.

We also follow this up by checking on admission that the patients understand exactly what will be involved. All potential donors are screened by checking their VDRL, the hepatitis B and C antibodies and HIV status. The screening process is completed in the operating theatre when the femoral head is swabbed and subsequently examined for contamination with micro-organisms. Finally, the patient's notes are tagged and at the three-month outpatient follow-up appointment the patient's HIV status is checked again. The expert advisory group on HIV infection at the Department of Health has stated that this is currently considered to be the best practice.

The femoral heads are retrieved during normal total hip replacement operations. When the femoral head is removed it is passed to the scrub nurse who trims off any remaining soft tissue attached to the femoral neck. It is then put into a sterile glass jar which is itself placed in a second outer polyethylene jar and then handed off the instrument table.

Accurate documentation is essential and it should be the surgeon's responsibility to make sure that the first part of the data sheet which confirms that there are no at-risk factors for transmission of disease is completed. The time of removal is noted and the head categorised roughly into size – small, medium or large.

The second part of the data sheet records that all the pre-operative blood tests are normal, including the 90-day HIV. Once these are completed, the femoral head can then be released for implantation.

Until recently we had been storing our femoral heads at −20 °C for a maximum period of one year. In practice, we have found that the four surgeons at Glenfield General Hospital have been able to provide enough femoral heads to service the needs of 12 orthopaedic surgeons practising in Leicester.

The vast majority of the femoral heads are used between three and six months after retrieval. The advantage of storing at −80 °C is that it is now possible to store the femoral heads for five years.

Occasionally, a three-month HIV test is missed during the normal patient follow-up. Extended storage allows us time to follow up these stragglers and to pick them up later.

The steps involved in the implantation process are clearly described on an instruction sheet which accompanies every femoral head sent out from the bank to the operating theatres. We consider it essential that the femoral head is swabbed and the specimen sent for culture and sensitivity prior to implantation. Should the recipient develop problems with sepsis, early identification of the organism is made possible. In practice, we have found that the positive results prior to implantation are extremely low.

We do not match for ABO blood groups but we do try wherever possible to provide Rhesus-negative bone for Rhesus-negative recipients. If this is not possible then the risk of producing Rhesus antibodies can, for practical purposes, be prevented by giving the recipient an obstetric-type dose of anti-D.

Lastly, it is mandatory to audit the bank's performance, not only to monitor the bacteriology aspects of the process but also to identify clearly where wastage is occurring and, hopefully, to educate your colleagues in order to reduce this, which is not easy.

Massive Allograft Bank

If you decide to establish a massive allograft bank, it is advisable to first cut your teeth on a femoral head bank! Increasing the scope of your bank to include retrieving and restoring massive allografts is a great deal more costly and complex to establish than a femoral head bank. Present and anticipated future demand clearly lies in the field of revision surgery, for which intercalcary rather than osteoarticular grafts are required. Our allografts are used locally for both revision and tumour surgery.

Educating people in your area should start before you set up the bank. The most critical personnel are the local transplant coordinators. You have to be able to convince them that retrieving bones is an important part of the whole process of organ retrieval. We have been very fortunate in the amount of support that our transplant coordinators have given us in Leicester. In the early stages we were very surprised at the degree of antipathy expressed concerning the process of bone retrieval. This largely stems from the fact that others involved in organ transplantation immediately have an image of an amoeba-like corpse following the retrieval of bone allografts.

The reconstruction of the donor's limb must be fully explained to all concerned. The donor's body must be seen to appear as near normal as possible following a bone donation. The limb with no bones in it has to be reconstructed.

The process of education includes medical colleagues, in particular those who work in the intensive care units, the theatre staff who will help you with the retrievals, and the potential members of the retrieval teams. It is absolutely essential that everyone behaves in a seemly manner because, unfortunately, we know of examples where retrieval teams have badly upset

Subject Index